Lecture Notes in Networks and Systems

Volume 123

The series "Lecture Notes in Networks and Systems" publishes the latest developments in Networks and Systems—quickly, informally and with high quality. Original research reported in proceedings and post-proceedings represents the core of LNNS.

Volumes published in LNNS embrace all aspects and subfields of, as well as new challenges in, Networks and Systems.

The series contains proceedings and edited volumes in systems and networks, spanning the areas of Cyber-Physical Systems, Autonomous Systems, Sensor Networks, Control Systems, Energy Systems, Automotive Systems, Biological Systems, Vehicular Networking and Connected Vehicles, Aerospace Systems, Automation, Manufacturing, Smart Grids, Nonlinear Systems, Power Systems, Robotics, Social Systems, Economic Systems and other. Of particular value to both the contributors and the readership are the short publication timeframe and the world-wide distribution and exposure which enable both a wide and rapid dissemination of research output.

The series covers the theory, applications, and perspectives on the state of the art and future developments relevant to systems and networks, decision making, control, complex processes and related areas, as embedded in the fields of interdisciplinary and applied sciences, engineering, computer science, physics, economics, social, and life sciences, as well as the paradigms and methodologies behind them.

**** Indexing: The books of this series are submitted to ISI Proceedings, SCOPUS, Google Scholar and Springerlink ****

More information about this series at http://www.springer.com/series/15179

Mehmet Zeki Sarıkaya ·
Hemen Dutta · Ahmet Ocak Akdemir ·
Hari M. Srivastava
Editors

Mathematical Methods and Modelling in Applied Sciences

Springer

Editors
Mehmet Zeki Sarıkaya
Department of Mathematics,
Faculty of Science and Letters
Düzce University
Düzce, Turkey

Ahmet Ocak Akdemir
Department of Mathematics,
Faculty of Science and Letters
University of Ağrı İbrahim Çeçen
Ağrı, Turkey

Hemen Dutta
Department of Mathematics
Gauhati University
Guwahati, Assam, India

Hari M. Srivastava
Department of Mathematics
and Statistics
University of Victoria
Victoria, BC, Canada

ISSN 2367-3370 ISSN 2367-3389 (electronic)
Lecture Notes in Networks and Systems
ISBN 978-3-030-43001-6 ISBN 978-3-030-43002-3 (eBook)
https://doi.org/10.1007/978-3-030-43002-3

This Springer imprint is published by the registered company Springer Nature Switzerland AG
The registered company address is: Gewerbestrasse 11, 6330 Cham, Switzerland

Contents

About the Editors

Mehmet Zeki Sarikaya is Professor at the Düzce University, Turkey. His research interests include aspects of operator theory, inequalities, special functions, especially selection principles, fractional calculus, and integral inequalities. In particular, he gave Bessel diamond operator in the literature. He supervised several doctoral and postgraduate students in the fields of applied mathematics and analysis. He carries out many courses at bachelor, graduate, and doctoral levels. He completed his PhD and focused on Bessel diamond operator, at the University of Afyon Kocatepe in 2007, under the supervision of Prof. Dr. Hüseyin Yildirim. He has published over 290 papers, two chapters on inequalities, and a lot of conference papers.

Hemen Dutta has been serving the Department of Mathematics at Gauhati University as a faculty member. He did his Master of Science in Mathematics, Postgraduate Diploma in Computer Application, Master of Philosophy in Mathematics, and Doctor of Philosophy in Mathematics. His research areas include functional analysis, mathematical modeling, etc. He has to credit over 100 items as research papers and chapters. He has published 10 books as textbooks, reference books, monographs, and edited books. He has delivered several talks at national and international levels and organized several academic events in different capacities. He has acted as a reviewer for journals and databases and associated with editing special issues in journals. He has published several articles in newspaper, popular books, magazines, and science portals.

Ahmet Ocak Akdemir is a faculty member in the Mathematics Department of Ağrı İbrahim Çeçen University, Turkey, since 2009. He completed his bachelor degree from Atatürk University, Turkey, in 2007 and started PhD analysis in the same year in Graduate School of Natural and Applied Sciences, Atatürk University. He received his PhD degree on the thesis titled "Integral inequalities for different kinds of convex functions on the coordinates" from Atatürk University in 2012. He has published several research articles in journals of repute on different types of integral inequalities, convex function classes, integral inequalities with the help of fractional integral operators, and new classes of convex functions and inequalities. He also worked as a director in several national research projects. He has been involved as a supervisor of a number of master's students and currently supervising several graduate students. He has received publication promotion awards from various public institutions and has participated in many conferences.

Hari Mohan Srivastava has held the position of Professor Emeritus in the Department of Mathematics and Statistics at the University of Victoria in Canada since 2006, having joined the faculty there in 1969, first as Associate Professor (1969–1974) and then as Full Professor (1974–2006). He earned his PhD degree in 1965 while he was a full-time member of the teaching faculty at the J. N. V. University of Jodhpur in India. He has held numerous visiting research and honorary chair positions at many universities and research institutes in different parts of the world. Having received several DSc (honoris causa) degrees as well as honorary memberships and honorary fellowships of many scientific academies and learned societies around the world, he is also actively associated editorially with numerous international scientific research journals. His current research interests include several areas of Pure and Applied Mathematical Sciences. He has published 33 books, monographs and edited volumes, 33 (and encyclopedia) chapters, 48 papers in international conference proceedings, and more than 1,250 scientific research articles in peer-reviewed international journals, as well as Forewords and Prefaces to many books and journals, and so on.

Fuzzy Kernel-Based Clustering and Support Vector Machine Algorithm in Analyzing Cerebral Infarction Dataset

Zuherman Rustam[1], Dea Aulia Utami[1(✉)], Jacub Pandelaki[2], Nadisa Karina Putri[1], and Sri Hartini[1]

[1] Department of Mathematics, FMIPA Universitas Indonesia, Kampus UI Depok, Depok 16424, Indonesia
deaauliautami@sci.ui.ac.id
[2] Medical Department of Radiology, RSUPN Dr. Cipto Mangunkusumo, Kota Jakarta Pusat, DKI, Jakarta 10430, Indonesia

Abstract. Ischemic stroke is a disease that occurs due to disruption of blood circulation to the brain due to blood clots in the brain. The blockage is called cerebral infarction. In diagnosing the presence of cerebral infarction in the brain, machine learning is used because it is not enough just to use a CT scan to diagnose. To deal with the problem of classification of cerebral infarction data obtained from Dr. Cipto Mangunkusumo's Hospital in Jakarta, this study proposes the use of Fuzzy C-Means Clustering (FCM), Fuzzy Possibilistic C-Means (FPCM), and Radial Base Function Fuzzy Possibilistic C-Means (RBFFPCM) method as a clustering method and a Support Vector Machine (SVM) method as a classification method. This method will be compared to the level of accuracy. The greatest level of accuracy is generated from the Radial Base Function Fuzzy Possibilistic C-Means (RBFFPCM) method with an accuracy value of 91%.

Keywords: Fuzzy C-Means Clustering (FCM) · Fuzzy Possibilistic C-Means (FPCM) · Radial Base Function Fuzzy Possibilistic C-Means (RBFFPCM) · Support Vector Machine (SVM) · Cerebral infarction · Ischemic stroke

1 Introduction

Stroke is the first leading cause of death after heart disease and diabetes [1]. There are 4.4 million people in the Southeast Asia region experiencing strokes [2]. This disease occurs due to blood circulation in the brain that is disrupted due to the presence of broken blood vessels and blockages in the brain (referred to as cerebral infarction) [3]. When a stroke occurs, tissue in the brain will die because of blood clots in the heart and other blood vessels, so that blood circulation that carries oxygen and nutrients to the body will stop [3]. Ischemic stroke is a stroke caused by a blockage or rupture of a blood vessel in the brain [4]. The blockage is called cerebral infarction.

Ischemic stroke can be known if in the patient's brain there is cerebral infarction [4]. CT scans can be used to determine the presence of cerebral infarction. However, it

M. Zeki Sarıkaya et al. (Eds.): ICMRS 2019, LNNS 123, pp. 1–11, 2020.
https://doi.org/10.1007/978-3-030-43002-3_1

is not enough just to use a CT scan to diagnose that there is infarction in the brain. In overcoming this, machine learning can be used.

There are several studies that have discussed the fuzzy clustering method, among them are, Havens et al. [5] compares the effectiveness of three different implementations that have the aim of improving the fuzzy c means clustering method. Among them are sampling with non-iterative extensions, incremental techniques and adding kernel functions to FCM. The results show that adding kernel functions to FCM is a good choice in data clustering. Pal et al. [6] proposes a new model called the possibilistic fuzzy c-means (PFCM) model.

This method generates membership functions, possibilistic functions simultaneously with the cluster center. But PFCM is less sensitive to outliers and can avoid the right clusters. H.Izakian et al. [7] proposed a hybrid fuzzy clustering method based on FCM and fuzzy PSO (FPSO) to streamline the clustering algorithm and produce quality solutions. To overcome the lack of fuzzy c-means that are sensitive to initialization and easily trapped in optimal local solutions, this paper is integrated with particle swarm algorithms.

Kannan et al. [8] uses the effective fuzzy method of the possibility of c-means by using norm distance which is induced by the kernel function for data on brain cancer and colon cancer in the basis of the microarray gene expression. This paper has reported the superiority of the proposed methods through cluster validation using silhouette accuracy, running time, number of iterations and well-separated clusters.

Dsouza et al. [9] provides a complete explanation of Support Vector Machines (SVM) that can be used for classification of uncertain data. SVM uses kernel configurations to produce better results in classification. In this paper, breast cancer data used for four types of SVM kernel methods, there are linear, polynomial, sigmoid and radial kernels.

This study proposes the use of clustering and classification methods. The clustering method used is Fuzzy C-Means Clustering (FCM), Fuzzy Possibilistic C-Means (FPCM), and Radial Base Function Fuzzy Possibistic C-Means (RBFFPCM). Meanwhile, the classification method used is Support Vector Machine (SVM). Validation on performance measurement used 5-fold cross validation. The end result is knowing how the proposed method influences the accuracy of infarct data prediction by calculating the accuracy of the model on the classifier.

2 Materials and Methods

2.1 Clustering Method

is an unsupervised learning method. In the process of grouping data, this study uses clustering methods. The concept of clustering method is to determine the value of a distance to know and measure the similarity of each object to be observed [10]. The commonly used distance measurement is Euclidean distance. The smaller the value of the Euclidean distance, the greater the similarity value of an object and the greater the value of the Euclidean distance, the lower the similarity value of an object [10]. After determining the size of the similarity of an object, grouping will be carried out. Some clustering methods used in this study include the following:

2.1.1 Fuzzy C-Means Clustering (FCM)

FCM is an unsupervised classification based on clustering. This method has a membership function that is defined based on the distance function, so that the degree of membership of a data pixel is defined based on how close the distance is to the center of the class [11]. The degree of membership will determine the existence of each data point in a group (cluster).

The concept are determine the cluster center to be marked as the average location for each cluster that exists [11]. For each cluster, each data has a degree of membership. In determining the cluster center will go to the right location, the membership value of each data and cluster center is repaired repeatedly [11]. This repetitive improvement is based on the objective function given in the equation below:

$$J(\boldsymbol{U}, \boldsymbol{V}, \boldsymbol{X}, c, m) = \sum\nolimits_{i=1=1}^{c} \sum\nolimits_{k=1}^{N} u_{ik}^{m} \|x_k - v_i\|^2 \tag{1}$$

with the constraint function $\sum_{i=1}^{c} u_{ik} = 1$ Where N is the amount of data, c is the number of clusters, $\boldsymbol{V}$ is the center of the cluster, $\boldsymbol{U}$ is the membership function, $\mathbf{X}$ is the data to be clustered, m is the fuzzy ($m > 1$) and $\|x_k - v_i\|^2$ is the distance between data point with the cluster center.

The value of the degree of membership in the FCM method is

$$u_{ik} = \frac{1}{\sum_{j=1}^{c} \left(\frac{\|x_k - v_i\|}{\|x_k - v_j\|} \right)^{2/(m-1)}}, u_{ik} \in U, k = 1, 2, \ldots, n \tag{2}$$

The updated i-cluster center is

$$v_i = \frac{\sum_{k=1}^{N} u_{ik}^{m} x_k}{\sum_{k=1}^{N} u_{ik}^{m}}, i = 1, 2, \ldots, c \tag{3}$$

With the iteration termination criteria as follows:

$$\Delta = \left\| \boldsymbol{V}^t - \boldsymbol{V}^{t-1} \right\| < \varepsilon \tag{4}$$

With $\boldsymbol{V}^t$ is the center of the cluster in the t-iteration and $\boldsymbol{V}^{t-1}$ is the center of the cluster in the previous iteration.

2.1.2 Fuzzy Possibilistic C-Means (FPCM)

This method defines clustering techniques by integrating membership functions in the Fuzzy C-Means (FCM) method and the function of the typicality of Positive C-Means (PCM) method to reduce the effect of data that has noisy and outliers. The objective functions of FPCM are as follows [12]:

$$J(\boldsymbol{U}, \boldsymbol{V}, \boldsymbol{X}, c, m, n, \tau) = \sum_{i=1}^{c} \sum_{k=1}^{n} \left(u_{ik}^{m} + \tau_{ik}^{\eta} \right) d^2(x_k, v_i), \quad m, \eta > 1 \tag{5}$$

With the constraints $\sum_{i=1}^{c} u_{ik} = 1, \forall k \in \{1, 2, \ldots, n\}, \sum_{k=1}^{\eta} \tau_{ik} = 1, \forall i \in \{1,2, \ldots, c\}$. Where n is the number of sample data, c is the number of clusters, m is the fuzzy degree,

η is the possibilistic degree, x_k is the k data, v_i is the center value in the i-cluster, u_{ik} is member value x_k in the i-cluster, and τ_{ik} is the typicality value x_k in the i cluster.

The value of the degree of membership in the FPCM method is

$$u_{ik} = \frac{1}{\sum_{j=1}^{c}\left(\frac{\|x_k - v_i\|}{\|x_k - v_j\|}\right)^{1/(m-1)}} \tag{6}$$

The value of the typicality of the FPCM method is

$$\tau_{ik} = \frac{\left(\frac{1}{\|x_k - v_i\|}\right)^{1/(\eta-1)}}{\sum_{l=1}^{n}\left(\frac{1}{\|x_l - v_i\|}\right)^{1/(\eta-1)}} \tag{7}$$

The updated *i*-cluster center is

$$v_i = \frac{\sum_{k=1}^{N}(u_{ik}^{m} + \tau_{ik}^{\eta})x_k}{\sum_{k=1}^{N}(u_{ik}^{m} + \tau_{ik}^{\eta})} \tag{8}$$

With the iteration termination criteria as follows:

$$\Delta = \left\| \boldsymbol{V}^{t} - \boldsymbol{V}^{t-1} \right\| < \varepsilon \tag{9}$$

With $\boldsymbol{V}^t$ is the center of the cluster in the t-iteration and $\boldsymbol{V}^{t-1}$ is the center of the cluster in the previous iteration.

2.1.3 Radial Base Function Fuzzy Possibilistic C-Means (RBFFPCM)

This method is a kernel-based clustering method. The clustering method used is Fuzzy Possible C-Means (FPCM) using the kernel Radial Base Function (RBF) [13]. Through nonlinear transformations, this kernel-based cluster method maps input data elements into higher-dimensional feature spaces [14]. In general, the kernel function is defined as follows [12]:

$$d^2(\emptyset(x_k), \emptyset(v_i)) = \|\emptyset(x_k) - \emptyset(v_i)\|^2 = 2(1 - K(x_k, v_i)) \tag{10}$$

With $K(x_k, v_i)$ is the Radial Base Function kernel function shown below

$$K\left(x_k, v_j\right) = exp\left(-\frac{\|x_k - v_j\|^2}{\sigma^2}\right) \tag{11}$$

The objective function of RBFFPCM is as follows:

$$J(U,V) = 2\sum_{i=1}^{c}\sum_{k=1}^{n}\left(u_{ik}^{m} + \tau_{ik}^{\eta}\right)\left(1 - K(x_k, v_i)\right), \quad m, \eta > 1 \tag{12}$$

With the constraints $\sum_{i=1}^{c} u_{ik} = 1, \forall k \in \{1, 2, \ldots, n\}, \sum_{k=1}^{\eta} \tau_{ik} = 1, \forall i \in \{1,2,\ldots,c\}$. Where n is the number of sample data, c is the number of clusters, m is the fuzzy degree,

$\eta$is the possibilistic degree, x_k is the k data, v_i is the center value in the i-cluster, u_{ik} is member value x_k in the i-cluster, and τ_{ik} is the typicality value x_k in the i cluster.

The value of the degree of membership in the RBFFPCM method is

$$u_{ik} = \frac{\frac{1}{(2-2K(x_k,v_i))}^{1/(m-1)}}{\sum_{j=1}^{c} \frac{1}{(2-2K(x_k,v_i))}^{1/(m-1)}} \tag{13}$$

The value of the typicality of the FPCM method is

$$\tau_{ik} = \frac{\frac{1}{(2-2K(x_k,v_i))}^{1/(\eta-1)}}{\sum_{l=1}^{n} \frac{1}{(2-2K(x_k,v_i))}^{1/(\eta-1)}} \tag{14}$$

The updated i-cluster center is

$$v_i = \frac{\sum_{j=1}^{n} u_{ik}^{m} K(x_k, v_i) x_k}{\sum_{j=1}^{n} u_{ik}^{m} K(x_k, v_i)} \tag{15}$$

With the iteration termination criteria as follows:

$$\Delta = \left\| \boldsymbol{V}^t - \boldsymbol{V}^{t-1} \right\| < \varepsilon \tag{16}$$

With $\boldsymbol{V}^t$ is the center of the cluster in the t-iteration and $\boldsymbol{V}^{t-1}$ is the center of the cluster in the previous iteration.

2.2 Classification Method

Is a grouping of data where the data used has a label or target class [15]. Algorithms to solve classification problems are categorized into supervised learning. The classification method used in this study is Support Vector Machine (SVM).

2.2.1 Support Vector Machine (SVM)

SVM was first introduced by Vapnik in 1995 [15]. This method is one of the supervised learning methods used for classification problems. The purpose of SVM is to classify it by finding the best hyperplane that separates data from defined classes [15]. The best hyperplane is a hyperplane that will maximize margins [16]. Where, margin is a perpendicular distance between the hyperplane and the closest data points of each class [16]. Meanwhile, the data closest to the hyperplane is called support vector [16].

Suppose there is a data point $\boldsymbol{x}_i, y_i$ where $i = 1, 2, \ldots, N$, with $\boldsymbol{x}_i \in R^d$ is a d dimension input vector that represents the number of features, and $y_i \in -1, 1$ with y_i are class labels of the infarction dataset, there are the infarct class and normal class. Optimal hyperplane is given which will separate data into two classes. The point $\boldsymbol{x}$ in the hyperplane satisfies the equation below

$$y(\boldsymbol{x}) = \boldsymbol{w}^T \boldsymbol{x} + b \tag{17}$$

Where $\boldsymbol{w}$ is a vector of the weight parameter values, and b is a bias that has a scalar value. The formed hyperplane will separate the data into two classes on the infarction dataset, there are the infarct class and the normal class, or the class SVM method that has positive and negative values.

The distance between the two hyperplane can be defined from the equation below:

$$\frac{\left|\boldsymbol{w}^T\boldsymbol{x}_i + b\right|}{\|\boldsymbol{w}\|} = \frac{1}{\|\boldsymbol{w}\|} \quad (18)$$

So that the total distance between the two hyperplanes is $\frac{2}{\|\boldsymbol{w}\|}$. The biggest margin is obtained by maximizing $\frac{2}{\|\boldsymbol{w}\|}$ which is equivalent to minimizing $\|\boldsymbol{w}\|^2$ [17]. So, the optimization problem from SVM is as follows:

$$\min \frac{1}{2}\|\boldsymbol{w}\|^2 \quad (19)$$

Subject to

$$y_i\left(\boldsymbol{w}^T\boldsymbol{x} + b\right) \geq 1, i = 1, 2, \ldots, n$$

If training data is not linearly separated, then a slack ε_i variable can be added which is used as a misclassification of the noisy example. Adding slack variables will change the formula to the following [8]:

$$\min \frac{1}{2}\|w\|^2 + C\sum \varepsilon_i \quad (20)$$

Subject to

$$y_i\left(\boldsymbol{w}^T\boldsymbol{x}_i + b\right) \geq 1 - \varepsilon_i \quad (21)$$

and

$$\varepsilon_i \geq 0 \ \forall i = 1, 2, \ldots, n$$

If $\varepsilon > 1$, there will be misclassification at that point. There is a parameter C that is used to avoid overfitting. This is referred to as the soft margin classification. This problem is an optimization problem with objective functions in the form of quadratic programming problems and linear inequality problems [17]. This problem can be solved by changing the SVM optimization problem to a dual problem using the Lagrange duality theorem [17]. The following is the formula:

$$g(x) = sgn\left(\sum_{i=1}^{r} y_i\sigma_i K(x_i, x) + a^*\right) \quad (22)$$

$$\text{s.t.}\ 0 \leq \sigma_i \leq C \quad (23)$$

where σ_i is the Lagrange duality solved by the quadratic optimization problem, a^* shows the optimum bias value, and $K(x_i, x)$ is the kernel function which is expressed as:

$$K\left(x_i, x_j\right) = exp\left(-\frac{\|x_i - x_j\|^2}{\sigma^2}\right) \quad (24)$$

Where, the kernel function used in this study is the kernel Radial Basis Function (RBF).

3 Data

Data on cerebral infarction was obtained from a database of ischemic stroke patients available at the Department of Radiology Dr. Cipto Mangunkusumo's Hospital, Jakarta, Indonesia. The data are from January to November 2018. The cerebral infarction data is 206 data with 7 features. Data validation is used with 5-fold cross validation, so the dataset will be divided into 5 parts. In sequence, data for each section will be used as 30% test data and 70% training data. Class data indicated by infarction is labeled '1' in the dataset, while the normal data class is labeled '0' in the dataset. Table 1 is part of the cerebral infarction data to be examined and Table 2 describes the definitions of the data features used.

Table 1. The example of cerebral infarction dataset

Area (cm^2)	Min	Max	Average	SD	Sum	Length (cm)	Target
0.2	−3	38	16.88	9.3	5166	2	1
0.1	15	44	30.64	7.37	7722	1.8	0
0.2	−5	51	19.19	12.44	4797	1.9	1
0.1	18	51	32.29	7.84	8136	1.8	0
0.1	−14	26	8.67	10.25	824	1.2	1
0.1	25	61	38.99	7.37	6122	1.5	0

Table 2. The features of cerebral infarction dataset

No	Feature	Definition of feature
1	Area	The size of the area from the infarction point
2	Min	Minimum value of infarction
3	Max	Maximum value of infarction
4	Average	Average value of infarction
5	SD	Standard error value of infarction
6	Sum	Sum value of infarction point
7	Length	Length of infarction point

4 Experiment Result

In evaluating the performance of the clustering method and classification in this study, evaluations were based on parameters such as accuracy, precision, recall and F1-measure. Where the formula in finding them is as follows.

$$Accuracy\ of\ Classification = \frac{(TN + TP)}{(FN + TP + FP + TN)} \times 100\% \quad (25)$$

$$Recall = \frac{TP}{(FN + TP)} \times 100\% \quad (26)$$

$$Precision = \frac{TP}{(FP + TP)} \times 100\% \quad (27)$$

$$F1 = 2 \times \frac{Precision \times Recall}{Precision + Recall} \times 100\% \quad (28)$$

The greater the value of accuracy, recall, precision and f1 score, the better the method used. And, the best classifiers have a f1 score that approaches 1. While the worst classifiers have a f1 score that approaches 0.

This research used a software program Python 3.6. Table 3 are the results of the accuracy of the overall methods used, there are FCM, FPCM, RBFPCM, and SVM.

Table 3. Accuracy of each method

	Accuracy	Data training	Data testing
FCM	0.841379310345	70%	30%
FPCM	0.88275862069	70%	30%
RBFPCM	0.91666767677	70%	30%
SVM	0.855172413793	70%	30%

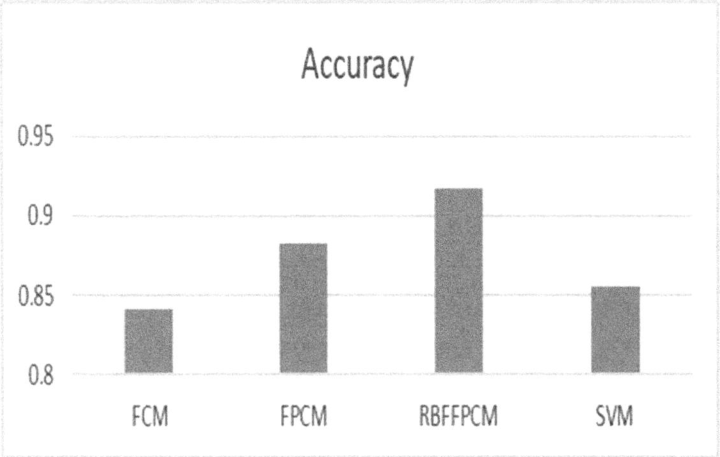

Fig. 1. Accuracy of each method

Based on the results of Table 3 and Fig. 1, the best accuracy obtained is 0.91666767677 which is when the RBFFPCM method is used. Meanwhile, the FCM method has the lowest level of accuracy among the others at 0.841379310345.

Table 4 shows the overall performance of the clustering model and the classification used.

Table 4. Classification report for each method

	Accuracy	Precision	Recall	f1 score
FCM	0.841379310345	0.84210526	0.85333333	0.84497041
FPCM	0.88275862069	0.85	0.93150685	0.88820225
RBFFPCM	0.91666767677	0.22480622	0.87878787	0.44846555
SVM	0.855172413793	0.92424242	0.79220779	0.85581395

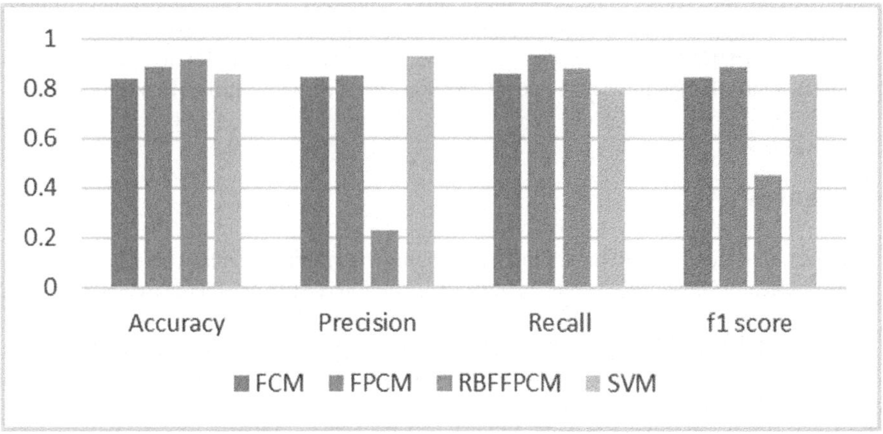

Fig. 2. Classification report for each method

Based on the Table 4 and Fig. 2, the FPCM method had better performance than the other methods used because it has a recall value of 0.93150685 followed by the RBFFPCM and FCM methods. However, based on the value of precision, the SVM method had better performance than the other methods used with a result of 0.92424242. Because the results of the recall and precision values get different the better methods, we can see the results of the f1 score. Based on the results of f1 score, the FPCM method had better performance than the other methods used with a result of 0.88820225.

5 Discussion

The results of the experiment show that the performance of the kernel based fuzzy clustering consisting of RBFFPCM and FPCM are the best method in the classification of cerebral infarction data in the human brain, can be seen from the high accuracy value. The method is compared with FCM and SVM. However, when viewed from the high recall, precision and f1 score values, the FPCM method is the best method.

6 Conclusion

In this study, we analyze the 70% training data and 30% testing data of cerebral infarction data used the kernel based fuzzy clustering method, there are radial base function fuzzy possibilistic c-means (RBFFPCM) and fuzzy possibilistic c-means (FPCM) which are generalizations of fuzzy c-means, possibilistic c-means method and additional kernel functions. The addition of kernel functions in the fuzzy clustering method can improve the accuracy of classification data without increasing too much computing costs. In addition, support vector machines are also used as a classification method. SVM is a greater classification method by finding the best hyperplane between data classes. In this paper, we compared the RBFFPCM with FPCM, FCM and SVM. The results shows that RBFFPCM is the best method than the other for analyze the cerebral infarction dataset with 91% accuracy value.

Acknowledgments. This work was financially supported by The Indonesian Ministry of Research and Higher Education, under Grant PDUPT 2019 (ID number NKB-1621/UN2.R3.1/HKP05.00/2019). This work supported by Department Radiology of Dr. Cipto Mangunkusumo's Hospital. We thank to all reviewers for the improvement of this article.

References

1. Kementrian Kesehatan Republik Indonesia. http://www.depkes.go.id. Accessed 7 Mar 2019
2. World Health Organization (WHO). https://who.int/topics/stroke/cerebrovascular/. Accessed 7 Mar 2019
3. Mentari, I.A., Naufalina, R., Rahmadi, M., Khotib, J.: Development of ischemic stroke model by right unilateral common carotid artery occlusion (RUCCAO) Method. Fol Med Indones **54**(3), 200–206 (2018)
4. Bay, V., Kjolby, B.F., Iversen, N.K., Mikkelsen, I.K., Ardalan, M., Nyengaard, J.R., Jespersen, S.N., Drasbek, K.R., Stergaard, L., Hansen, B.: Stroke infarct volume estimation in fixed tissue : comparison of diffusion kurtosis imaging to diffusion weighted imaging and histology in a rodent MCAO model. PLoS ONE **13**(4), e0196161 (2018)
5. Havens, T.C., Bezdek, J.C., Leckie, C., Palaniswami, M.: Fuzzy c-means algorithms for very large data. IEEE Trans. Fuzzy Syst. **20**(6), 1130–1146 (2012)
6. Pal, N.R., Pal, K., Keller, J.M., Bezdek, J.C.: A possibilistic fuzzy c-means clustering algorithm. IEEE Trans. Fuzzy Syst. **13**(4), 517–530 (2005)
7. Izakian, H., Abraham, A.: Fuzzy C-means and fuzzy swarm for fuzzy clustering problem. Expert Syst. Appl. **38**, 1835–1838 (2011)
8. Kannan, S.R., Devi, R., Ramathilagam, S., Hong, T.P.: Effective fuzzy possibilistic c-means: an analyzing cancer medical database. Soft. Comput. **21**, 2835–2845 (2017)
9. Dsouza, K.J., Ansari, Z.A.: Experimental exploration of support vector machine for cancer cell classification. In: IEEE International Conference on Cloud Computing in Emerging Markets (2017)
10. Saad, M.F., Salah, M., Lee, J., Kwon, O.: A modified fuzzy possibilistic C-means for context data clustering toward efficient context prediction. In: New Challenges for Intelligent Information SCI, vol. 351, pp. 157–165 (2011)
11. Zhang, C., Zhou, Y., Martin, T.: Similarity based fuzzy and possibilistic c-means algorithm. In: Proceedings of the 11th Joint Conference on Information Sciences (2008)

12. Kannan, S.R., Devi, R., Ramathilagam, S., Hong, T.P., Ravikumar, A.: Robust fuzzy clustering algorithms in analyzing high dimensional cancer databases. J. Appl. Soft Comput. **35**, 199–213 (2015)
13. Rustam, Z., Talita, A.S.: Fuzzy Kernel C-means algorithm for intrusion detection systems. J. Theor. Appl. Inf Technol. **81**(1), 161 (2015)
14. Rustam, Z., Talita, A.S.: Fuzzy Kernel K-medoids algorithm for multiclass multidimensional data classification. J. Theor. Appl. Inf Technol. **80**(1), 147 (2015)
15. Liu, J., Zio, E.: Integration of feature vector selection and support vector machine for classification of imbalanced data. Appl. Soft Comput. J. **75**, 702–711 (2017)
16. Mathew, J., Pang, C.K., Luo, M., Leong, W.H.: Classification of imbalanced data by oversampling in kernel space of support vector machine. IEEE Trans. Neural Netw. Learn. Syst. **29**(9), 4065–4076 (2018)
17. Wang, H., Zheng, B., Yoon, S.W., Ko, H.S.: A support vector machine-based ensemble algorithm for breast cancer diagnosis. Eur. J. Oper. Res. **267**, 687–699 (2018)

A Predator-Prey Model with Fear Factor, Allee Effect and Periodic Harvesting

Dipo Aldila(✉) and Padma Sindura Adhyarini

Department of Mathematics, Universitas Indonesia,
Kampus UI Depok, Depok 16424, Indonesia
aldiladipo@sci.ui.ac.id

Abstract. A two-dimensional predator-prey model constructed in this article to study how the fear of predator on prey will reduce the intrinsic growth rate of prey. The Allee effect in prey population is included to describe the minimum requirement of the individual to achieve the positive growth of the prey population. Anti-predation and harvesting on the predator population are also considered in the model. We non-dimensionalized the model first before analyzing the existence and the local stability criteria of all equilibrium points. The analytical results showed how the harvesting intervention would determine the local stability criteria of the coexistence equilibrium point. We found that the time scale separation does not impact the stability of equilibrium points. Some numerical simulation is given to show how the system behaves depend on the periodic harvesting interventions.

Keywords: Predator · Prey · Allee effect · Fear factor · Anti-predation · Periodic harvesting

1 Introduction

There are many kinds of interaction between two species, such as commensalism symbiosis, mutualism symbiosis, parasitism symbiosis, prey-predator interaction, and many more. In the predator-prey interactions, prey behavior is one of many important factors in the predator environment. The predator may die when there's no prey as their foods. Since the first time introduced by Alfred J. Lotka in 1910 [1] and Vito Volterra in 1926 [2], many kinds of a mathematical model for predator-prey are studied, for instance in [3–5].

Several factors should be considered in developing the predator-prey model, for instance, the fear of predator on prey [6,7]. When the predator existence becomes a threat for the prey, they will develop their self-defense, alert their traits, and so on. The increasing of the fear of predator on prey will lead to a decreasing in the growth rate of the prey population.

Supported by Universitas Indonesia with PITTA Research Grant, ID number: NKB-0627/UN2.R3.1/HKP.05.00/2019.

M. Zeki Sarıkaya et al. (Eds.): ICMRS 2019, LNNS 123, pp. 12–23, 2020.
https://doi.org/10.1007/978-3-030-43002-3_2

Another critical factor is the Allee effect phenomena. This is a phenomenon in biology/ecology where the population need a minimum density to growth and known as the Allee coefficient. Several mathematical models have been studied the effect of this phenomena in dynamical population models [8–10].

Another factor that needs to be considered is the anti-predation of the prey population. The anti-predation behavior might appear from at least two points of view, i.e., morphological change of behavior [11,12] or when the prey attack the predator [13,14]. In the anti-predation model, the adult prey not only hunts the adult predator but sometimes they also hunt to kill the juveniles predators.

Based on the previous facts, here in this paper, we construct a predator-prey model involving the above factors: fear-factor, Allee-effect, anti-predation, and also harvesting rate in the predator population. To make the research closer to real-life situations, the harvesting rate will be simulated as a time-dependent periodic function.

This paper is organized as follows. After a general explanation of the paper in this section, in Sect. 2, we present the construction of the mathematical model. Section 3 is devoted to analyzing the model concerning the existence and local stability of equilibrium points. The phase portrait simulation also is given in this section. In Sect. 4, some numerical simulations will be given to see how fear factor, periodic harvesting rate, and the time scale separation affect the dynamic of the model. In Sect. 5, some conclusion will be given.

2 Mathematical Model

In this section, we consider a time-dependent of harvesting rate to control the dynamic of our predator-prey interaction model. Let X and Y be the prey and the predator population, respectively. We assume that the prey population grows according to the logistic population with an intrinsic growth rate of r and carrying capacity K. Therefore, we have

$$\frac{dX}{dt} = rX\left(1 - \frac{X}{K}\right).$$

Next, we assume that there is a fear of predator on prey, which can reduce the per capita growth rate of the prey population. This fear factor is modeled as a reduction factor of the logistic growth rate given by $F(f) = \frac{1}{1+fY}$, where f is the level of fear. We also assume that there are Allee effects in the prey population. It implies the requirement of a minimum number of the prey population to guarantee the increasing of the prey population. If we assume that the Allee coefficient as θ, then the dynamic of the prey population without the predator population will be given by:

$$\frac{dX}{dt} = rX\left(1 - \frac{X}{K}\right)(X - \theta)\frac{1}{1+fY}.$$

The predator population dynamic without the existence of the prey population is considered by following the negative exponential growth model and decreasing by the time-dependent harvesting intervention, i.e.

$$\frac{dY}{dt} = -mY - h(t)Y,$$

where m and $h(t)$ are natural death rate and periodic harvesting rate, respectively. We model the predation term with a first type functional response with a as the maximum uptake of predation for predator population. We also assume that there is an anti predation from the prey population which might reduce the growth of predator population, with the rate of β. With the above assumptions, the model is given by:

$$\frac{dX}{dy} = rX\left(1 - \frac{X}{K}\right)(X - \theta)\left(\frac{1}{1 + fY}\right) - aXY \tag{1a}$$

$$\frac{dY}{dt} = a\alpha XY - \beta XY - mY - h(t)Y, \tag{1b}$$

where α is the energy conversion parameter from predation.

In the next section, we will study the model 1 with the assumption of a constant harvesting rate. We non-dimensionalized the model 1 to analyze the existence and local stability criteria of all equilibrium points.

3 Model Analysis

In this section, we simplify our model in system 1 to reduce the number of parameters, and then followed with equilibrium existence and local stability analysis in the next section.

3.1 Non-dimensionalization of the Model

Assume that $x = \frac{X}{K}$, $y = \frac{a}{rK}Y$ as the non dimensional population size of prey and predator, respectively. Also, let us assume that $\theta = \frac{\theta}{K}$, $m = \frac{m}{a\alpha K}$, $f = \frac{rKf}{a}$, and $t = a\alpha Kt$ and substitute it into the model 1 to obtain:

$$\frac{dx}{dt} = \frac{1}{\epsilon}\left(x(1 - x)(x - \theta)\frac{1}{1 + fy} - xy\right) \tag{2a}$$

$$\frac{dy}{dt} = xy(1 - \beta) - (m + h)y, \tag{2b}$$

where ϵ is the time separation parameter between prey and predator. In the next section, we will analyze this model instead of a model in Eq. 1.

3.2 Equilibrium Points and Their Stability

Next, we discuss the equilibrium point of model 2. Taking the right-hand side of model 2, we have four different equilibrium point.

1. Trivial equilibrium $\Gamma_0 = (x_0, y_0) = (0, 0)$.
   The first equilibrium is the trivial equilibrium which describe the extinction of predator and prey population. To analyzed the local stability, we evaluate Γ_0 in their Jacobian matrix which given by:

$$\mathcal{J}_0 = \begin{bmatrix} -\frac{\theta}{\epsilon} & 0 \\ 0 & -m-h \end{bmatrix}. \tag{3}$$

   Since all the diagonal elements are negative, it is trivial that this equilibrium will always locally asymptotically stable for all condition of parameters. This result written in the following theorem.

Theorem 1. *The predator-prey model 2 has a trivial equilibrium Γ_0 which always locally asymptotically stable (LAS).*

2. Equilibrium $\Gamma_1 = (x_1, y_1) = (1, 0)$.
   The second equilibrium point is the equilibrium when only prey population exist, while the extinction occur in predator population. Similar with previous analysis, we analyze the local stability of Γ_3 which give us:

$$\mathcal{J}_1 = \begin{bmatrix} \frac{-1+\theta}{\epsilon} & -\epsilon^{-1} \\ 0 & 1-\beta-h-m \end{bmatrix}. \tag{4}$$

   Since we have $\theta < 1$, then $-1+\theta < 0$. The other eigenvalue, i.e $\lambda = 1-\beta-h-m$ will be negative if $\beta+h+m \geq 1$. The stability criteria of Γ_1 is given in the following theorem.

Theorem 2. *The equilibrium point Γ_1 of the predator prey model 2 is:*

(a) Saddle if $\beta+h+m < 1$,
(b) Stable node if $\beta+h+m \geq 1$.

Proof. The eigenvalues of $\mathcal{J}_1$ are $\lambda_1 = \frac{-1+\theta}{\epsilon}$ which always negative, and $\lambda_2 = 1-\beta-h-m$ which can be positive or negative. If we have $\beta+h+m < 1$, we have $\lambda_2 > 0$ which make Γ_1 become saddle. In the other hand, if $\beta+h+m \geq 1$, we have $\lambda_2 \leq 0$ which make Γ_1 LAS.

3. Equilibrium $\Gamma_2 = (x_2, y_2) = (\theta, 0)$.
   The third equilibrium point is the Allee equilibrium point, when predator population extinct and prey population is in the Alley coefficient size. The Jacobian matrix of Γ_2 is given by:

$$\mathcal{J}_2 = \begin{bmatrix} \frac{\theta(1-\theta)}{\epsilon} & -\frac{\theta}{\epsilon} \\ 0 & \theta(1-\beta)-m-h \end{bmatrix}. \tag{5}$$

Since we have $\theta \in (0, 1)$, we have one positive eigenvalue, i.e $\frac{\theta(1-\theta)}{\epsilon}$. The other eigenvalue i.e $\lambda_2 = \theta(1-\beta)-m-h$ will determine the stability of Γ_2. This result is concluded in the following theorem.

Theorem 3. *The equilibrium point Γ_2 of the predator prey model 2 is always unstable, specifically*

(a) Saddle if $\frac{\theta(1-\beta)}{h+m} < 1$,
(b) Unstable node if $\frac{\theta(1-\beta)}{h+m} \geq 1$.

Proof. Similar with the proof of Theorem 2, the stability type of Γ_2 fully determined by the second eigenvalue, since the first eigenvalue already positive. The second eigenvalue, $\lambda_2 = \theta(1-\beta) - m - h$ will negative if $\frac{\theta(1-\beta)}{h+m} < 1$. Therefore, we have that Γ_2 is a saddle equilibrium point if $\frac{\theta(1-\beta)}{h+m} < 1$. In the other hand, $\lambda_2 > 0$ if $\frac{\theta(1-\beta)}{h+m} \geq 1$. Therefore, we have that Γ_2 is an unstable node equilibrium point if $\frac{\theta(1-\beta)}{h+m} < 1$.

4. Equilibrium $\Gamma_3 = (x_3, y_3) = (x_3^*, y_3^*)$.
   The last equilibrium point is the interior equilibrium point which given by $\Gamma_3 = (x_3, y_3) = \left(\frac{h+m}{1-\beta}, y_3^*\right)$, where y_3^* is the positive root of the second degree polynomial given by:

$$y_3^2 + \frac{1}{f}y_3 + \frac{1}{f(1-\beta)^2}(\beta + h + m - 1)(\theta(\beta - 1) + h + m) = 0. \qquad (6)$$

   Since we have the coefficient of y_3^2 and y_3 is already positive, we only could have a unique positive root of Eq. 6 i.e when $\frac{1}{f(1-\beta)^2}(\beta + h + m - 1)(\theta(\beta - 1) + h + m) < 0$. This result is given in the following theorem.

Theorem 4. *The predator-prey model in 2 has a unique interior equilibrium point Γ_3 whenever $\theta < \frac{h+m}{1-\beta} < 1$.*

Proof. The proof is directly from another algebraic form of $(\beta+h+m-1)(\theta(\beta-1) + h + m) < 0$.

Next we will analyze the stability criteria of this equilibrium. First, to simplify the form of this equilibrium, let $m_1 = \frac{h+m}{1-\beta}$, $x_3 = \frac{h+m}{1-\beta}$, and $y_3 = \frac{-1+(1-\beta)\sqrt{1+4f\left(1-\frac{h+m}{1-\beta}\right)\left(\frac{h+m}{1-\beta}-\theta\right)}}{2f}$. Evaluate $\Gamma_3 = (x_3, y_3)$ in the Jacobian matrix gave us:

$$\mathcal{J}_3 = \begin{bmatrix} A & B \\ C & 0 \end{bmatrix} = \begin{bmatrix} -\frac{f{y_3}^2 - 2\,\theta\,x_3 + 3\,{x_3}^2 + \theta - 2\,x_3 + y_3}{\epsilon\,(f y_3 + 1)} & \frac{x_3\left(-f^2{y_3}^2 - f\theta\,x_3 + f{x_3}^2 + f\theta - f x_3 - 2\,f y_3 - 1\right)}{\epsilon\,(f y_3 + 1)^2} \\ -y_3\,(\beta - 1) & 0 \end{bmatrix}$$

with the characteristic polynomial of the eigenvalues is given by:

$$\lambda^2 + A\lambda + BC = 0. \qquad (7)$$

Therefore, the stability criteria of Γ_3 determined by the sign of A, BC and the sign of the discriminant $A^2 - 4BC$. We have that Γ_3 is:

(a) Stable spiral if $A > 0$ and $A^2 - 4BC < 0$
(b) Stable node if $A > 0$ and $A^2 - 4BC \geq 0$
(c) Unstable node if $A < 0$ and $BC > 0$
(d) Unstable spiral if $A < 0$ and $A^2 - 4BC < 0$
(e) Stable focus if $A = 0$ and $BC > 0$.

Proof. The real part of the eigenvalues of $\mathcal{J}_3$ is given by the sign of A, that is it will positive when $A < 0$, negative when $A > 0$ and zero when $A = 0$. Therefore, Γ_3 is asymptotically stable when $A > 0$, unstable when $A < 0$ and stable focus if $A = 0$. The kind of behavior of the stability, whether it is node or not is determined by the sign of the discriminant.

3.3 Phase Portrait of the Model

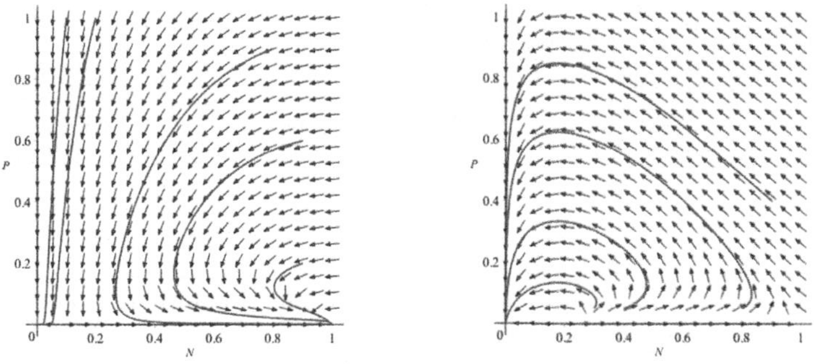

(a) (b)

Fig. 1. The phase portrait of system 2 when the predator extinction is stable (a) when $\beta = 0.2, h = 0.7, m = 0.2$ and unstable (b) when $\beta = 0.1, h = 0.01, m = 0.15$.

From the previous subsection, it can be seen that the predator population will extinct when $\beta + h + m \geq 1$ for every initial condition in $\mathbb{R}^2_+$. Please see Fig. 1(a) for this situation. On the other hand, Fig. 1(b) depicts how the extinction of the predator population will follow by the extinction of the prey population, when $\beta + h + m < 1$. This condition shows that the uncontrolled harvesting rate in predator population may cause the extinction of predator and prey population. It also tells us, when the predator is getting easier to be killed in the predation process, their extinction became a certain condition.

Next, in Fig. 2, we show three different dynamics around the coexistence equilibrium. Figure 2(a) and (b) show the stable and unstable spiral of the coexistence equilibrium. Figure 2(b) shows that all initial condition in $\mathbb{R}^2_+$ will be pushed away from all equilibrium points to the extinction of all population equilibrium point. On the other hand, Fig. 2(a) shows a bi-stability condition, that is, the convergence point of the trajectories of system 2 will depend on the initial condition, whether it converges to the coexistence equilibrium point, or in the extinction of all population. When the predator population in the initial condition is much larger than the tolerance of the number of prey population, both population will tend to extinction condition. Figure 2(c) shows a stable limit cycle when $\beta = 0.01, h = 0.01, m = 0.57, E_* = (0.5858, 0.1401)$. This situation means that all initial conditions that are exterior of the stable limit cycle will be attracted to the stable limit cycle. On the other hand, the initial condition that is interior of the limit cycle will behave as a stable center equilibrium. Therefore we have the Hopf-bifurcation appear in this condition, and the Hopf-bifurcation is sub-critical.

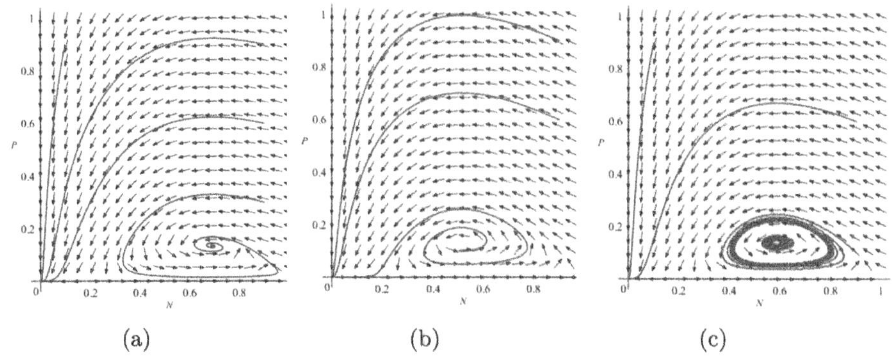

Fig. 2. The existence of coexist equilibrium. (a) stable equilibrium when $\beta = 0.2, h = 0.3, m = 0.25$, (b) unstable equilibrium when $\beta = 0.1, h = 0.31, m = 0.15$, and (c) stable limit cycle when $\beta = 0.01, h = 0.01, m = 0.57$.

4 Autonomous Simulations

In this section, some numerical simulations will be performed based on the analytical result in the previous article. We also investigate the effect of the time scale, fear factor, and harvesting intervention on the dynamical behavior of the system 2.

4.1 Effect of Fear Factor

From the previous analytical results, we found that the cost of the fear parameter (f) does change the size of the equilibrium point in the predator population, but not in the prey population. In the following simulation, we choose $\epsilon = 1, \theta = 0.2, m = 0.64, h = 0.01, beta = 0.1$ and varying f with $0, 10$ and 50. We have that the prey population is 0.72 while the predator population is 0.14, 0.08 and 0.045 for $f = 0, 10$ and 50, respectively. The larger f will reduce the predator size in the equilibrium. In the Fig. 3, we show how the fear factor reduces the equilibrium size of the predator. The blue, red and black curve indicate $f = 0, 10$ and 50, respectively.

4.2 Effect of Harvesting Rate

In this section, we will see how harvesting the prey population may change the dynamic of our model. First, we use same parameters value: $\epsilon = 1, \theta = 0.2, m = 0.64, f = 0.1, \beta = 0.1$ with varying h with $0, 0.06$ and 0.8. The coexistence equilibrium is exist when $h = 0$ or $h = 0.06$ while the predator population extinct when the harvesting intervention is uncontrolled ($h = 0.8$). The dynamic of this scenario can be seen in Fig. 4.

Our next simulation aims to see how the periodic harvesting rate affects the dynamic of our model. All parameters value are the same as in Fig. 4, except for

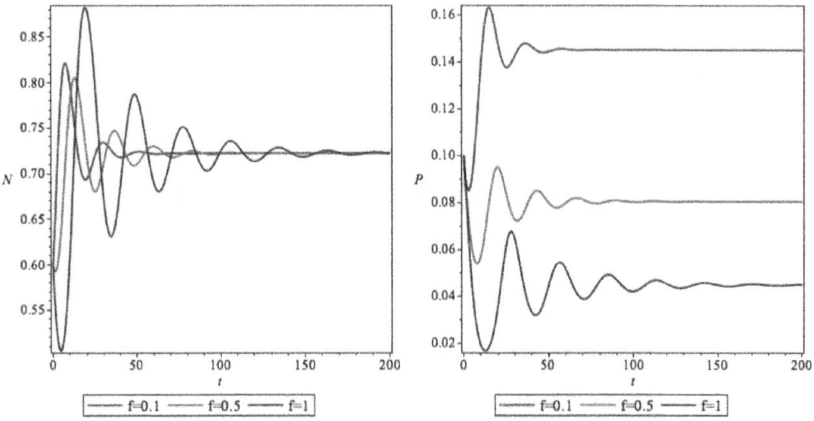

(a) (b)

Fig. 3. The dynamic of prey (a) and predator (b) population with various f

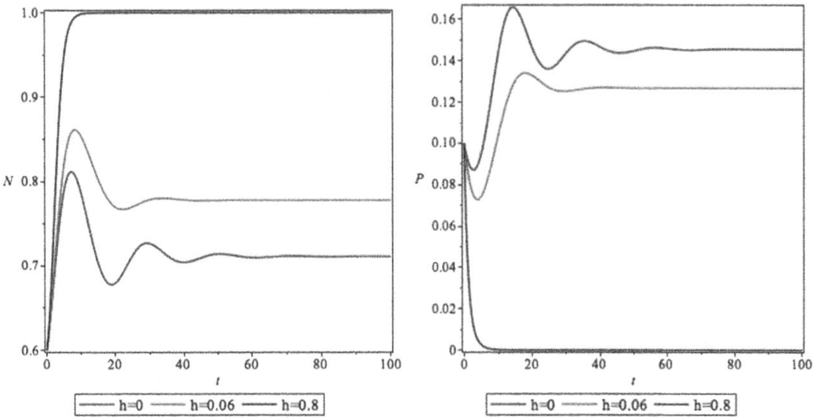

(a) (b)

Fig. 4. The dynamic of prey (a) and predator (b) population with various h

the harvesting rate. First scenario is when we set $h = 0.066$ constant all the time, $h = 0.24$ every 25 time unit and 0 in the next 25 time interval, and $h = 0.12$ every 50 time unit and 0 in the next 50 time interval. We use Fourier series to transform the piecewise harvesting rate in the 2nd and 3rd scenario, which gave us the following equations:

$$h_2 = 0.13 + 0.13 * \sin(\frac{1}{25} * \pi * t) + 0.04 * \sin(\frac{3}{25} * \pi * t) + 0.02 * \sin(\frac{5}{25} * \pi * t) + 0.019 * \sin(\frac{7}{25} * \pi * t)$$

$$h_3 = 0.06 + 0.06 * \sin(\frac{1}{50} * \pi * t) + 0.02 * \sin(\frac{3}{50} * \pi * t) + 0.01 * \sin(\frac{1}{10} * \pi * t) + 0.009 * \sin(\frac{7}{50} * \pi * t).$$

From Fig. 5, a longer interval for harvesting intervention gives a better result since the 3rd scenario provides more time for the predator population to reach its initial equilibrium point.

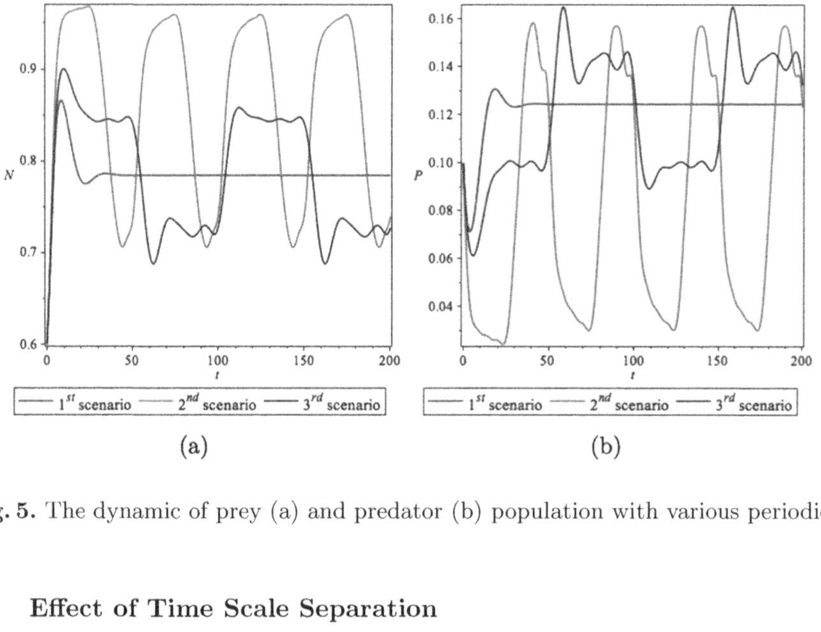

(a) (b)

Fig. 5. The dynamic of prey (a) and predator (b) population with various periodic h

4.3 Effect of Time Scale Separation

From the previous analysis, the non-dimensionalization process generates the new parameter, ϵ, as the time scale parameter. The parameter ϵ describes how predator and prey have a different life cycle with separate life spans. $\epsilon = 1$ indicates the predator and the prey have the same life cycle. If $\epsilon < 1$, then prey have a faster life cycle than predator. $\epsilon > 1$ is defined conversely. We have shown analytically that ϵ does not change the equilibrium and their local stability condition. Although, the smaller ϵ will accelerate the system 2 to reach the stability of each equilibrium points. To see this result, we assume the constant parameters $\theta = 0.2, f = 1, m = 0.64, h = 0.01, \beta = 0.1$ while ϵ varying. With these parameters, the system 2 has a stable interior-point $\Gamma_3 = (0.722, 0.128)$. Please see Fig. 6 for this condition. When ϵ is getting smaller, the prey and predator population reach the stable Γ_3 faster.

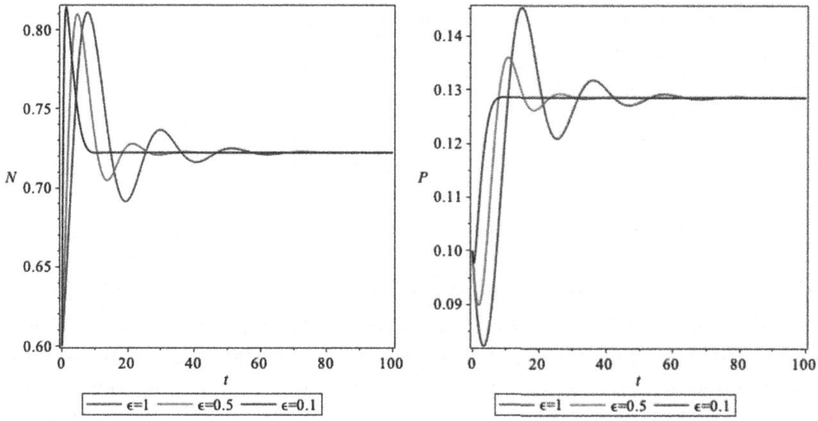

(a) (b)

Fig. 6. We use $\theta = 0.2, f = 1, m = 0.64, h = 0.01, \beta = 0.1, (N(0), P(0)) = (0.6, 0.1)$ while ϵ vary. It can be seen varying ϵ does not change the stability of the interior point both for prey (a) or predator (b) population.

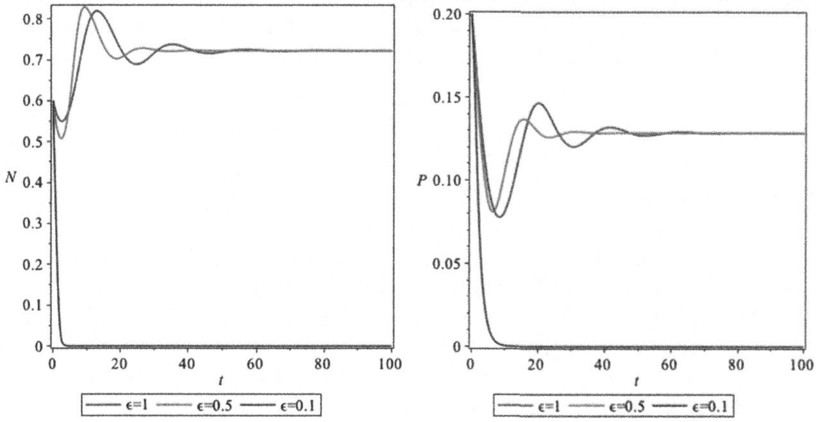

(a) (b)

Fig. 7. We use same parameter value as in Fig. 6 but change the initial condition, i.e $(N(0), P(0)) = (0.6, 0.2)$. It can be seen that even varying ϵ does not change the stability of the interior point, but the neighborhood of local stability of the interior point changed in prey (a) and predator (b) population.

Next, we use same parameter value as in Fig. 6, except the initial condition $(N(0), P(0)) = (0.6, 0.2)$ as in Fig. 7. With the same initial condition for each scenario and set $\epsilon = 0.1$, the dynamic of predator and prey population does not tend to the interior equilibrium point. On the contrary, the system 2 goes to the extinction equilibrium point Γ_0.

With these result, we can conclude that although ϵ does not change the condition of the local stability of all equilibrium point(s), it does change the quickness of system 2 to reach the equilibrium point and also shift the equilibrium points neighborhood which can make the initial condition goes to which stable equilibrium point.

5 Conclusions

Here in this paper, we investigated how fear-factor, anti-predation, Allee-effect, and harvesting intervention affect the dynamic of a predator-prey model. We examined the existence and local stability criteria of the equilibrium points. We find that the fear-factor in prey population does not change the coexistence size of prey population, but only in the predator population. We also find that harvesting in an uncontrolled situation might lead to the extinction of the predator population. From time scale separation analysis, we find that even though time scale does not change the existence and local stability of coexistence equilibrium, but it does change the neighborhood of each equilibrium to be locally stable.

Acknowledgments. We thank to all reviewers for their valuable comments. This research is funded by Universitas Indonesia with PITTA research grant Scheme, 2019 (ID number: NKB-0627/UN2.R3.1/HKP.05.00/2019).

References

1. Lotka, A.J.: Contribution to the theory of periodic reaction. J. Phys. Chem. **14**(3), 271–274 (1910). https://doi.org/10.1021/j150111a004
2. Volterra, V.: Variazioni e fluttuazioni del numero d'individui in specie animali conviventi. Mem. Acad. Lincei Roma. **2**, 31–113 (1926)
3. Putri, A., Aldila, D.: Prey-predator interaction model between herring and seals with phocine distemper virus (PDV) infection in seals. In: AIP Conference Proceedings 2023, vol. 020239 (2018). https://doi.org/10.1063/1.5064236
4. Triharyuni, S., Aldila, D.: A mathematical model of predator-prey interaction between seal-herring and steelhead trout. In: AIP Conference Proceedings 1862, vol. 030152 (2017). https://doi.org/10.1063/1.4991256
5. Sayekti, I.M., Malik, M., Aldila, D.: One-prey two-predator model with prey harvesting in a food chain interaction. In: AIP Conference Proceedings 1862, vol. 030124 (2017). https://doi.org/10.1063/1.4991228
6. Wang, X., Zanette, L., Zou, X.: Modelling the fear factor in predator-prey interactions. J. Math. Biol. **73**(5), 1179–1204 (2016). https://doi.org/10.1007/s00285-016-0989-1
7. Zanette, L.Y., White, A.F., Allen, M.C., Clinchy, M.: Perceived predation risk reduces the number of offspring songbirds produce per year. Science **334**, 1398–1401 (2011). https://doi.org/10.1126/science.1210908
8. Gonzalez-Oliviars, E., Mena-Lorca, J., Rojas-Palma, A., Flores, J.D.: Dynamical complexities in the Leslie-Gower predator-prey model as consequences of the Allee effect on prey. Appl. Math. Model. **35**, 366–381 (2011). https://doi.org/10.1016/j.apm.2010.07.001

9. Rojas-Palma, A., Gonzales-Olivares, E.: Optimal harvesting in a predator-prey model with Allee effect and sigmoid functional response. Appl. Math. Model. **36**, 1864–1874 (2012). https://doi.org/10.1016/j.apm.2011.07.081
10. Li, T., Huang, X., Xie, X.: Stability of a stage-structured predator-prey model with Allee effect and harvesting. Commun. Math. Biol. Neurosci. **2019**, 13 (2019)
11. Lima, S.L.: Stress and decision making under the risk of predation: recent developments from behavioral, reproductive, and ecological perspectives. Adv. Study Behav. **27**, 215–290 (1998)
12. Relyea, R.A.: How prey respond to combined predators: a review and an empirical test. Ecology **84**, 1827–1839 (2003)
13. Saito, Y.: Prey kills predator: counter-attack success of a spider mite against its specific phytoseiid predator. Exper. Appl. Acarol. **2**, 47–62 (1986)
14. Tang, B., Xiao, Y.: Bifurcation analysis of a predator-prey model with anti predator behaviour Chaos. Chaos, Solitons Fractals **70**, 58–68 (2015)

Mathematical Modeling of Rock Massif Dynamics Under Explosive Sources of Disturbances

G. K. Zakir'yanova(✉)

Institute of Mechanics and Mechanical Engineering, Almaty, Kazakhstan
gulmzak@mail.ru

Abstract. Mathematical models are developed for the dynamics of the rock mass during wave propagation using the model of anisotropic elastic media. For such media by use of the apparatus of generalized functions theory Green's tensor are received. Based on the constructed Green's tensor the fundamental stress tensor are obtained. For action any mass forces distributed in the medium the generalized solutions are also obtained. The influence of the degree of anisotropy of the medium on the character of the stress-strain state of the rock mass in the vicinity of various types of disturbance sources is studied. Some results of numerical calculations of the fundamental solutions, the fundamental stress tensor for various anisotropic elastic media are given. As acting loads, sources of perturbations of various types are considered, including impulse ones.

Keywords: Elastic medium · Anisotropy · Green's tensor · Stress-strain state · Generalized functions

1 Introduction

Investigation of the propagation and diffraction of waves in a rock massif under the influence of various sources of disturbances are the actual problems of mechanics and mathematical physics. The most economical and effective for solving of such problems are methods of mathematical modeling. The development of mathematical models of wave processes in bodies and medium for the purpose of the most adequate description of real physical processes is related to the construction of solutions of boundary value problems for hyperbolic systems of equations the solutions of which can have characteristic surfaces on which the solutions themselves, or their derivatives, are discontinuous [1, 2]. In physical processes, they describe shock waves, on the fronts of which the studied characteristics of the process (velocities, stresses, pressure, temperature, etc.) can have jumps.

In order to take into account the real properties of the rock massif, we consider an anisotropic elastic medium whose equations of motion are described by a strictly hyperbolic system of second-order equations with constant coefficients. The velocities of wave's propagation in this case depend on the type of deformation, which are determined

M. Zeki Sarıkaya et al. (Eds.): ICMRS 2019, LNNS 123, pp. 24–32, 2020.
https://doi.org/10.1007/978-3-030-43002-3_3

by the type of the applied load, and may be also from the direction of motion of the wave front. Green's tensors -fundamental solutions that satisfy the radiation conditions, and the stress tensor are constructed. The influence of the degree of anisotropy on the character of diffraction processes in anisotropic media is studied using real rock as an example of numerical experiments with the action of concentrated sources of various types that are characteristic of explosive influences. It is shown that the waveguide properties of the medium depend essentially on the elastic parameters. For weak anisotropy, the fronts of shock waves from the concentrated sources are ellipsoidal, similar to spherical in an isotropic elastic medium. For strong anisotropy, there are lacunae where the medium is at quiescence [3]. In the calculations, elastic media with different numbers of lacunae and different locations of there are considered. Based on the constructed Green's tensor, solutions of the equations for any mass forces distributed in the medium are presented.

2 Motion Equations

Consider an anisotropic elastic medium, the motion equations of which are described by a system of hyperbolic equations with derivatives of the second order of the form:

$$L_{ij}(\partial_x, \partial_t)u_j(x,t) + G_i(x,t) = 0,\ (x,t) \in R^{N+1} \tag{1}$$

$$L_{ij}(\partial x, \partial t) = C_{ij}^{ml}\partial_m\partial_l \ -\ \rho\delta_{ij}\partial_t^2,\quad i,j,m,l = \overline{1,N} \tag{2}$$

$$C_{ij}^{ml} = C_{ij}^{lm} = C_{ji}^{ml} = C_{ml}^{ij} \tag{3}$$

where $x = (x_1, \ldots, x_N) \in R^N$ ($N = 2$ corresponds to the plane deformation, $N = 3$ corresponds to the spatial case), $\partial_x = (\partial_1, \ldots, \partial_N)$, $\partial_i = \partial/\partial x_i$, $\partial_t = \partial/\partial t$, δ_{ij} is Kronecker symbol, G_i are the components of mass force vector, ρ is the density of the medium, $t \in R^1$. The matrix of elastic constants C_{ij}^{ml} has symmetry properties with respect to permutation of the indices (3) and satisfies the strict hyperbolicity condition

$$W(n,v) = C_{ij}^{ml}n_m n_l v^i v^j > 0\ \forall n \neq 0,\quad v \neq 0.$$

Assuming the summation over the repeated indexes in the above-mentioned limits of their variation in the product (similar to the tensor convolution), we omit the sum sign.

Solutions of Eq. (1) can have characteristic surfaces on which the solutions themselves or their derivatives are discontinuous. In physical problems, they describe shock waves, which is characteristic of external influences that have an impact character and are represented by discontinuous or singular functions. Conditions on the shock wave fronts are [4]

$$[u_i(x,t)]_{F_t} = 0 \ \Rightarrow [u_{i,t}\, n_l + c u_{i,l}]_{F_t} = 0 \tag{4}$$

$$\left[\sigma_{ij}n_j\right]_{F_t} = -[c u_{i,t}]_{F_t} \tag{5}$$

Here $\sigma_{ij} = C_{ij}^{ml}u_{m,l}$, $[f]_{F_t}$ is the jump of function f on F_t. The velocity of wave front motion c is defined by solving the characteristic equation of the system (1):

$$\det\left\{C_{ij}^{ml}\nu_m\nu_l - \rho c^2\delta_{ij}\right\} = 0, \tag{6}$$

where $(\nu, \nu_t) = (\nu_{1,\dots}\nu_N, \nu_t)$ is vector of characteristic normal connected with velocity c by relation $c = -\nu_t/\|\nu\|_{R^N}$, $\|\nu\| = \sqrt{\nu_j\nu_j}$. Equation (6) has (taking into account multiplicity) the $2N$ real roots: $c = \pm c_k(\nu)$, $0 < c_k \le c_{k+1}$, $k = \overline{1, N-1}$, having the meaning of the phase velocities for the harmonic analysis of the system (1) and depending on the direction of wave propagation in the general case. The first condition from (4) means continuity condition; condition (7) corresponds to the momentum conservation condition on the wave fronts, which relates the velocity jump on a wave front to the stress jump. Therefore, such a surface is called the shock wave front. In [5] conditions (4)–(5) were received by using the generalized functions theory.

3 Dynamical Green's Tensor and Fundamental Stress Tensor

For problems of mechanics, the problem of construction of fundamental solutions is complicated, since media movements are not described by a single equation, but by a system of equations. Therefore, it is necessary to build fundamental solutions (Green's tensors) for systems of equations of media, the order of which depends on the dimension of the solution space. Green's tensors can be used to receive the solution for various mass forces. And they are used to build of kernels of singular boundary integral equations for solving boundary value problems [6, 7]. The fundamental solutions for anisotropic elastic media have been studied in [8–16] and by other authors. Fundamental solutions are determined up to solutions of a homogeneous system of equations. A special place among them is occupied by the Green's tensor, which satisfies the conditions: $U_j^k(x,t) = 0\ for\ t \le 0$, $\|x\| \ \ge c_{\max} t$, c are roots of the characteristic equation.

To construct the Green's tensor, it is convenient to use the Fourier transform with parameters $(\xi, \omega) = (\xi_1, \dots, \xi_N, \omega)$, which brings the system (1) to a system of linear algebraic equations of the form

$$L_{ij}(i\xi, i\omega)\,\overline{U}_j^k(\xi, \omega) + \delta_i^k = 0, \tag{7}$$

Here $L_{ij}(\xi, \omega)$ are homogeneous second-order polynomials corresponding to the differential operators (2). Solving the system (7), we obtain the transform of the Green's tensor, which, by virtue of the homogeneity of the differential polynomials, has the form

$$U_{jk}(i\xi, i\omega) = -Q_{jk}(i\xi, i\omega)\,Q^{-1}(i\xi, i\omega) = Q_{jk}(i\xi, i\omega)\,Q^{-1}(i\xi, i\omega)$$

where $Q_{jk}(\cdot)$ are the cofactor of the element with the index (k, j) of $\{L(i\xi, i\omega)\}$, $Q(\cdot)$ is the symbol of L (2): $Q(i\xi, i\omega) = (-1)^N \det\{L_{ij}(\xi, \omega)\}$.

In [5] the Green tensor for any dimension N was constructed.

The symmetry relations (3) allow the tensor of elastic constants C_{ij}^{ml} represented in the form of a square matrix $C_{\alpha\beta}$ $(\alpha, \beta = \overline{1, 6})$. The correspondence between the pairs of indexes (ij), (ml) and the indexes α, β, established by the scheme $(11) \leftrightarrow 1$, $(22) \leftrightarrow 2$, $(33) \leftrightarrow 3$, $(23) = (32) \leftrightarrow 4$, $(31) = (13) \leftrightarrow 5$, $(12) = (21) \leftrightarrow 6$. Hooke's law for an anisotropic (orthotropic) elastic medium, which is under the conditions of plane deformation, will be written in the form $\sigma_{11} = C_{11}u_{1,1} + C_{12}u_{2,2}$, $\sigma_{12} = C_{66}u_{1,2}$,

$\sigma_{22} = C_{21}u_{1,1} + C_{22}u_{2,2}$. For such a medium, the Green's tensor are [9, 16]:

$$U_j^k(x,t) = \frac{1}{\pi t}\mathrm{Im} \sum_{\substack{q=1 \\ \mathrm{Im}\,\zeta q > 0}}^{2} \frac{Q_{jk}\left(\zeta_q, 1, (x_1\zeta_q + x_2)/t\right)}{Q_{,\zeta}\left(\zeta_q, 1, (x_1\zeta_q + x_2)/t\right)} \tag{8}$$

Here $Q_{11}(\xi_1, \xi_2, \omega\cdot) = C_{66}\xi_1^2 + C_{22}\xi_2^2 + \rho\omega^2$, $Q_{22}(\xi_1, \xi_2, \omega\cdot) = C_{11}\xi_1^2 + C_{66}\xi_2^2 + \rho\omega^2$, $Q_{12}(\xi_1, \xi_2, \omega\cdot) = -(C_{12} + C_{66})\xi_1\xi_2$ are polynomials of ζ_q are the roots of equation $Q(\zeta, 1, x_1\zeta + x_2) = 0$, $Q = Q_{11}Q_{22} - Q_{12}^2$. In the expression (8), the residues of the fractional-rational functions in the upper half-plane are summed, which requires knowledge of the values of the roots of the polynomial $Q : Q(\zeta, 1, x_1\zeta + x_2) = 0$. The roots of this equation of the fourth degree are complex conjugate; therefore, we always have two roots satisfying the condition $Im\,\zeta \geq 0$. In the case of an isotropic medium $C_{ij}^{ml} = \lambda\delta_{ij}\delta_{lm} + \mu\left(\delta_{im}\delta_{jl} + \delta_{jm}\delta_{il}\right)$ (λ, μ are elastic constants of Lame) we have $\zeta_1 = -\left(x_1x_2 + c_1t\sqrt{r - c_1^2t^2}\right)/(x_1^2 - c_1^2t^2)$, $\zeta_2 = -\left(x_1x_2 + c_2t\sqrt{r - c_2^2t^2}\right)/(x_2^2 - c_2^2t^2)$ where $r = \sqrt{x_ix_i}$, $c_1 = \sqrt{\lambda + 2\mu/\rho}$, $c_2 = \sqrt{\mu/\rho}$.

The Green's tensor allows to construct a fundamental stress tensor, the components of which, according to Hooke's law are

$$S_{ik}^m(x,t) = C_{ij}^{ml}\, U_j^k{}_{,l}\,(x,t)$$

For an orthotropic medium, it will have the form

$$S_{ij}^k(x,t) = \frac{H(t)}{\pi t}C_{ij}^{ml}\mathrm{Im} \sum_{\substack{q=1 \\ \mathrm{Im}\,\zeta_k > 0}}^{2} \frac{Q_{mk,x_l}Q_{,\zeta} - Q_{mk}\left(Q_{,\zeta}\right)_{,x_l}}{\left(Q_{,\zeta}\right)^2}$$

We denote

$$\Gamma_i^k(x,t,n) = S_{ik}^m(x,t)n_m(x)$$

and introduce a tensor T:

$$T_k^i(x,t,n) = -\Gamma_i^k(x,t,n) = -C_{ij}^{ml}n_m(x)U_j^k{}_{,l}\,(x,t)$$

then the system of Eq. (1) for tensor S can be written in the form:

$$S_{ik,m}^m(x,t) - \rho U_i^k(x,t) + \delta_i^k\delta(x)\delta(t)$$

Note the properties of tensors of fundamental solutions

$$U_i^k(x,t) = U_i^k(-x,t), \qquad U_i^k(x,t) = U_k^i(x,t),$$
$$S_{ik}^m(x,t) = -S_{ik}^m(-x,t),$$
$$T_i^k(x,t,n) = -T_i^k(-x,t,n) = -T_i^k(x,t,-n)$$

At the fixed k and n a tensor T_i^k is the generalized solution of system (1), corresponding to the multipole

$$G_i(x,t) = C_{ik}^{ml} n_m(x)\delta_{,l}(x)\delta(t)$$

These and other properties were proved in [7].

We will call T_i^k as the multipole matrix, since it describes the fundamental solutions of system (1) generated by concentrated sources of the multipole type [17].

4 Generalized Solution

A generalized solution of equations for arbitrary mass forces can be represented as a convolution. For example, the study of the processes of wave propagation from foci of earthquakes is associated with the investigation of the stress-strain state of the medium under the action of distributed mass forces. For regular functions of mass forces $G_k(x,t)$ the components of the displacement field are the following integral representations:

$$\hat{u}_i(x,t) = \int_0^\infty d\tau \int_{R^N} U_{ik}(x-y, t-\tau) G_k(y,\tau) dV(y)$$

For a distant source of an earthquake, the distance to which substantially exceeds its dimensions, the models of concentrated sources in the form of singular generalized functions with point support (poles, dipoles, multipoles, etc.) are used [17]. Thus, the displacement field in the vicinity of the earthquake is well described by a concentrated load with axial symmetry, is a flat center expansion: $G = -D[u\, grad\, \delta(r - r_0)]\, u$, D is the magnitude of the dipole moment. In the case of a concentrated moment of the force represented as $G = -MF^0[u\, grad\, \delta(r - r_0)]$. The displacement field then has the form of a convolution U_j^k with the corresponding G_k:

$$u_j = U_{jk} * G_k, \quad j,k = \overline{1,N}$$

For concentrated source such singular generalized functions with point support it is necessary the convolution to take by rules of convolution definition in the theory of generalized functions.

5 Wave Propagation in Anisotropic Media

Anisotropic medium with characteristics closest to the real environment, in particular rocks. Wave propagation in such medium is subject to more complex laws than in an isotropic medium, and the stress-strain state of the medium depends strongly on the degree of anisotropy. In media with weak anisotropy of elastic properties, the wave propagation pattern is similar to the wave propagation pattern in an isotropic medium, but wave fronts representing closed smooth curves differ slightly from concentric circles. In environments with a strong anisotropy of elastic properties, lacunae arise. The coordinates of such regions satisfy the conditions $\mathrm{Im}\zeta_q(x_1, x_2, t) = 0, \quad q = 1, 2$. This phenomenon is associated with the waveguide properties of a highly anisotropic medium, which are sharply expressed in directions with predominant rigidity and are weakened in those where the rigidity is small. The existence of lacunas for hyperbolic equations with constant coefficients, which include the equations of motion of the anisotropic elastic medium, was detected by Petrovsky [3]. He introduced the necessary and sufficient conditions for the existence of lacunas, components of the complement to the surface of the wavefront in which the fundamental solution vanishes (strong lacunas). An example of strong lacunas gives, in particular, the system of Eq. (1) in an even-dimensional space. Lacunas whose coordinates satisfy the conditions $\mathrm{Im}\zeta_q(x_1, x_2, t) = 0, \quad q = 1, 2$ arise for certain constants of Eq. (1) corresponding to strongly anisotropic media. For such media, the wave front patterns differ sharply from the classical front as in the case of isotropic media and have a complex non-smooth form. So, for strongly anisotropic topaz [19, 20] there are lacunas (they are represented by triangular regions), which located on both axes simultaneously (Figs. 1 and 2):

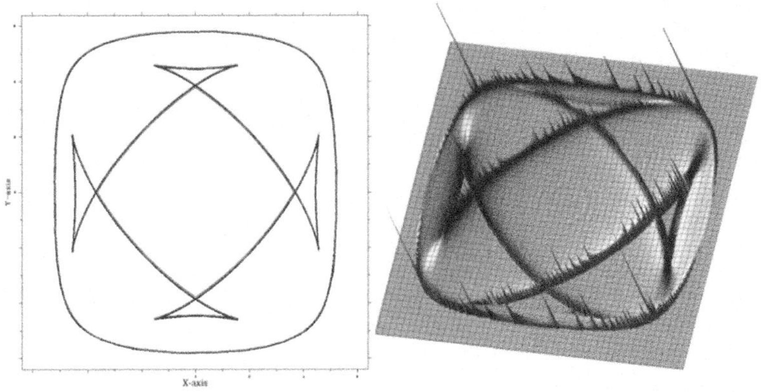

a) **b)**

Fig. 1. Picture of wave fronts (**a**) and the amplitude of U_{11} Green tensor component (**b**) for topaz at $t = 5.8$ ms under the action of a concentrated impulse force For potassium pentaborate there are lacunas too. In this case lacunas located between the axes:

In [18] presented the results of calculations of fundamental solutions corresponding to the action of concentrated forces, the amplitude of the movement, as well as pictures of wave fronts for various medium. The calculations were performed for of isotropic siltstone, and anisotropic media: aragonite, zinc, and topaz and potassium pentaborate. Below are the results of calculations of the stress-strain state in the perturbed zone. The figures show the distribution pattern of displacements along the axis for the media under consideration in cases of concentrated force, dipole, flat expansion center, rotation center.

Figure 3 shows the distribution of the displacement tensor components along the axis (inclination angle) for potassium pentaborate under the action of the concentrated force (I), the concentrated moment (III), the center of rotation (V).

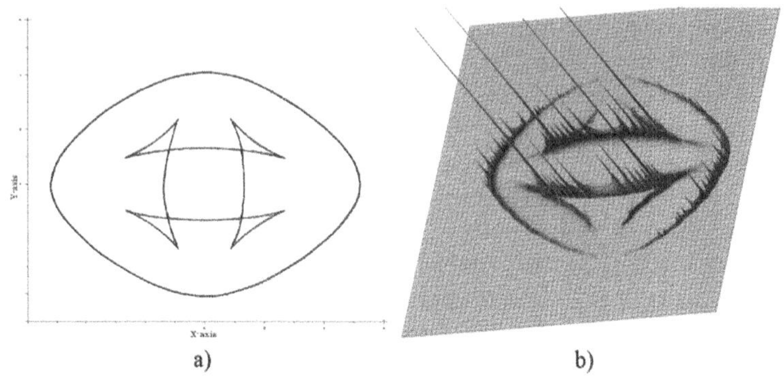

Fig. 2. Picture of wave fronts (**a**) and the amplitude of U_{11} Green tensor component (**b**) for potassium pentaborate at $t = 5.8$ ms under the action of a concentrated impulse force

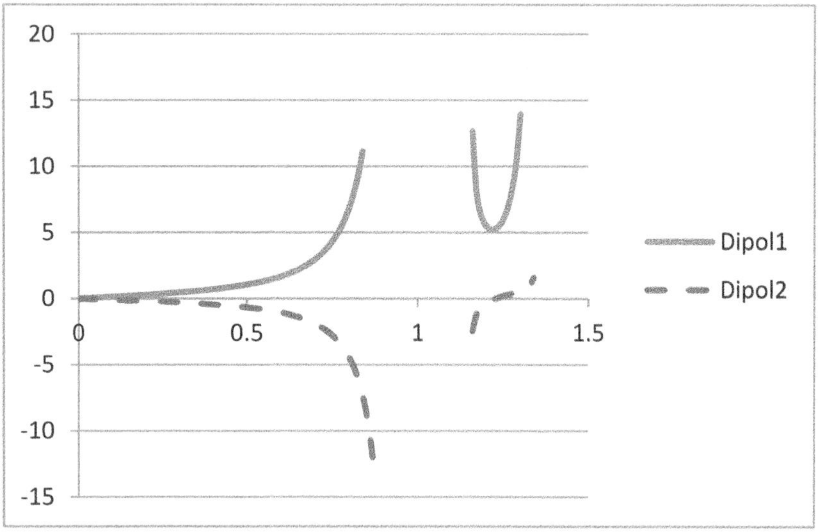

Fig. 3. Components of the Green's tensor for topaz under the action of a dipole

Figure 4 shows the distribution of the components of the displacement tensor along an axis located at an angle $\theta = \pi/4$ to the axis x_1 for potassium pentaborate under the action of a concentrated force (I), a concentrated moment (III), the center of rotation (V). In the calculations it was assumed that $D = 1, M = 1$, in the case of a dipole $e = (1, 0)$, for a concentrated moment $G^0 = (1, 0), e = (0, 1)$.

Fundamental solutions obtained in this can be used to receive the solution for various mass forces also they are needed to build kernels of singular boundary integral equations for solving boundary value problems (Fig. 5).

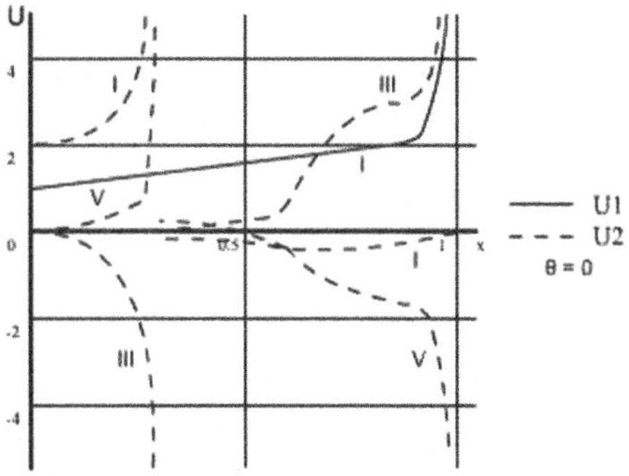

Fig. 4. Components of the Green's tensor for potassium pentaborate under the action of concentrated forces

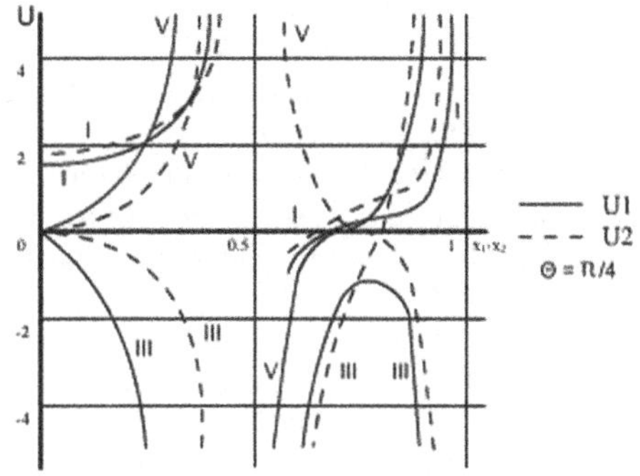

Fig. 5. Components of the Green's tensor for potassium pentaborate under the action of concentrated forces

6 Conclusion

The results of these studies will be useful in developing mathematical models and methods for studying the processes of wave propagation and diffraction in a rock mass if are used the models of mechanics of a deformable solid in order to most adequately describe real physical processes that accompany, in particular, explosive and seismic effects on the array and ground and underground facilities also.

Acknowledgments. This research is financially supported by a grant from the Ministry of Science and Education of the Republic of Kazakhstan (Grant No. AP05135494).

References

1. Hormander, L.: Linear Partial Differential Operators. Springer, Berlin (1963)
2. Vladimirov, V.S.: Equations of Mathematical Physics. Marcel Dekker, New York (1971)
3. Petrovsky, I.G.: Lectures on Partial Differential Equations, State Publishing House of the physical and mathematical literature, Moscow (1961)

4. Petrashen, G.I.: Waves in Anisotropic Media. Nauka, Leningrag (1973)
5. Alekseeva, L.A., Zakir'yanova, G.K.: The green matrix for strictly hyperbolic systems with second-order derivatives. Differ. Equ. **37**(4), 517–523 (2001)
6. Alekseyeva, L.A., Dildabayev, S., Zakir'yanova, G.K., Zhanbyrbayev, A.B.: Boundary integral equation method in two and three dimensional problems of elastodynamics. Comput. Mech. **18**(2), 147–157 (1996)
7. Alexeyeva, L.A., Zakir'yanova, G.K.: Generalized solutions of initial-boundary value problems for second-order hyperbolic systems. Comput. Math. Math. Phys. **51**(7), 1194–1207 (2011)
8. Payton, R.G.: Wave propagation in a restricted transversely isotropic solid whose slowness surface contains conical points. Q. J. Mech. Appl. Math. **45**, 183–197 (1992)
9. Payton, R.G.: Two-dimensional anisotropic elastic waves emanating from a point source. Proc. Camb. Philol. Soc. **70**, 191–210 (1971)
10. Burridge, R.: The singularity on the plane lids of the wave surface of elastic media with cubic symmetry. Q. J. Mech. Appl. Math. **20**, 41–56 (1967)
11. Payton, R.G.: Elastic Wave Propagation in Transversely Isotropic Media. Martinus Nijhoff, Dordrecht (1983)
12. Budreck, D.E.: An eigenfunction expansion of elastic wave Green's function for anisotropic media. Q. J. Mech. Appl. Math. **46**, 1–26 (1993)
13. Zhu, H.: A method to evaluate three-dimensional time-harmonic elastodynamic Green's functions in transversely isotropic media. J. Appl. Mech. Trans. ASME **59**, 587–590 (1992)
14. Wang, C.Y., Achenbach, J.D.: Elastodynamic fundamental solutions for anisotropic solids. Geophys. J. Int. **118**, 384–392 (1994)
15. Wang, C.Y., Achenbach, J.D.: Three-dimensional time-harmonic elastodynamic Green's functions for anisotropic solids. Proc. R. Soc. Lond. Ser. A **449**, 441–458 (1995)
16. Zakir'yanova, G.K: Boundary Integral Equations for the Nonstationary Dynamics of Anisotropic Media, Cand. Sci. (Phys.-Math.) Dissertation, Almaty (1990)
17. Ketch, V., Teodorescu, P.: Introduction to the Theory of Generalized Functions with Applications in Engineering. Wiley, New York (1978)
18. Zakir'yanova, G.K.: New Trends in Analysis and Interdisciplinary Applications Trends in Mathematics. Springer, Cham (2017). https://doi.org/10.1007/978-3-319-48812-7_52
19. Fedorov, F.I.: Theory of Elastic Waves in Crystals. Plenum Press, New York (1968)
20. Clark, Jr.: Handbook of Physical Constants Geological Society of America, vol. 97 (1966)

Residual Power Series Approach for Solving Linear Fractional Swift-Hohenberg Problems

Shatha Hasan[1], Mohammed Al-Smadi[1(✉)], Shaher Momani[2,3], and Omar Abu Arqub[3]

[1] Applied Science Department, Ajloun College,
Al-Balqa Applied University, Ajloun 26816, Jordan
{shatha,mhm.smadi}@bau.edu.jo

[2] Department of Mathematics and Sciences, College of Humanities and Sciences, Ajman University, Ajman, UAE

[3] Department of Mathematics, The University of Jordan, Amman 11942, Jordan

Abstract. In this paper, a modified residual fractional power series technique is applied to provide an analytic-numeric approximated solution for linear time-fractional Swift-Hohenberg equation. The proposed algorithm relies on minimizing the residual error that results when approximating the solution by a truncated fractional Taylor series. The fractional derivative is computed by Caputo time-fractional derivative. The approximate solution obtained by the proposed approach is coinciding well with the exact one. To test the potentially, accuracy and reliability of the proposed technique, the linear time-fractional Swift-Hohenberg equation is considered. The numerical results indicate that the residual fractional power series method is simple, accurate, efficient and suitable for solving various types of differential equations with fractional order.

Keywords: Caputo fractional derivative · Multiple fractional power series · Residual function · Swift-Hohenberg equations

1 Introduction

The theory of fractional calculus has been gained in the last years a great attention in various fields related to science and engineering because of the enormous range of its applications and the critical role which it plays to describe a complex dynamical behavior of real-world problems such as the fluid flow model, traffic flow model, and diffusion models [1–6]. As well as, the topic of fractional calculus aids to simplify the controlling design with no any shortage of hereditary. The derivatives of fractional order are powerful tools for interpreting several physical problems, for instance, electrical circuits, damping laws, and rheology. Fractional partial differential equations (FPDEs) are constructed since they have attracted the attention of the scientists and engineers for their ability in providing exact explanation of linear and nonlinear phenomena which appear, for example, in fluid mechanics wherever continuum assumption is not well. Therefore, fractional model can be deemed to be the best operator [7–11].

In the current paper, we aim to employ the residual fractional power series (RFPS) algorithm in constructing an analytic-numeric approximate solution for the linear time-fractional Swift-Hohenberg equation (TF-SHE) in the form:

M. Zeki Sarıkaya et al. (Eds.): ICMRS 2019, LNNS 123, pp. 33–43, 2020.
https://doi.org/10.1007/978-3-030-43002-3_4

$$D_{\tau_0}^{\beta} v(x,t) + (1-r)v(x,t) + 2\frac{\partial^2 v(x,t)}{\partial x^2} + \frac{\partial^3 v(x,t)}{\partial x^3} = 0, 0 < \beta \leq 1, \quad (1)$$

subject to the initial condition

$$v(x,0) = \psi(x), \quad (2)$$

where r is a real bifurcation parameter, $v(x,t)$ is a scalar valued function defined on the line or the plane, and $D_{\tau_0}^{\beta}$ denotes the Caputo fractional derivative of order β. In the case of $\beta = 1$, the Swift-Hohenberg equation of fractional order in (1) reduces to the classical Swift-Hohenberg equation.

The Swift-Hohenberg equation was proposed and studied by Jack Swift and Pierre Hohenberg [12]. This equation is derived from thermal convection equations and plays a vital role in describing the temperature and fluid velocity of thermal convection [12], as well as, in describing pattern formation theory [13]. Many experts have recently been interested in studying the time fractional Swift-Hohenberg equation because of their important applications in thermal physics and engineering [14–17].

In fact, obtaining the exact solutions for fractional differential equations is not easy in general. So, numerical techniques are required. The residual fractional power series (RFPS) method is an analytic-approximate technique that was proposed and developed in [18] to solve some kinds of fuzzy differential equations. It is an efficient optimization method that produces the solution of a given linear or nonlinear equations in the form of a truncated power series without a need for linearization, discretization or being exposed to perturbation [19–23]. The RFPS algorithm does not depend on making a comparison between the corresponding coefficients to obtain a recurrence relation for them. Instead, it defines a residual error function and uses some theorems to produce a chain of equations that evaluate the power series coefficients. The RFPS solution results in a form of a rapidly convergent power series especially when the exact solution is polynomial [24–27]. The proposed method is suitable for various differential equations so that it can be applied to other issues such as problems in [28–30].

The rest of the current work is structured as follows: in the next section, we give a summary of some necessary definitions and preliminary results concerning fractional operators. In Sect. 3, the analysis of RFPS method is presented. In Sect. 4, numerical example is given to illustrate the capability of the proposed method. The end of this paper is a conclusion in Sect. 5.

2 Preliminaries

In this section, we revisit the Riemann-Liouville fractional integral and Caputo fractional derivative in addition to some basic properties of them.

Definition 1. The integral operator for Riemann-Liouville of order $\beta \geq 0$ of $v(x,t)$ is defined by:

$$J_t^{\beta} v(x,t) = \begin{cases} \frac{1}{\Gamma(\beta)} \int_0^t (t-\xi)^{\beta-1} v(\xi,x) d\xi, \ x \in I, & t > \xi \geq 0, \ \beta > 0, \\ v(x,t), & \beta = 0. \end{cases} \quad (3)$$

Definition 2. For a positive integer η with $\eta - 1 < \beta \leq \eta$, the derivative operator for Caputo of time-fractional order β of $\upsilon(x,t)$ is defined by

$$D_t^{\beta}\upsilon(x,t) = \begin{cases} \frac{1}{\Gamma(\eta-\beta)}\int_0^t (t-\xi)^{\eta-\beta-1}\frac{\partial^{\eta}\upsilon(x,\xi)}{\partial\xi^{\eta}}d\xi, & \eta-1<\beta<\eta, \\ \frac{\partial^{\eta}\upsilon(x,t)}{\partial t^{\eta}}, & \beta=\eta\in \mathrm{N}. \end{cases} \quad (4)$$

Theorem 1. If $\eta - 1 < \beta < \eta$ and $\eta \in \mathrm{N}$, then $\mathcal{D}_t^{\beta} J_t^{\beta}\upsilon(x,t) = \upsilon(x,t)$ and $J_t^{\beta} D_t^{\beta}\upsilon(x,t) = \upsilon(x,t) - \sum_{k=0}^{\eta-1}\frac{\partial^k \upsilon(x,0^+)}{\partial t^k k!}t^k$, where $t > 0$.

3 Analysis of the RPS Method

In this section, we introduce some basic definitions and results concerning the fractional power series (FPS) representation to be able to construct a series expansion with respect to time-fractional derivatives.

Definition 3. A power series representation of the form

$$\sum_{k=0}^{\infty} a_k(t-\tau_0)^{k\beta} = a_0 + a_1(t-\tau_0)^{\beta} + a_2(t-\tau_0)^{2\beta} + \ldots,$$

where $0 \leq \eta - 1 < \beta \leq \eta$ and $t \geq \tau_0$ is referred to as the fractional power series (FPS) about τ_0.

Theorem 2 [31]. Let $\upsilon(t)$ be a function with the FPS representation at τ_0 of the form $\upsilon(t) = \sum_{k=0}^{\infty}\upsilon_k(t-\tau_0)^{k\beta}$. If $\upsilon(t) \in C[\tau_0, \tau_0+\rho)$, and $D_t^{k\beta}\upsilon(t) \in C(\tau_0, \tau_0+\rho)$, for $k = 0, 1, 2, \ldots$, then the coefficients υ_k will take the form $\upsilon_k = \frac{D_t^{k\beta}\upsilon(\tau_0)}{\Gamma(k\beta+1)}$, where $D_t^{k\beta} = D_t^{\beta}D_t^{\beta}\ldots D_t^{\beta}$ (k-times) while ρ denotes the radius of convergence.

Definition 4. For $\eta - 1 < \beta \leq \eta$ a power series of the form, where $x \in I, t \geq \tau_0$,

$$\sum_{k=0}^{\infty}\mu_k(x)(t-\tau_0)^{k\beta} = \mu_0(x) + \mu_1(x)(t-\tau_0)^{\beta} + \mu_2(x)(t-\tau_0)^{2\beta} + \ldots,$$

is referred to as the multiple fractional power series (MFPS) about $t = \tau_0$. Here, $\mu_k(x)$ are functions of the variable x which we call the coefficients of the series.

Theorem 3. Suppose that $\upsilon(x,t)$ has a MFPS representation at $t = \tau_0$ of the form

$$\upsilon(x,t) = \sum_{k=0}^{\infty}\upsilon_k(x)(t-\tau_0)^{k\beta},$$

where $0 \leq \eta - 1 < \beta \leq \eta$, $x \in I$, and $\tau_0 \leq t < \tau_0 + \rho$. If $D_t^{k\beta}\upsilon(x,t)$ are continuous on $I \times (\tau_0, \tau_0+\rho)$, $k = 0, 1, 2, \cdots$, then $\upsilon_k = \frac{D_t^{k\beta}\upsilon(x,\tau_0)}{\Gamma(k\beta+1)}$, $k = 0, 1, 2, \cdots$.

Currently, the RFPS algorithm proposed that the solution of linear TF-SHE (1) and (2) about $\tau_0 = 0$ has the following MFPS:

$$v(x,t) = \sum_{n=0}^{\infty} \psi_n(x) t^{n\beta}, \tag{5}$$

where $v_0(x) = \psi(x)$. So, the analytic solution can be rewritten as

$$v(x,t) = \psi(x) + \sum_{n=1}^{\infty} \psi_n(x) t^{n\beta}. \tag{6}$$

The approximate series form of (6) can be given by the following j^{th}-truncated MFPS:

$$v^j(x,t) = \psi(x) + \sum_{n=1}^{j} \psi_n(x) t^{n\beta}. \tag{7}$$

Now, define the so-called the jth-residual function as follows

$$Res^j(x,t) = D_0^\beta v^j(x,t) + (1-r)v^j(x,t) + 2\frac{\partial^2 v_j(x,t)}{\partial x^2} + \frac{\partial^3 v_j(x,t)}{\partial x^3}, \tag{8}$$

and the residual function as follows

$$Res(x,t) = \lim_{j\to\infty} Res^j(x,t) = D_0^\beta v(x,t) + (1-r)v(x,t) + 2\frac{\partial^2 v(x,t)}{\partial x^2} + \frac{\partial^3 v(x,t)}{\partial x^3}. \tag{9}$$

From [16–20], the following results can be easily established:

(a) $Res(x,t) = \lim_{j\to\infty} Res^j(x,t) = 0$ for each $t \geq 0$ and $x \in I$.

(b) $D_0^{(n-1)\beta} \mathrm{Res}(x,0) = D_0^{(n-1)\beta} Res^j(x,0) = 0,\ n = 1, 2, 3, \cdots, j$.

Therefore, in order to determine the $\psi_n(x), n = 1, 2, 3, \cdots, j$, the following fractional differential equation must be solved

$$D_0^{(n-1)\beta} Res^j(x,0) = 0,\ \ j = 1, 2, 3, \cdots. \tag{10}$$

Now, to evaluate the first unknown coefficient, $\psi_1(x)$, of (9), substitute $v^1(x,t) = \psi(x) + \psi_1(x)t^\beta$ into the 1st-residual function $Res^1(x,t)$ to get:

$$\begin{aligned} Res^1(x,t) &= D_0^\beta\big(\psi(x) + \psi_1(x)t^\beta\big) + (1-r)\big(\psi(x) + \psi_1(x)t^\beta\big) \\ &\quad + 2\frac{\partial^2\big(\psi(x) + \psi_1(x)t^\beta\big)}{\partial x^2} + \frac{\partial^3\big(\psi(x) + \psi_1(x)t^\beta\big)}{\partial x^3} \\ &= \psi_1(x) + (1-r)\big(\psi(x) + \psi_1(x)t^\beta\big) + 2\Big(\psi''(x) + \psi_1''(x)t^\beta\Big) \\ &\quad + \Big(\psi^{(3)}(x) + \psi_1^{(3)}(x)t^\beta\Big). \end{aligned}$$

Thus, depending on the fact $Res^j(x,0) = 0$, we get $\psi_1(x) = \frac{(r-1)\psi(x)-2\psi''(x)-\psi^{(3)}(x)}{\Gamma(\beta+1)}$. Hence, the 1st-RFPS approximated solution for (1) and (2) can be expressed as

$$v^1(x,t) = \psi(x) + \frac{(r-1)\psi(x) - 2\psi''(x) - \psi^{(3)}(x)}{\Gamma(\beta+1)} t^\beta. \tag{11}$$

Likewise, to evaluate the second unknown coefficient $\psi_2(x)$, write $v^2(x,t) = \psi(x) + \psi_1(x)t^\beta + \psi_2(x)t^{2\beta}$ into $Res^2(x,t)$ of (9) and then find new discretized form of this residual function as follows:

$$\begin{aligned}
Res^2(x,t) &= D_0^\beta\Big(\psi(x) + \psi_1(x)t^\beta + \psi_2(x)t^{2\beta}\Big) \\
&+ (1-r)\Big(\psi(x) + \psi_1(x)t^\beta + \psi_2(x)t^{2\beta}\Big) \\
&+ 2\frac{\partial^2}{\partial x^2}\Big(\psi(x) + \psi_1(x)t^\beta + \psi_2(x)t^{2\beta}\Big) \\
&+ \frac{\partial^3}{\partial x^3}\Big(\psi(x) + \psi_1(x)t^\beta + \psi_2(x)t^{2\beta}\Big) \\
&= \Gamma(\beta+1)\psi_1(x) + \frac{\Gamma(2\beta+1)}{\Gamma(\beta+1)}\psi_2(x)t^{2\beta} \\
&+ (1-r)\Big(\psi(x) + \psi_1(x)t^\beta + \psi_2(x)t^{2\beta}\Big) \\
&+ 2\Big(\psi''(x) + \psi_1''(x)t^\beta + \psi_2''(x)t^{2\beta}\Big) \\
&+ \Big(\psi^{(3)}(x) + \psi_1^{(3)}(x)t^\beta + \psi_2^{(3)}(x)t^{2\beta}\Big).
\end{aligned}$$

Considering (10) for $j = 2$, we get

$$D_0^\beta Res^2(x,t) = \Gamma(2\beta+1)\psi_2(x) + (1-r)\psi_1(x) + \psi_1''(x) + \psi_1^{(3)}(x).$$

By using the fact that $D_0^\beta Res^2(x,0) = 0$, and based on the previous result of $\psi_1(x)$, it yields that

$$\psi_2(x) = -\frac{(1-r)\psi_1(x) + \psi_1''(x) + \psi_1^{(3)}(x)}{\Gamma(2\beta+1)}.$$

Hence, the 2nd-RFPS approximated solution for (1) and (2) can be written as

$$\begin{aligned}
v^2(x,t) = \psi(x) &+ \frac{(r-1)\psi(x) - 2\psi''(x) - \psi^{(3)}(x)}{\Gamma(\beta+1)} t^\beta \\
&- \frac{(1-r)\psi_1(x) + \psi_1''(x) + \psi_1^{(3)}(x)}{\Gamma(2\beta+1)} t^{2\beta}.
\end{aligned}$$

This technique can be repeated till the arbitrary order coefficients of MFPS solution for (1) and (2) are determined.

4 Numerical Applications

In this section, we consider the application of RPS method for solving linear TF-SHE to demonstrate the performance and efficiency of the present technique by using Mathematica 10.0 software package.

Example 1: Consider Swift-Hohenberg equation in the form:

$$D_0^\beta v(x,t) + (1-r)v(x,t) + 2\frac{\partial^2 v(x,t)}{\partial x^2} + \frac{\partial^3 v(x,t)}{\partial x^3} = 0, 0 < \beta \le 1, \tag{12}$$

with the initial conditions

$$v(x,0) = e^x. \tag{13}$$

According to apply RFPS algorithm, we start with $v(x,0) = e^x$, assume that the jth-truncated MFPS solution of (12) and (13), has the following form:

$$v^j(x,t) = e^x + \sum\nolimits_{n=1}^{j} \psi_n(x) t^{n\beta}, \tag{14}$$

Consequently, the jth-residual function $Res^j(x,t)$ of (12) can be given by

$$\begin{aligned} Res^j(x,t) &= D_0^\beta v^j(x,t) + (1-r)v^j(x,t) + 2\frac{\partial^2 v_j(x,t)}{\partial x^2} + \frac{\partial^3 v_j(x,t)}{\partial x^3} \\ &= D_0^\beta\left(e^x + \sum_{n=1}^{j}\psi_n(x)t^{n\beta}\right) + (1-r)\left(e^x + \sum_{n=1}^{j}\psi_n(x)t^{n\beta}\right) \\ &+ 2\frac{\partial^2}{\partial x^2}\left(e^x + \sum_{n=1}^{j}\psi_n(x)t^{n\beta}\right) + \frac{\partial^3}{\partial x^3}\left(e^x + \sum_{n=1}^{j}\psi_n(x)t^{n\beta}\right), \end{aligned} \tag{15}$$

where the unknown coefficients $\psi_n(x)$ for $n = 1, 2, 3, \cdots, j$ can be determined by doing the following manner: For $j = 1$, we have

$$\begin{aligned} Res^1(x,t) &= D_0^\beta\left(e^x + \psi_1(x)t^\beta\right) + (1-r)\left(e^x + \psi_1(x)t^\beta\right) + 2\left(e^x + \psi_1''(x)t^\beta\right) \\ &+ \left(e^x + \psi_1^{(3)}(x)t^\beta\right) \\ &= \Gamma(\beta+1)\psi_1(x) + (1-r)\left(e^x + \psi_1(x)t^\beta\right) + 2\left(e^x + \psi_1''(x)t^\beta\right) \\ &+ \left(e^x + \psi_1^{(3)}(x)t^\beta\right). \end{aligned}$$

Then, use the fact $Res^1(x,0) = 0$ to get $\psi_1(x) = \frac{(r-4)e^x}{\Gamma(\beta+1)}$.

Again, to find out the value of $\psi_2(x)$ put $j = 2$, into the jth-residual function of (15) to get

$$Res^2(x,t) = D_0^\beta\left(e^x + \psi_1(x)t^\beta + \psi_2(x)t^{2\beta}\right) + (1-r)(e^x + \psi_1(x)t^\beta + \psi_2(x)$$

$$+2\left(e^x+\psi_1''(x)t^\beta+\psi_2''(x)t^{2\beta}\right)+\left(e^x+v_1^{(3)}(x)t^\beta\right.$$
$$\left.+\psi_2^{(3)}(x)t^{2\beta}\right)$$
$$=\Gamma(\beta+1)\psi_1(x)+\frac{\Gamma(2\beta+1)}{\Gamma(\beta+1)}\psi_2(x)t^{2\beta}+(1-r)\left(e^x\right.$$
$$\left.+\psi_1(x)t^\beta+\psi_2(x)t^{2\beta}\right)+2\left(e^x+\psi_1''(x)t^\beta+\psi_2''(x)t^{2\beta}\right)+\left(e^x\right.$$
$$\left.+\psi_1^{(3)}(x)t^\beta+\psi_2^{(3)}(x)t^{2\beta}\right),$$

and then by applying D_0^β on both sides of the resulting equation, we have the following:

$$D_0^\beta Res^2(x,t)=\Gamma(2\beta+1)\psi_2(x)$$
$$+\Gamma(\beta+1)\Big((1-r)\psi_1(x)+2\psi_1''(x)+\psi_1^{(3)}(x)\Big).$$

Based on the fact $D_0^\beta Res^2(x,0)=0$ and the previous result of $\psi_1(x)$, we obtain $\psi_2(x)=\frac{(r-4)^2e^x}{\Gamma(2\beta+1)}$.

Similarly, for $j=3$, the 3rd-residual function $Res^3(x,t)$, can be written by

$$Res^3(x,t)=D_0^\beta\left(e^x+\psi_1(x)t^\beta+\psi_2(x)t^{2\beta}+\psi_3(x)t^{3\beta}\right)$$
$$+(1-r)\left(e^x+\psi_1(x)t^\beta+\psi_2(x)t^{2\beta}+\psi_3(x)t^{3\beta}\right)$$
$$+2\left(e^x+\psi_1''(x)t^\beta+\psi_2''(x)t^{2\beta}+\psi_3''(x)t^{3\beta}\right)$$
$$+\left(e^x+\psi_1^{(3)}(x)t^\beta+\psi_2^{(3)}(x)t^{2\beta}+\psi_3^{(3)}(x)t^{3\beta}\right)$$
$$=\Gamma(\beta+1)\psi_1(x)+\frac{\Gamma(2\beta+1)}{\Gamma(\beta+1)}\psi_2(x)t^\beta+\frac{\Gamma(3\beta+1)}{\Gamma(2\beta+1)}\psi_3(x)t^{2\beta}$$
$$+(1-r)\left(e^x+\psi_1(x)t^\beta+\psi_2(x)t^{2\beta}+\psi_3(x)t^{3\beta}\right)$$
$$+2\left(e^x+\psi_1''(x)t^\beta+\psi_2''(x)t^{2\beta}+\psi_3''(x)t^{3\beta}\right)$$
$$+\left(e^x+\psi_1^{(3)}(x)t^\beta+\psi_2^{(3)}(x)t^{2\beta}+\psi_3^{(3)}(x)t^{3\beta}\right),$$

and then by utilizing the fact $D_0^{(j-1)\beta}Res^j(x,t)$ at $j=3$, we have that

$$D_0^{2\beta}Res^3(x,t)=\Gamma(3\beta+1)\psi_3(x)$$
$$+\Gamma(2\beta+1)\Big((1-r)\psi_2(x)+2\psi_2''(x)+\psi_2^{(3)}(x)\Big).$$

Now, by solving $D_0^{2\beta}Res^3(x,t)=0$ at $t=0$, taking into account the values of $\psi_1(x)$ and $\psi_2(x)$, one can get $\psi_3(x)=\frac{(r-4)^3e^x}{\Gamma(3\beta+1)}$.

Using the same fashion and the fact $D_0^{3\beta}Res^4(x,0)=0$ for $j=4$, one can obtain that $\psi_4(x)=\frac{(r-4)^4e^x}{\Gamma(4\beta+1)}$.

Moreover, depend on the fact that $D_0^{(j-1)\beta} Res^j(x,0) = 0$ for $j = 5, 6, 7, 8$, the 8th-RFPS approximated solution for TF-SHE (12) and (13) is given by

$$\begin{aligned} v^8(x,t) = {} & e^x + \frac{(r-4)e^x}{\Gamma(\beta+1)}t^\beta + \frac{(r-4)^2e^x}{\Gamma(2\beta+1)}t^{2\beta} + \frac{(r-4)^3e^x}{\Gamma(3\beta+1)}t^{3\beta} + \frac{(r-4)^4e^x}{\Gamma(4\beta+1)}t^{4\beta} \\ & + \frac{(r-4)^5e^x}{\Gamma(5\beta+1)}t^{5\beta} + \frac{(r-4)^6e^x}{\Gamma(6\beta+1)}t^{6\beta} + \frac{(r-4)^7e^x}{\Gamma(7\beta+1)}t^{7\beta} \\ & + \frac{(r-4)^8e^x}{\Gamma(8\beta+1)}t^{8\beta}. \end{aligned}$$

Therefore, the approximated solution of (1) and (2) has general form which is coinciding well with the exact solution for $\beta = 1$, such that

$$v(x,t) = e^x\left(1 + (r-4)t + \frac{(r-4)^2}{2}t^2 + \frac{(r-4)^3}{3!}t^3 + \ldots\right) = e^{x+(r-4)t}.$$

For numerical simulation, Table 1 shows the absolute errors of approximated solution, $v^8(x,t)$, for TF-SHE (12) and (13) at $\beta = 1$ and $r = 3$ for fixed values of x and some selected grid of t with step size 0.2.

Table 1. The absolute error of Example 1.

x	t_i	Absolute errors
−2	0.2	$1.872113575 \times 10^{-13}$
	0.4	$9.399298007 \times 10^{-11}$
	0.6	$3.544659497 \times 10^{-9}$
	0.8	$4.632465531 \times 10^{-8}$
0	0.2	$1.383337889 \times 10^{-12}$
	0.4	$6.945193309 \times 10^{-10}$
	0.6	$2.619168793 \times 10^{-8}$
	0.8	$3.422954769 \times 10^{-7}$
2	0.2	$1.022026908 \times 10^{-11}$
	0.4	$5.131843395 \times 10^{-9}$
	0.6	$1.935318509 \times 10^{-7}$
	0.8	$2.529240481 \times 10^{-7}$

Also, Table 2 shows approximated solution, $v^8(x,t)$ of Swift-Hohenberg Eq. (1) and (2) for some values of the time-fractional derivative β. In this table, we take $\beta \in \{0.75, 0.85, 0.95, 1\}$. with a fixed value of x at some selected grid of t and step size 0.2. To show the accuracy of the RFPS-algorithm, the effect of different values of the fractional derivative order $0 < \beta \le 1$ for the 8th-RFPS approximate solution are plotted in Fig. 1 for TF-SHE (12) and (13) with different values of β and $r = 3$ for all $t \in [0, 1]$ and $x \in [-2, 2]$.

Table 2. Numerical results for Example 1 using the RFPS method.

x	t_i	$\beta = 1$	$\beta = 0.95$	$\beta = 0.85$	$\beta = 0.75$
-2	0.1	0.149568619	0.151803880	0.157413659	0.165063119
	0.3	0.182683524	0.187820819	0.200120958	0.216001808
	0.5	0.223130159	0.231051931	0.249777899	0.273609578
	0.7	0.272531777	0.283504869	0.309314225	0.341977026
	0.9	0.332870925	0.347357704	0.381358908	0.424277875
2	0.1	8.166169913	8.288211024	8.594494546	9.0121409788
	0.3	9.974182454	10.254669268	10.926923409	11.793299130
	0.5	12.18249392	12.615007995	13.637411209	14.938576717
	0.7	14.87973084	15.478841403	16.887984458	18.671312959
	0.9	18.17413671	18.965088059	20.821490852	23.164787083

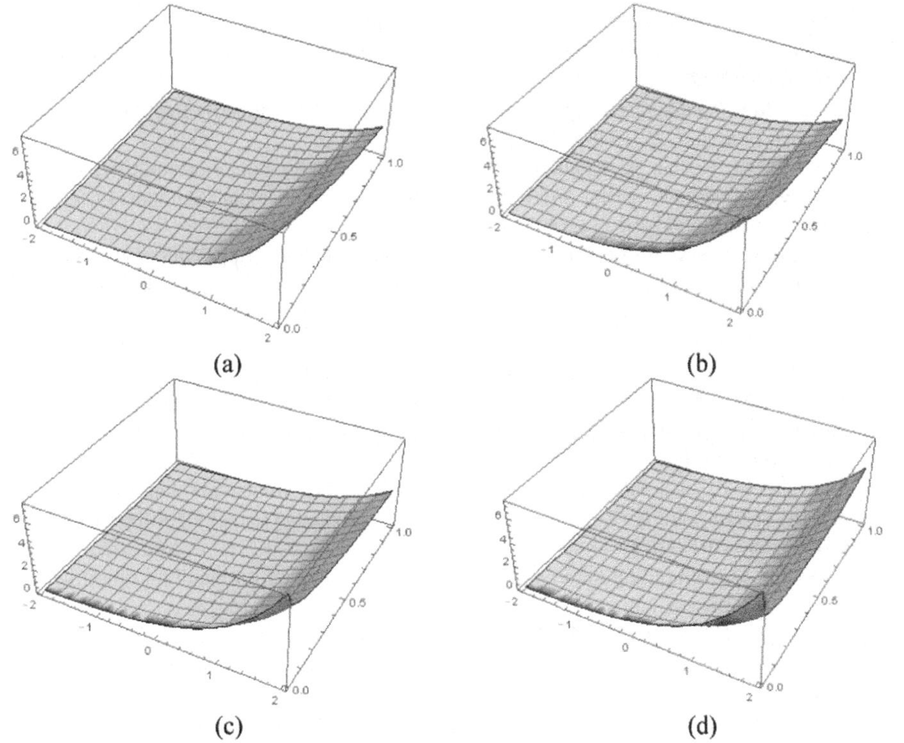

Fig. 1. Surface plot of the 8th-RFPS approximated solution of Example1 for $r = 3$ with: (a) $\beta = 1$ (b) $\beta = 0.75$ (c) $\beta = 0.5$ and (d) $\beta = 0.25$.

5 Conclusion

In the present paper, the analytic-numeric solution of linear TF-SHE has been constructed and analyzed by utilizing an efficient accurate method, named the residual fractional power series algorithm. The solution methodology depends on constructing a residual function as well as applying the generalized Taylor formula in sense of Caputo time-fractional derivative. The proposed approach provides the solutions in the form of rapidly convergent series with easily computable coefficients. Graphical and numerical results are performed by Mathematica 10. The results demonstrate the accuracy, efficiency and the capability of the present method. Therefore, the residual fractional power series algorithm is reliable, effective, simple, straightforward tool for handling a wide range of linear time-fractional differential equations.

References

1. Kilbas, A.A., Srivastava, H.M., Trujillo, J.J.: Theory and Applications of Fractional Differential Equations. North-Holland Mathematics Studies, vol. 204. Elsevier Science B.V., Amsterdam (2006)
2. Klimek, M.: Fractional sequential mechanics—models with symmetric fractional derivative. Czech. J. Phys. **51**(12), 1348–1354 (2001)
3. Al-Smadi, M., Freihat, A., Khalil, H., Momani, S., Khan, R.A.: Numerical multistep approach for solving fractional partial differential equations. Int. J. Comput. Methods **14**, 1750029 (2017). https://doi.org/10.1142/S0219876217500293
4. Momani, S., Abu Arqub, O., Freihat, A., Al-Smadi, M.: Analytical approximations for Fokker-Planck equations of fractional order in multistep schemes. Appl. Comput. Math. **15**(3), 319–330 (2016)
5. Baleanu, D., Golmankhaneh, A.K., Baleanu, M.C.: Fractional electromagnetic equations using fractional forms. Int. J. Theor. Phys. **48**, 3114–3123 (2009)
6. Al-Smadi, M., Abu Arqub, O.: Computational algorithm for solving Fredholm time-fractional partial integrodifferential equations of Dirichlet functions type with error estimates. Appl. Math. Comput. **342**, 280–294 (2019)
7. Altawallbeh, Z., Al-Smadi, M., Komashynska, I., Ateiwi, A.: Numerical solutions of fractional systems of two-point BVPs by using the iterative reproducing kernel algorithm. Ukr. Math. J. **70**(5), 687–701 (2018)
8. Al-Smadi, M.: Simplified iterative reproducing kernel method for handling time-fractional BVPs with error estimation. Ain Shams Eng. J. **9**(4), 2517–2525 (2018)
9. Abu Arqub, O., Al-Smadi, M.: Atangana-Baleanu fractional approach to the solutions of Bagley-Torvik and Painlevé equations in Hilbert space. Chaos Solitons Fractals **117**, 161–167 (2018)
10. Abu Arqub, O., Al-Smadi, M.: Numerical algorithm for solving time-fractional partial integrodifferential equations subject to initial and Dirichlet boundary conditions. Numer. Methods Partial Differ. Equ. **34**(5), 1577–1597 (2017)
11. Abu Arqub, O., Odibat, Z., Al-Smadi, M.: Numerical solutions of time-fractional partial integrodifferential equations of Robin functions types in Hilbert space with error bounds and error estimates. Nonlinear Dyn. **94**(3), 1819–1834 (2018)
12. Swift, J., Hohenberg, P.C.: Hydrodynamic fluctuations at the convective instability. Phys. Rev. **15**(1), 319 (1977)

13. Ryabov, P.N., Kudryashov, N.A.: Nonlinear waves described by the generalized Swift Hohenberg equation. J. Phys. Conf. Ser. **788** (2017). Article No. 012032
14. Khan, N.A., Khan, N., Ayaz, M., Mahmood, A.: Analytical methods for solving the time-fractional Swift-Hohenberg (S-H) equation. Comput. Math. Appl. **61**, 2182–2185 (2011)
15. Merdan, M.: A numeric-analytic method for time-fractional Swift-Hohenberg (S-H) equation with modified Riemann-Liouville derivative. Appl. Math. Model. **37**(6), 4224–4231 (2013)
16. Vishal, K., Kumar, S., Das, S.: Application of homotopy analysis method for fractional Swift Hohenberg equation–revisited. Appl. Math. Model. **36**, 3630–3637 (2012)
17. Li, W., Pang, Y.: An Iterative Method for Time-Fractional Swift-Hohenberg Equation. Adv. Math. Phys. **2018** (2018). Article ID 2405432, 13 pages
18. Abu Arqub, O.: Series solution of fuzzy differential equations under strongly generalized differentiability. J. Adv. Res. Appl. Math. **5**(1), 31–52 (2013)
19. Hasan, S., Al-Smadi, M., Freihet, A., Momani, S.: Two computational approaches for solving a fractional obstacle system in Hilbert space. Adv. Differ. Eq. **2019**, 55 (2019)
20. Moaddy, K., Al-Smadi, M., Hashim, I.: A novel representation of the exact solution for differential algebraic equations system using residual power-series method. Discrete Dyn. Nat. Soc. **2015** (2015). Article ID 205207, 12 pages
21. Komashynska, I., Al-Smadi, M., Ateiwi, A., Al-Obaidy, S.: Approximate analytical solution by residual power series method for system of Fredholm integral equations. Appl. Math. Inf. Sci. **10**(3), 975–985 (2016)
22. Komashynska, I., Al-Smadi, M., Abu Arqub, O., Momani, S.: An efficient analytical method for solving singular initial value problems of nonlinear systems. Appl. Math. Inf. Sci. **10**(2), 647–656 (2016)
23. Freihet, A., Hasan, S., Al-Smadi, M., Gaith, M., Momani, S.: Construction of fractional power series solutions to fractional stiff system using residual functions algorithm. Adv. Differ. Eq. **2019**, 95 (2019). https://doi.org/10.1186/s13662-019-2042-3
24. Alaroud, M., Al-Smadi, M., Ahmad, R.R., Salma Din, U.K.: Computational optimization of residual power series algorithm for certain classes of fuzzy fractional differential equations. Int. J. Diff. Eq. **2018** (2018). Article ID 8686502, 11 pages
25. Abu-Gdairi, R., Al-Smadi, M., Gumah, G.: An expansion iterative technique for handling fractional differential equations using fractional power series scheme. J. Math. Stat. **11**(2), 29–38 (2015)
26. El-Ajou, A., Abu Arqub, O., Al-Smadi, M.: A general form of the generalized Taylor's formula with some applications. Appl. Math. Comput. **256**, 851–859 (2015)
27. Alaroud, M., Al-Smadi, M., Ahmad, R.R., Salma Din, U.K.: An analytical numerical method for solving fuzzy fractional Volterra integro-differential equations. Symmetry **11**(2), 205 (2019)
28. Al-Smadi, M.: Reliable numerical algorithm for handling fuzzy integral equations of second kind in Hilbert spaces. Filomat **33**(2), 583–597 (2019)
29. Area, I., Losada, J., Nieto, J.J.: A note on the fractional logistic equation. Phys. A Stat. Mech. Appl. **444**, 182–187 (2016)
30. Al-Smadi, M., Abu Arqub, O., Shawagfeh, N., Momani, S.: Numerical investigations for systems of second-order periodic boundary value problems using reproducing kernel method. Appl. Math. Comput. **291**, 137–148 (2016)
31. El-Ajou, A., Abu Arqub, O., Al, Zhour Z., Momani, S.: New results on fractional power series: theories and applications. Entropy **15**(12), 5305–5323 (2013)

Kernel-Based Fuzzy Clustering for Sinusitis Dataset

Zuherman Rustam[1](✉), Nadisa Karina Putri[1], Jacub Pandelaki[2], Widyo Ari Nugroho[2], Dea Aulia Utami[1], and Sri Hartini[1]

[1] Department of Mathematics, University of Indonesia, 16424 Depok, Indonesia
rustam@sci.ui.ac.id, putrinadisa@gmail.com
[2] Department of Radiology, Dr. Cipto Mangunkusumo Hospital, DKI, 10430 Jakarta, Indonesia

Abstract. Sinusitis is a condition resulting from inflammation of sinus walls. In handling the disease, machine learning method is often used to find more precise and accurate treatment plan for patients. For instance, fuzzy clustering is widely used for pattern recognition and data mining. Due to uncertainty and ambiguity, this method is used to overcome the non-linearity of medical dataset. In this study, fuzzy clustering was provided with kernel methods. We used some Kernel methods such as, Kernelized Fuzzy c-Means (KFCM), Kernelized Possibilistic c-Means (KPCM), Kernelized Fuzzy Possibilistic c-Means (KFPCM), and Kernelized Possibilistic Fuzzy c-Means (KPFCM) for clustering sinusitis dataset. The dataset was retrieved from Cipto Mangunkusumo Hospital Jakarta, Indonesia, which contains 4 features and 200 instances of this condition. These level of accuracy and model performance are used to compare these approaches. The result showed that KFCM has the highest accuracy for categorizing sinusitis dataset with accuracy of 96.97% and running time of 0.01 s.

Keywords: Kernelized Fuzzy c-Means (KFCM) · Kernelized Possibilistic c-Means (KPCM) · Kernelized Fuzzy Possibilistic c-Means (KFPCM) · Kernelized Possibilistic Fuzzy c-Means (KPFCM) · Sinusitis

1 Introduction

Sinusitis is a condition resulting from inflammation of lining tissues of sinus. The inflammation is caused by the blocking of the tissue which is filled with fluids as a result [1]. Sinus is an air cavity in the nose interconnected by air ducts inside the skull [2]. This inflammation leads to infections in the tissues, resulting to difficulties in breathing. There are several types of sinusitis depending on the period of time a patient has been affected. For example, acute sinusitis emerges only 2–4 weeks of the infection, sub-acute inflammation between the 4–12 weeks, chronic sinusitis 12 weeks or longer, and recurrent sinusitis which happens several times across the year [1]. This study work, focused on the acute type of sinusitis.

In many cases, most patients try to approach the disease using local remedies which often do not yield positive results. If not treated properly, sinusitis may last for a long time and lead to much severe diseases such as meningitis, bone abscess, osteomyelitis,

M. Zeki Sarıkaya et al. (Eds.): ICMRS 2019, LNNS 123, pp. 44–54, 2020.
https://doi.org/10.1007/978-3-030-43002-3_5

or the patients may lose their sense of smell permanently [2]. It is important for patients to have a radiology examination in the hospital in order to understand the type infection they may be struggling with. This check is usually carried out through a Computed Tomography Scanning (CT scan) or a Magnetic Resonance Imaging (MRI) [3].

Clustering refers to a categorization method which assign data into groups [4, 5, 17]. In this process, data with high similarities are placed in the same cluster while dissimilar ones are in a different bunch [5, 17]. There are two types of clustering, hard and soft. In hard clustering the data either completely belong to a cluster or not. Contrastingly, the data may belong to more than one cluster in soft clustering [4, 17]. In this study, we will focus on fuzzy-clustering, this is a part of soft clustering.

Fuzzy c-means (FCM) is a clustering method that has been extensively used in data mining [6]. It has a membership function ranging from 0 to 1, where 0 indicates the data being the farthest from the cluster's center while 1 represent the closest. This membership constraint is used to prevent trivial solution for the value being 0 [6]. Unfortunately, FCM cannot handle noise data or outlier well enough. To overcome this problem, a Possibilistic Fuzzy c-Means (PCM) method was developed in 1993 [6]. This method neglects the constraint presented in FCM and build a more objective function. As the result, PCM handle the outlier better than the FCM, though it has a propensity to produce coincidental clusters [8]. This is a weakness yet to be addressed.

In 1997, Pal et al. proposed an integrated method of FCM and PCM and named Fuzzy Possibilistic c-Means (FPCM), the approach which combines the benefits of each of the methods [6, 14]. In order to address the weakness in FCM and PCM, FPCM defines not only membership function but also typicalities. The method calculates membership and possibilities at the same time [8]. According to Huber's statistic, this model is strong for outlier and noise data. In 2005, Bezdek et al. designed a new method called Possibilistic Fuzzy c-Means (PFCM) which was intended to subdue FCM's sensitivity to noise data and PCM's subtleness to coincidental clusters [9]. However, the aforementioned methods use Euclidean function but cannot address the nonlinearity separable problems along with high dimensional data.

In medical field, the diagnosis of disease is often based on the experience and knowledge of practitioners. To support the process of diagnosing as well as for more accuracy, clustering method is always used [10]. For instance, fuzzy clustering is efficient in handling uncertainty and ambiguity in medical data [11], including Sinusitis dataset. In this study, the Euclidean distance function was changed to Kernel high dimensional which obtain higher accuracy. Furthermore, Kernelized Fuzzy c-Means (KFCM), Kernelized Possibilistic c-Means (KPCM), Kernelized Fuzzy Possibilistic c-Means (KFPCM), and Kernelized Possibilistic Fuzzy c-Means (KPFCM) were compared for sinusitis dataset. By using the kernel-based distance, it occurs that the results of clustering are better.

2 Materials and Methods

2.1 Kernel Function

Since Euclidean function is not able to deal with complex problems, kernel function is used to ensure input X is mapped to a high dimensional feature space G. The distance formula is then transformed to kernel in the feature space [12]

$$X = (x_1, x_2, \ldots, x_n) \rightarrow \Phi(\mathrm{X}) = (\phi(x_1), \ldots, \phi(x_N)), \quad (1)$$

$$\left\|x_j - v_i\right\| \rightarrow \left\|\phi\left(x_j\right) - \phi(v_i)\right\|, \quad (2)$$

where Kernel in high dimensional feature space was

$$K\left(x_i, x_j\right) = \left\langle\phi(x_i)\cdot\phi\left(x_j\right)\right\rangle. \quad (3)$$

Even though there are other functions in kernel, we used the Gaussian Radial Basis Function (GRBF) Kernel [12]:

$$K\left(x_i, x_j\right) = exp\left(\frac{-\left\|x_i - x_j\right\|^2}{2\sigma^2}\right) \quad (4)$$

where σ is a parameter which defines the width of features, ranges from zero to infinite [12]. The squared distance becomes:

$$\begin{aligned}\|\phi(x_k) - \phi(v_i)\|^2 &= \langle[\phi(x_k) - \phi(v_i)]\cdot[\phi(x_k) - \phi(v_i)]\rangle \\ &= K(x_k, x_k) + K(v_i, v_i) - 2K(x_k, v_i) = 2 - 2K(x_k, v_i)\end{aligned} \quad (5)$$

We changed all the Euclidean distance formula with the high dimensional kernel function for FCM, PCM, FPCM, and PFCM.

2.2 Kernelized Fuzzy C-Means (KFCM)

Fuzzy c-Means was identical with the later fuzzy k-Means [9]. Given dataset in p-dimensional vector space, $\mathrm{X} = \{\mathrm{x}_1, \mathrm{x}_2, \ldots, \mathrm{x}_\mathrm{n}\} \subset \Re^p$, the objective function of FCM is [6, 8]:

$$J_{FCM}(U, V) = \sum\nolimits_{k=1}^{n}\sum\nolimits_{i=1}^{c}(u_{ik})^m\|x_k - v_i\|^2, \quad (6)$$

where c is the number of clusters, $2 \leq c \leq n$, p is the number of data items, m is weight exponent to discover the amount of fuzziness, $1 \leq m \leq \infty$, n is number of data points, u_{ik} is the membership of x_k in class i, and v_i is the vector cluster center of class i. The membership function and cluster center formula are as follows:

$$u_{ik} = \left(\sum\nolimits_{j=1}^{c}\left(\frac{\|x_k - v_i\|}{\left\|x_k - v_j\right\|}\right)^{\frac{2}{m-1}}\right)^{-1} \quad i = 1, 2, \ldots, c,\ j = 1, 2, .., n \quad (7)$$

$$v_i = \frac{\sum_{j=1}^{n}(u_ik)^m x_k}{\sum_{j=1}^{n}(u_ik)^m}. \quad (8)$$

As stated in earlier, the Euclidean distance in Eq. (6) was changed to kernel function. The objective function becomes as follows [9]:

$$J_{KFCM}(U, V) = \sum\nolimits_{i=1}^{c}\sum\nolimits_{k=1}^{n}(u_{ik})^m\|\phi(x_k) - \phi(v_i)\|^2$$

$$= \sum_{i=1}^{c} \sum_{k=1}^{n} (u_{ik})^m (2 - 2K(x_k, v_i)). \tag{9}$$

Minimize Eq. (9) in order under the constraint of u_{ik}, the membership function and cluster center become as follows:

$$u_{ik} = \frac{(1 - K(x_k, v_i))^{-\frac{1}{(m-1)}}}{\sum_{j=1}^{c} \left(1 - K\left(x_k, v_j\right)\right)^{-\frac{1}{(m-1)}}}, \tag{10}$$

$$v_i = \frac{\sum_{k=1}^{n} (u_{ik})^m K(x_k, v_i) x_k}{\sum_{k=1}^{n} (u_{ik})^m K(x_k, v_i)}. \tag{11}$$

2.3 Kernelized Possibilistic C-Means (KPCM)

Given the dataset $X = \{x_1, x_2, \ldots, x_n\} \subset \Re^p$, the objective function of PCM is [7–9]:

$$J_{PCM}(U, V) = \sum_{i=1}^{c} \sum_{k=1}^{n} u_{ik}^m \|x_k - v_i\|^2 + \sum_{i=1}^{c} \eta_i \sum_{k=1}^{n} (1 - u_{ik})^m, \tag{12}$$

$$\eta_i = \frac{\sum_{k=1}^{n} u_{ik}^m \|x_k - v_i\|^2}{\sum_{k=1}^{n} u_{ik}^m}, \tag{13}$$

where η_i are positive numbers. In high dimensional feature space G, the objective function is

$$J_{KPCM}(U, V) = \sum_{k=1}^{n} \sum_{i=1}^{c} u_{ik}^m \|\phi(x_k) - \phi(v_i)\|^2 + \sum_{i=1}^{c} \eta_i \sum_{k=1}^{n} (1 - u_{ik})^m. \tag{14}$$

In order to find the partition of x into the subset, minimize Eq. (14), then we obtain

$$u_{ik} = \left[1 + \left(\frac{2 - 2K(x_k, v_i)}{\eta_i}\right)^{\frac{1}{m-1}}\right]^{-1}, \forall i, k \tag{15}$$

$$\phi(\mathrm{v_i}) = \frac{\sum_{k=1}^{n} u_{ik}^m \phi(x_k)}{\sum_{k=1}^{n} u_{ik}^m}, \forall i \tag{16}$$

Equation (16) cannot be computed directly, both sides of the equation are multiplied by $\phi\left(x_j\right)^T$.The equation takes the form:

$$K\left(x_j, v_i\right) = \frac{\sum_{k=1}^{n} u_{ik}^m K\left(x_k, x_j\right)}{\sum_{k=1}^{n} u_{ik}^m}, \forall i, j \tag{17}$$

$$\eta_i = \beta \frac{\sum_{k=1}^{n} u_{ik}^m (2 - 2K(x_k, v_i))}{\sum_{k=1}^{n} u_{ik}^m}. \tag{18}$$

where β is a positive real number and it is commonly equal to one.

2.4 Kernelized Fuzzy Possibilistic C-Means (KFPCM)

Fuzzy Possibilistic c-Means (FPCM) is a combination of both Fuzzy and Possibilistic c-means. It is not only contained the membership function but also the typicality (t) in order to find the accurate characteristic for the dataset [6, 8, 13, 14]. The objective function of FPCM is shown in Eq. (19)

$$J_{FPCM}(U,T,V)=\sum_{i=1}^{c}\sum_{k=1}^{n}\left(u_{ik}^{m}+t_{ik}^{m}\right)\|x_k-v_i\|^2, \tag{19}$$

with constraints of u and t

$$\sum_{i=1}^{c}u_{ik}=1,\forall j\in\{1,\ldots,n\}, \tag{20}$$

$$\sum_{i=1}^{c}t_{ik}=1,\forall i\in\{1,\ldots,n\}. \tag{21}$$

By using Kernel function, we change the Euclidean distance with Eq. (19), the objective function of KFPCM is [8]:

$$\begin{aligned}J_{KFPCM}(U,T,V)&=\sum_{i=1}^{c}\sum_{k=1}^{n}\left(u_{ik}^{m}+t_{ik}^{m}\right)\|\phi(x_k)-\phi(v_i)\|^2\\&=\sum_{i=1}^{c}\sum_{k=1}^{n}\left(u_{ik}^{m}+t_{ik}^{m}\right)(2-2K(x_k,v_i))\end{aligned} \tag{22}$$

To minimize the objective function, the membership function u and v become

$$u_{ik}=\frac{(2-2K(x_k,v_i))^{-\frac{1}{m-1}}}{\sum_{j=1}^{c}(2-2K(x_k,v_j))^{-\frac{1}{m-1}}}, \tag{23}$$

$$v_i=\frac{\sum_{k=1}^{n}u_{ik}^{m}K(x_k,v_i)\mathrm{x_k}}{\sum_{k=1}^{n}u_{ik}^{m}K(x_k,v_i)}. \tag{24}$$

2.5 Kernelized Possibilistic Fuzzy C-Means (KPFCM)

Possibilistic Fuzzy c-Means (PFCM) is also a mixture of FCM and PCM formed to overcome the weaknesses of both FCM and PCM. Given dataset $\mathrm{X}=\{\mathrm{x}_1,\mathrm{x}_2,\ldots,\mathrm{x}_\mathrm{n}\}\subset\Re^p$, the objective function is as follows [14–16]:

$$\begin{aligned}J_{PFCM}(\mathbf{U},\mathbf{T},\mathbf{V})&=\sum_{k=1}^{n}\sum_{i=1}^{c}\left(au_{ik}^{m}+bt_{ik}^{\lambda}\right)\|x_k-v_i\|^2\\&+\sum_{i=1}^{c}\eta_i\sum_{k=1}^{n}(1-t_{ik})^{\lambda},\end{aligned} \tag{25}$$

with constraints

$$\sum_{i=1}^{c}u_{ik}=1,\forall j\in\{1,\ldots,n\}, \tag{26}$$

$$t_{ik}=\left(1+\left(\frac{b\|x_k-v_i\|^2}{\eta_i}\right)\right)^{-1}, \tag{27}$$

$$v_i = \frac{\sum_{k=1}^{n}\left(au_{ik}^{m} + bt_{ik}^{\lambda}\right)x_k}{\sum_{k=1}^{n}\left(au_{ik}^{m} + bt_{ik}^{\lambda}\right)}, \tag{28}$$

where u is the membership of x, $0 \leq u_{ij} \leq 1$, v is the cluster center, t is the typicality values depending on all data, $0 \leq t_{ij} \leq 1$, m and λ are the weighting exponent, $m, \lambda > 0$, a and b are constants defining the relative important for the membership and typicality, $a > 0, b > 0$.

By changing the Euclidean distance with high dimensional kernel space, the objective function is as follows [15]:

$$J_{KPFCM}(U,T,V) = \sum_{k=1}^{n}\sum_{i=1}^{c}\left(au_{ik}^{m} + bt_{ik}^{\lambda}\right)\|\phi(x_k) - \phi(v_i)\|^2 + \sum_{i=1}^{c}\eta_i\sum_{k=1}^{n}(1 - t_{ik})^{\lambda}. \tag{29}$$

To minimize Eq. (29) using Lagrangian multipliers, the function of membership and typicality becomes:

$$u_{ik} = \left[\sum_{j=1}^{c}\left(\frac{1 - K(\mathrm{x_k}, \mathrm{v_i})}{1 - K(\mathrm{x_k}, \mathrm{v_j})}\right)^{-\frac{1}{m-1}}\right]^{-1}, \forall i, k \tag{30}$$

$$t_{ik} = \left[1 + \left(\frac{b}{\eta_i}(2 - 2K(\mathrm{x_k}, \mathrm{v_i}))\right)^{-\frac{1}{\lambda-1}}\right]^{-1}, \forall i, k \tag{31}$$

$$\phi(\mathrm{v_i}) = \frac{\sum_{k=1}^{n}\left(au_{ik}^{m} + bt_{ik}^{\lambda}\right)\phi(\mathrm{x}_k)}{\sum_{k=1}^{n}\left(au_{ik}^{m} + bt_{ik}^{\lambda}\right)}, \forall i. \tag{32}$$

By multiplying each side of Eq. (32) by $\phi(x_j)^T$, the equation becomes as follows:

$$K(\mathrm{x_j}, \mathrm{v_i}) = \frac{\sum_{k=1}^{n}\left(au_{ik}^{m} + bt_{ik}^{\lambda}\right)K(\mathrm{x}_k, \mathrm{x}_j)}{\sum_{k=1}^{n}\left(au_{ik}^{m} + bt_{ik}^{\lambda}\right)}, \forall i, j \tag{33}$$

$$\eta_i = \beta\frac{\sum_{k=1}^{n} u_{ik}^{m}(2 - 2K(\mathrm{x_k}, \mathrm{v_i}))}{\sum_{k=1}^{n} u_{ik}^{m}}. \tag{34}$$

where β is a positive real number and it is commonly equal to one.

2.6 Statistical Measurement

In this study, we have evaluated the proposed method with several measurements described in Tables 1 and 2.

Table 1. Confusion matrix [17]

Actual vs. Predicted	Positive	Negative
Positive	TP	FP
Negative	FN	TN

Table 2. Terminology of statistical measurement [17]

Name	Formula	Function
Accuracy (A)	$A = \frac{TP+TN}{TP+TN+FP+FN}$	Measurement of the algorithm in prediction
Precision (P)	$P = \frac{TP}{TP+FP}$	Measure classifier correctness
Recall (R)	$R = \frac{TP}{TP+FN}$	Measure classifier sensitivity or completeness
F1 Score (F1)	$F1 = 2 \times \frac{P \times R}{P+R}$	Measure the weighted average of the precision and recall

3 Dataset

This dataset was obtained from RSCM Hospital Jakarta, Indonesia, which contains 4 features, 2 classes, and 200 instances. The description for each feature is explained in Table 3. Data authentication uses 10-fold cross validation.

Table 3. Description of features in sinusitis data

No.	Feature name	Feature description
1.	Gender (G)	L for man and W for woman
2.	Age	The patient's age when being checked
3.	Hounsfield Units (HU)	The sinusitis area, where the number is attained from one point of the area
4.	Air Cavity (AC)	Part of the bones which are normal and not affected by sinusitis
5.	Diagnosis (T)	K is chronic sinusitis and A is acute sinusitis

4 Experiment and Result

The performance for each model, KFCM, KPCM, KFPCM, and KPFCM was tested using sinusitis dataset. The performance was then compared based on their average accuracy, sensitivity, precision, and iteration time. Each of the model used 10-fold cross validation and Gaussian RBF Kernel with $\sigma = 1000$. The parameters used in the models are $m = 2$, $\lambda = 2$, $a = 0.5$, and $b = 0.5$.

The accuracy result for the methods used is shown in Table 4 and Fig. 1. Evidently, the highest accuracy was obtained by KFCM and KPFCM, where each attained 96,67%. Because the accuracy is the same, the model performance for overall methods was compared as shown in Table 5 and Fig. 2. Due to the similar results posted by KFCM and KPFCM, we compared the running time for each method as shown in Fig. 3. Finally, it is evident the fastest running time was achieved by FKCM, this is only 0,0109375 s.

Table 4. Accuracy for each method

Methods	Accuracy (%)
KFCM	96,66666667
KPCM	94,76190476
KFPCM	96,19047619
KPFCM	96,66666667

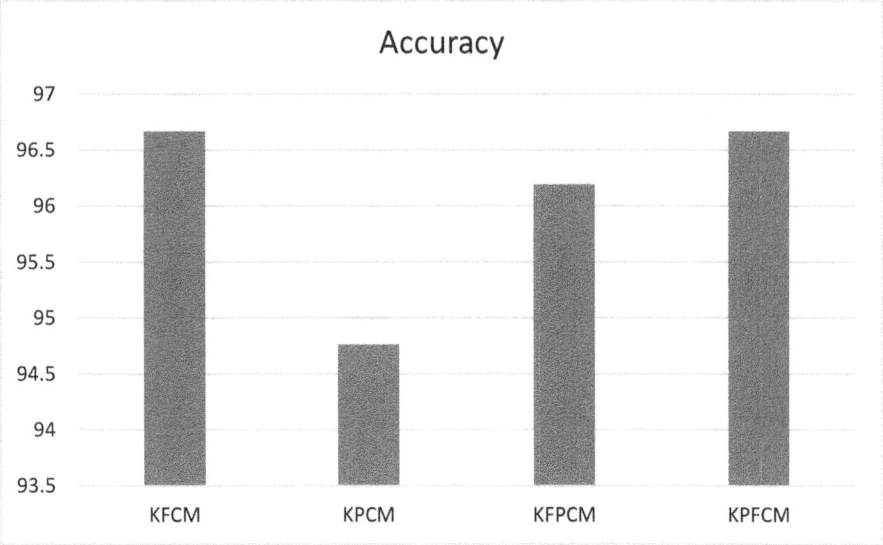

Fig. 1. Accuracy for each method

Table 5. Model performance

Methods	Accuracy (%)	Sensitivity (%)	Precision (%)	F-score (%)
KFCM	96,66666667	97,33333333	72,03599153	72,8696786
KPCM	94,76190476	96,07070707	53,24629499	56,0160967
KFPCM	96,19047619	97,22222222	71,59491884	72,09440075
KPFCM	96,66666667	97,33333333	72,03599153	72,8696786

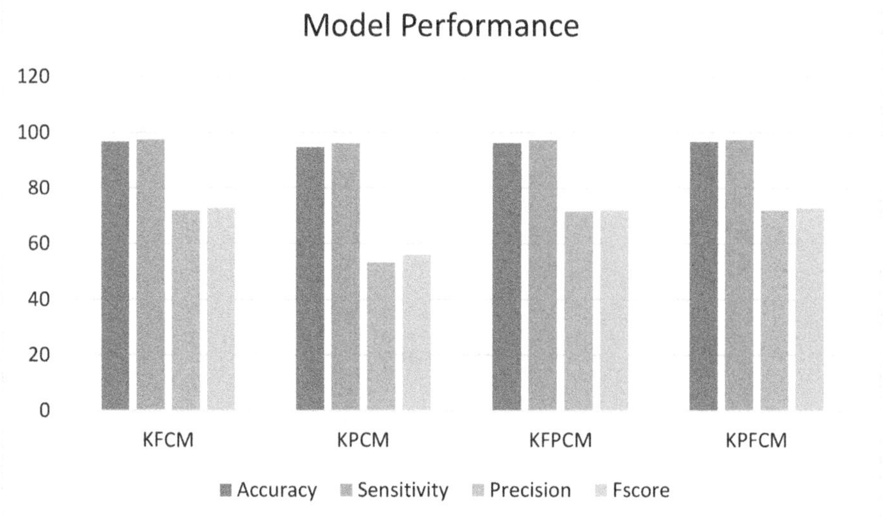

Fig. 2. Model performance

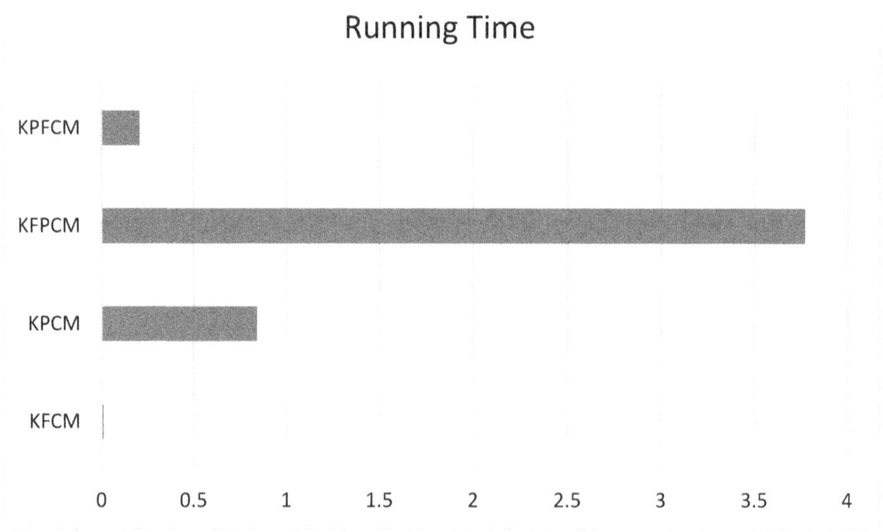

Fig. 3. Running time for each method

5 Discussion

The performance of kernel based fuzzy clustering was best with KFCM and KPFCM for sinusitis dataset, accuracy value. The other two methods, KPCM and KFPCM, were also compared using the accuracy, precision, recall, and f1 score. However, when viewed based on the running time, KFCM was the best.

6 Conclusion

This study analyzed sinusitis dataset using fuzzy clustering modified by changing the Euclidean distance with kernel function. The clustering methods used include Kernelized Fuzzy c-Means (KFCM), Kernelized Possibilistic c-Means (KPCM), Kernelized Fuzzy Possibilistic c-Means (KFPCM), and Kernelized Possibilistic Fuzzy c-Means (KPFCM). The addition of kernel functions was expected to improve the performance of the methods and reduce running time and cost. The results show that KFCM is better than the other methods which categorizes sinusitis dataset with accuracy of 96,97% and running time of 0,01 s.

Acknowledgement. We wish to express our gratitude to the University of Indonesia and PIT.9 2019 research grant scheme (ID number NKB-0039/UN2.R3.1/HKP.05.00/2019) for facilitating this study. This work was also supported by the Department Radiology of Dr. Cipto Mangunkusumo's Hospital, and we are so grateful to them. We also thank all the reviewers for the improvements made.

References

1. WebMD. https://www.webmd.com/allergies/sinusitis-and-sinus-infection#1
2. MediciNet. https://www.medicinenet.com/sinusitis/article.htm
3. Alodokter. https://www.alodokter.com/sinusitis
4. Suganya, R., Shanthi, R.: Fuzzy C-means algorithm – a review. Int. J. Sci. Res. Publ. **2**, 440–442 (2012)
5. Bezdek, J.C.: Pattern Recognition with Fuzzy Function Algorithms. Plenum Press, New York (1981)
6. Pal, N.R., Pal, K., Bezdek, J.C.: A mixed c-means clustering model. IEEE Trans. Fuzzy Syst. **1**, 11–21 (1997). https://doi.org/10.1109/fuzzy.1997.616338. IEEE
7. Krishnapuram, R., Keller, M.: A possibilistic approach to clustering. IEEE Trans. Fuzzy Syst. **1**(2), 98–110 (1993). https://doi.org/10.1109/91.227387
8. Shanmugapriya, B., Punithvalli, M.: A new kernelized fuzzy possibilistic C-means for high dimensional data clustering based on kernel-induced distance measure. In: International Conference on Computer Communication and Informatics, pp. 1–5, IEEE (2013). https://doi.org/10.1109/iccci.2013.6466319
9. Kalam, R., Thomas, C., Rahiman, M.A.: Gaussian kernel based fuzzy c-means clustering algorithm for image segmentation. In: The Second International Conference on Computer Science, Engineering, and Information Technology, pp. 47–46 (2016). https://doi.org/10.5121/csit.2016.60405

10. Rustam, Z., Wulan, A., Jannati, M.V., Fauzan, A.A.: Application kernel modified fuzzy c-means for glumatosis cerebri. In: Proceedings, 2016 12th International Conference on Mathematics, Statistics, and Their Applications, ICMSA: In Conjunction with the 6th Annual International Conference of Syiah Kuala University, pp. 33–38 (2017). https://doi.org/10.1109/icmsa.2016.7954303
11. Vijayarani, S., Sudha, S.: An efficient clustering algorithm for prediction diseases from hemogram blood test samples. Indian Journal of Science and Technology **18**, 1 (2015). https://doi.org/10.17485/ijst/2015/v8i17/52123
12. Wu, X.H.: A possibilistic C-means clustering algorithm based on kernel methods. In: International Conference on Communications, Circuits, and Systems, pp. 2062–2063. IEEE (2006). https://doi.org/10.1109/cimca.2005.1631512
13. Yong, Z., Yue'e, L., Shixiong, X.: Robust fuzzy-possibilistic C-means algorithm. In: Second International Symposium on Intelligent Information Technology Application, pp. 669–673. IEEE (2008). https://doi.org/10.1109/iita.2008.146
14. Pal, N.R., Pal, K., Keller, J.M., Bezdek, J.C.: A possibilistic fuzzy c-means clustering algorithm. IEEE Trans. Fuzzy Syst. **13**(5), 517–530 (2005). https://doi.org/10.1109/tfuzz.2004.840099. IEEE
15. Wu, X.H., Zhou, J.J.: Possibilistic fuzzy c-Means clustering model using kernel methods. In: Proceedings of the 2005 International Conference on Computational Intelligence for Modelling, Control and Automation, and International Conference Intelligent Agents, Web Technologies, and Internet Commerce, IEEE (2005). https://doi.org/10.1109/cimca.2005.1631512
16. Magaina, B.O., Ruelas, R., Corona-Nakamura, M.A., Andina, D.: An improvement to the possibilistic fuzzy c-Means clustering algorithm. In: World Automation Congress, Budapest, Hungary, IEEE (2006). https://doi.org/10.1109/wac.2006.376056
17. Burkov, A.: The Hundred-Page Machine Learning Book. Andriy Burkov, Canada (2019)

Mathematical Modelling for Complex Biochemical Networks and Identification of Fast and Slow Reactions

Sarbaz H. A. Khoshnaw[1(✉)] and Hemn M. Rasool[2]

[1] Department of Mathematics, College of Basic Education, University of Raparin, Main Road, Ranya, Kurdistan Region, Iraq
sarbaz.hamza@uor.edu.krd

[2] Department of Mathematics, Faculty of Science and Health, Koya University, Daniel Mitterrand Boulevard, Koya KOY45, Kurdistan Region, Iraq
hemn.muhammad@koyauniversity.org

Abstract. This paper reviews some mathematical models of biochemical reaction networks. The models contain a large number species and reactions. This is a difficult task and it requires some effective computational tools. Techniques of model reduction are important approaches for minimizing the number of elements. One of the classical and common techniques of model reduction is quasi equilibrium approximation (QEA). According to this approach, the fast reactions simply reach equilibrium very fast. It allows one to classify the model reaction rates into slow and fast terms. This study suggest QEA technique to simplify and calculate analytical approximate solutions for non–competitive inhibition enzymatic reactions in different cases. On the other hand, the suggested method may not work very well analytically with higher dimensional biochemical networks. As a result, we propose an algorithm to identify slow and fast reactions in complex systems. The proposed algorithm provides a great step further in developing QEA technique. This is applied to dihydrofolate reductase (DHFR) cell signaling pathways. The algorithm would be easily applied by biologists and chemists for various purposes such as identifying slow–fast reactions and critical model elements. Finally, computational simulations show that many cell signalling pathways can reach equilibrium in a short interval of time. Interestingly, all reactions mainly become slow for a long range of time.

Keywords: Mathematical modelling · Quasi–equilibrium approximation · Slow and fast reactions · Computational simulations · Cell signalling pathways

1 Introduction

There are a variety of kinetic reactions with non-linear terms. The majority of such models include high–dimensional species and reactions. This becomes a

Supported by University of Raparin and Koya University.

M. Zeki Sarıkaya et al. (Eds.): ICMRS 2019, LNNS 123, pp. 55–69, 2020.
https://doi.org/10.1007/978-3-030-43002-3_6

difficult task in modelling process because such models sometimes need a huge array of computational tools for simulations and identifications. Thus, mathematical methods are effective tools to solve such problems. One modern way to provide a reasonable answer for the issue is minimizing the number of elements. This is sometimes called model reduction in systems biology. According to this approach, the dynamics of original and simplified models should be similar in appearance, but contain fewer variables and parameters. There are some well known techniques of model reduction for complex biochemical networks. The geometric singular perturbation method is a classical method of model reduction. This is applied for kinetic models with slow and fast variables [1]. Another common method of model reduction is quasi steady state approximation (QSSA), this is also used when the models equations divided into slow and fast subsystems [2]. Accordingly, quasi equilibrium approximation (QEA) that is sometimes called rapid equilibrium approximation is well known approach for minimizing the number of species and parameters in cell signalling pathways [3,4]. Recently, the method has been used for minimizing the number of elements in biochemical cell signalling pathways [5,6]. More recently, the method was further improved and applied in enzymatic reactions [7]. Further details and understanding about the method can be seen in these studies [8–10]. The problem in this study is identifying the slow and fast reactions for complex biochemical reaction networks. Therefore, we study quasi equilibrium approximation to divide model reactions into slow and fast reactions. Then, we analyse the fast subsystems and calculate some analytical approximate solutions. We apply QEA method to non–competitive inhibition enzymatic reaction network for two different cases. When the model equations have higher dimensional elements then applying QEA may not be easy analytically. This is because such models may have many possibilities to identify slow and fast reactions. Therefore, a computational procedure is required to separate fast and slow reactions for complex networks. Consequently, we propose a simple algorithm to identifying slow–fast subsystems for complex biochemical reaction networks. The suggested algorithm applies on dihydrofolate reductase (DHFR) cell signalling pathways. Results show that the majority of reversible reactions for complex cell signalling pathways get equilibrium very fast on short interval of time. Interestingly, the main contribution in this study is dividing slow and fast subsystems for complex biochemical reaction networks based on some suggested steps.

2 Methods

We consider n reversible reactions which are given below

$$\sum_{j=1}^{m} \alpha_{ij} A_j \underset{k_i^-}{\overset{k_i^+}{\rightleftharpoons}} \sum_{j=1}^{m} \beta_{ij} A_j, \quad i = 1, 2, ..., n. \tag{1}$$

where $A_j, j = 1, 2, ..., m$ are chemical components, α_{ij} and β_{ij} are non–negative integers, $k_i^+ > 0$ and $k_i^- \geq 0$ are the reaction rate coefficients [11,12]. The reaction rates are expressed as follows

$$v_i = k_i^+ \prod_{j=1}^{m} c_j^{\alpha_{ij}}(t) - k_i^- \prod_{j=1}^{m} c_j^{\beta_{ij}}(t). \tag{2}$$

Therefore, the model equations can be given

$$\frac{dC}{dt} = \sum_i \gamma_i v_i. \tag{3}$$

where $\gamma_{ij} = \beta_{ij} - \alpha_{ij}$, for $i = 1, 2, ..., n$ and $j = 1, 2, .., m$.

The QEA method was introduced as a model reduction technique. This is for minimizing the number of elements for such systems. Recently, Gorban and Karlin developed the idea of QEA for invariant manifolds in physical and chemical kinetics [13]. The main idea of QEA is related to divide the model reaction rates into slow and fast. Then, Eq. (3) takes the form:

$$\frac{dC}{dt} = \sum_{s,slow} \gamma^s V^s(C, K, t) + \frac{1}{\epsilon} \sum_{f,fast} \gamma^f V^f(C, K, t), \tag{4}$$

where ϵ is a small parameter ($0 < \epsilon \ll 1$), V^s and V^f are slow and fast reaction rates respectively, γ^s and γ^f are slow and fast stoichiometric vectors respectively. The fast subsystem becomes

$$\frac{dC}{dt} = \frac{1}{\epsilon} \sum_{f,fast} \gamma^f V^f(C, K, t). \tag{5}$$

The reader can see further details about the method and its applications in [5,7]. The fast subsystems can be analysed in order to calculate some analytical approximate solutions. Therefore, the quasi–equilibrium manifold is given by the following algebraic equations

$$\begin{aligned} &\sum_{f,fast} \gamma^f V^f(C, K, t) = 0, \\ &h^i(C) = b_i, \end{aligned} \tag{6}$$

where $h^i(C)$ for $i = 1, 2, ..., p$ are linear independent conservation laws for the fast subsystems and b_i are constants functions. For small parameter ϵ, the equations $\sum_{f,fast} \gamma^f V^f(C, K, t) = 0$ gives some approximate solutions. In this study, we have applied the above method to simplify and analyse some enzymatic chemical reactions. The reader can see further details and applications of model reductions in systems biology in [14–24].

3 Non-competitive Inhibition Enzymatic Reactions

An important branch of studying enzymatic reactions is enzyme inhibition. This is occurred in any organic or inorganic chemical compound that shared their molecules with enzyme active sites. Almost all enzyme inhibitors have less molecular compounds. They can be combined with enzymes in order to have enzyme–inhibitor complex. Inhibitors play important roles in biological and biochemical models such as mechanism controlling, regulating metabolic activities and identifying active side for amino acids. For understanding how inhibitors work with enzymes [25]; see Fig. 1.

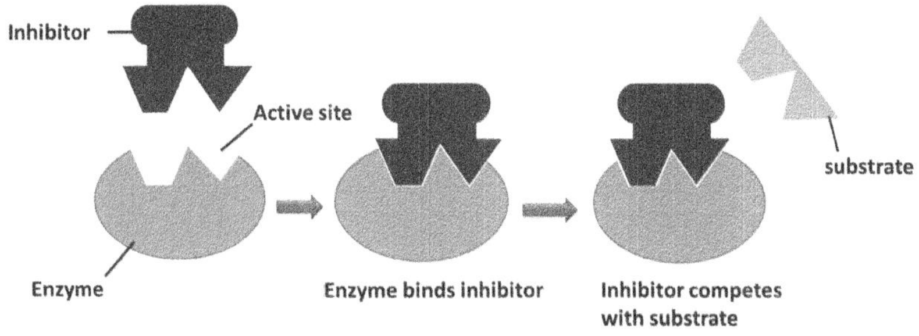

Fig. 1. Demonstrating how competitive inhibition works with enzymes [25].

The chemical reaction networks for non–competitive inhibition enzymatic reactions are defined below

$$\begin{array}{ccccccc} & E & + & S & \underset{k_{-1}}{\overset{k_1}{\rightleftharpoons}} & ES & \overset{k_2}{\longrightarrow} & E+P \\ & + & & & & + & & \\ & I & & & & I & & \\ k_{-3} & \upharpoonleft\downharpoonright & k_3 & & k_{-4} & \upharpoonleft\downharpoonright & k_4 & \\ & EI & + & S & \underset{k_{-5}}{\overset{k_5}{\rightleftharpoons}} & ESI & \overset{k_6}{\longrightarrow} & EI+P \end{array} \quad (7)$$

where E, S, I, ES, EI, ESI and P are enzyme, substrate, inhibitor, enzyme substrate complex, enzyme inhibitor complex, enzyme substrate inhibitor complex and product, respectively. The parameters $k_1, k_{-1}, k_2, k_3, k_{-3}, k_4, k_{-4}, k_5, k_{-5}$ and k_6 are kinetic constants [25]. The concentrations of species are $e = [E], s = [S], i = [I], p = [P], c_1 = [ES], c_2 = [EI]$ and $c_3 = [ESI]$. The model equations are

$$
\begin{aligned}
\frac{ds}{dt} &= -k_1es + k_{-1}c_1 - k_5c_2 + k_{-5}c_3,\\
\frac{de}{dt} &= -k_1es + k_{-1}c_1 - k_3ei + k_{-3}c_2 + k_2c_1,\\
\frac{di}{dt} &= -k_3ei + k_{-3}c_2 - k_4c_1i + k_{-4}c_3,\\
\frac{dc_1}{dt} &= k_1es - k_{-1}c_1 - k_4c_1i + k_{-4}c_3 - k_2c_1,\\
\frac{dc_2}{dt} &= k_3ei - k_{-3}c_2 - k_5c_2s + k_{-5}c_3 + k_6c_3,\\
\frac{dc_3}{dt} &= k_5c_2s - k_{-5}c_3 + k_4c_1i - k_{-4}c_3 - k_6c_3,\\
\frac{dp}{dt} &= k_2c_1 + k_6c_3,
\end{aligned}
\tag{8}
$$

with the initial conditions

$$
e(0) = e_0, s(0) = s_0, i(0) = i_0, c_1(0) = c_2(0) = c_3(0) = p(0) = 0. \tag{9}
$$

There are there independent stoichiometric conservation laws:

$$
e + c_1 + c_2 + c_3 = e_0, s + p + c_1 + c_3 = s_0, i + c_2 + c_3 = i_0. \tag{10}
$$

In order to simplify model equations and calculate some analytical solutions, we assume two different cases for fast reactions as they are given below.

3.1 Case A: One Fast Reaction

We assume that the model network has only one fast reversible reaction. By applying QEA method for chemical reactions (7), if possible suppose that the first reaction ($E + S \underset{k_{-1}}{\overset{k_1}{\rightleftharpoons}} ES$) becomes quasi–equilibrium. Let $k_1 = \frac{k_1^+}{\epsilon}$ and $k_{-1} = \frac{k_1^-}{\epsilon}$ where $\epsilon = \frac{e_0}{s_0}$ then $k_1^+ = \frac{k_1 e_0}{s_0}$ and $k_1^- = \frac{k_{-1} e_0}{s_0}$. Then, there are slow and fast reaction rates

$$
\begin{aligned}
\frac{ds}{dt} &= \frac{1}{\epsilon} g^f(s, e, c_1, t) + g_1^s(s, c_2, c_3, t),\\
\frac{de}{dt} &= \frac{1}{\epsilon} g^f(s, e, c_1, t) + g_2^s(e, i, c_2, t) + g_3^s(c_1, t),\\
\frac{di}{dt} &= g_2^s(e, i, c_2, t) + g_4^s(i, c_1, c_3, t),\\
\frac{dc_1}{dt} &= \frac{-1}{\epsilon} g^f(s, e, c_1, t) + g_4^s(i, c_1, c_3, t) - g_3^s(c_1, t),\\
\frac{dc_2}{dt} &= -g_2^s(e, i, c_2, t) + g_1^s(s, c_2, c_3, t) + g_5^s(c_3, t),\\
\frac{dc_3}{dt} &= -g_1^s(s, c_2, c_3, t) - g_4^s(i, c_1, c_3, t) - g_5^s(c_3, t),\\
\frac{dp}{dt} &= g_3^s(c_1, t) + g_5^s(c_3, t),
\end{aligned}
\tag{11}
$$

where $g^f(s,e,c_1,t) = -k_1^+ es + k_1^- c_1$, $g_1^s(s,c_2,c_3,t) = -k_5 c_2 s + k_{-5} c_3$, $g_2^s(e,i,c_2,t) = -k_3 ei + k_{-3} c_2, g_3^s(c_1,t) = k_2 c_1, g_4^s(i,c_1,c_3,t) = -k_4 c_1 i + k_{-4} c_3$ and $g_5^s(c_3,t) = k_6 c_3$. For $\epsilon \longrightarrow 0$, the quasi–equilibrium approximation can be applied. Then, the fast subsystems given

$$\begin{aligned} \frac{ds}{dt} &= \frac{1}{\epsilon}(-k_1^+ es + k_1^- c_1), \\ \frac{de}{dt} &= \frac{1}{\epsilon}(-k_1^+ es + k_1^- c_1), \\ \frac{dc_1}{dt} &= \frac{-1}{\epsilon}(-k_1^+ es + k_1^- c_1). \end{aligned} \tag{12}$$

Therefore, we obtain two slow variables $b_1(s,c_1) = s + c_1$ and $b_2(e,c_1) = e + c_1$. From the equation $g^f(s,e,c_1,t) = 0$, the slow manifold is calculated

$$M_0^* = \left\{ (s,e,c_1) \in R^3 : s = \frac{k_1^- c_1}{k_1^+ e} \right\}. \tag{13}$$

We fix slow variables (b_1 and b_2), we obtain algebraic equations

$$k_1^+ es - k_1^- c_1 = 0, s + c_1 = b_1, e + c_1 = b_2. \tag{14}$$

Thus, a quadratic equation for c_1 is

$$k_1^+ c_1^2 - (k_1^+ b_1 + k_1^+ b_2 + k_1^-) c_1 + k_1^+ b_1 b_2 = 0. \tag{15}$$

We can solve Eq. (15) for c_1

$$c_1(b_1,b_2) = \frac{1}{2}\left[\left(b_1 + b_2 + \frac{k_1^-}{k_1^+}\right) \pm \sqrt{\left(b_1 + b_2 + \frac{k_1^-}{k_1^+}\right)^2 - 4b_1 b_2} \right].$$

The solution for s and e are

$$s(b_1,b_2) = b_1 - \frac{1}{2}\left[\left(b_1 + b_2 + \frac{k_1^-}{k_1^+}\right) - \sqrt{\left(b_1 + b_2 + \frac{k_1^-}{k_1^+}\right)^2 - 4b_1 b_2} \right],$$

$$e(b_1,b_2) = b_2 - \frac{1}{2}\left[\left(b_1 + b_2 + \frac{k_1^-}{k_1^+}\right) - \sqrt{\left(b_1 + b_2 + \frac{k_1^-}{k_1^+}\right)^2 - 4b_1 b_2} \right].$$

To simplify the above solutions, it can be assumed that the concentration of substrate is much large than the total concentration of enzyme.

$$[S] \gg [ES] \quad i.e. \quad b_1 \gg c. \tag{16}$$

Then, Eq. (15) takes the form

$$\left(1 + \frac{b_2}{b_1} + \frac{k_1^-}{k_1^+ b_1}\right) c_1 = b_2 + O\left(\frac{c_1}{b_1}\right). \tag{17}$$

The approximate solutions are

$$c_1 \approx \frac{b_1 b_2}{b_1 + b_2 + k^*}, s(b_1,b_2) = \frac{b_1(b_1 + k^*)}{b_1 + b_2 + k^*}, e(b_1,b_2) = \frac{b_2(b_2 + k^*)}{b_1 + b_2 + k^*}, \tag{18}$$

where $k^* = k_1^- / k_1^+$.

3.2 Case B: Two Fast Reactions

We assume that the model network has only two fast reversible reactions. If possible suppose that the first and the third reactions ($E+S \underset{k_{-1}}{\overset{k_1}{\rightleftharpoons}} ES$ and $E+I \underset{k_{-3}}{\overset{k_3}{\rightleftharpoons}} EI$) become quasi–equilibrium. Let $k_1 = \frac{k_1^+}{\epsilon}, k_{-1} = \frac{k_1^-}{\epsilon}$ and $k_3 = \frac{k_3^+}{\epsilon}, k_{-3} = \frac{k_3^-}{\epsilon}$ where $\epsilon = \frac{e_0}{s_0}$ then $k_1^+ = \frac{k_1 e_0}{s_0}, k_1^- = \frac{k_{-1} e_0}{s_0}$ and $k_3^+ = \frac{k_3 e_0}{s_0}, k_3^- = \frac{k_{-3} e_0}{s_0}$. Then, the model equations become

$$\begin{aligned}
\frac{ds}{dt} &= \frac{1}{\epsilon} g_1^f(s, e, c_1, t) + g_1^s(s, c_2, c_3, t),\\
\frac{de}{dt} &= \frac{1}{\epsilon} g_1^f(s, e, c_1, t) + \frac{1}{\epsilon} g_2^f(e, i, c_2, t) + g_2^s(c_1, t),\\
\frac{di}{dt} &= \frac{1}{\epsilon} g_2^f(e, i, c_2, t) + g_3^s(i, c_1, c_3, t),\\
\frac{dc_1}{dt} &= \frac{-1}{\epsilon} g_1^f(s, e, c_1, t) + g_3^s(i, c_1, c_3, t) - g_2^s(c_1, t),\\
\frac{dc_2}{dt} &= \frac{-1}{\epsilon} g_2^{f_2}(e, i, c_2, t) + g_1^s(s, c_2, c_3, t) + g_4^s(c_3, t),\\
\frac{dc_3}{dt} &= -g_1^s(s, c_2, c_3, t) - g_3^s(i, c_1, c_3, t) - g_4^s(c_3, t),\\
\frac{dp}{dt} &= g_2^s(c_1, t) + g_4^s(c_3, t),
\end{aligned} \tag{19}$$

where $g_1^f(s, e, c_1, t) = -k_1^+ es + k_1^- c_1, g_2^f(e, i, c_2, t) = -k_3^+ ei + k_{-3}^- c_2$, $g_1^s(s, c_2, c_3, t) = -k_5 c_2 s + k_{-5} c_3, g_2^s(c_1, t) = k_2 c_1, g_3^s(i, c_1, c_3, t) = -k_4 c_1 i + k_{-4} c_3$ and $g_4^s(c_3, t) = k_6 c_3$. For $\epsilon \longrightarrow 0$, the quasi–equilibrium approximation can be applied. The fast subsystems are

$$\begin{aligned}
\frac{ds}{dt} &= \frac{1}{\epsilon}(-k_1^+ es + k_1^- c_1),\\
\frac{de}{dt} &= \frac{1}{\epsilon}(-k_1^+ es + k_1^- c_1) + \frac{1}{\epsilon}(-k_3^+ ei + k_3^- c_2),\\
\frac{di}{dt} &= \frac{1}{\epsilon}(-k_3^+ ei + k_3^- c_2),\\
\frac{dc_1}{dt} &= \frac{-1}{\epsilon}(-k_1^+ es + k_1^- c_1),\\
\frac{dc_2}{dt} &= \frac{-1}{\epsilon}(-k_3^+ ei + k_3^- c_2).
\end{aligned} \tag{20}$$

Thus, there model has three slow variables $b_1(s, c_1) = s + c_1$, $b_2(e, c_1, c_2) = e + c_1 + c_2$ and $b_3(i, c_2) = i + c_2$. After solving $g_1^f(s, e, c_1, t) = 0$ and $g_2^f(e, i, c_2) = 0$, the slow manifold is given

$$M_0^* = \left\{ (s, e, c_1) \in R^3 : s = \frac{k_1^- c_1}{k_1^+ e}, i = \frac{k_3^- c_2}{k_3^+ e} \right\}. \tag{21}$$

Again, we have some algebraic equations

$$\begin{aligned}
&k_1^+ es - k_1^- c_1 = 0, k_3^+ ei - k_3^- c_2 = 0,\\
&s + c_1 = b_1, e + c_1 + c_2 = b_2, i + c_2 = b_3.
\end{aligned} \tag{22}$$

Then, there are two non-linear algebraic equations

$$\begin{aligned} k_1^+ c_1^2 + k_1^+ c_1 c_2 - (k_1^+ b_1 + k_1^+ b_2 + k_1^-) c_1 - k_1^+ b_1 c_2 + k_1^+ b_1 b_2 = 0, \\ k_3^+ c_2^2 + k_3^+ c_1 c_2 - (k_3^+ b_2 + k_3^+ b_3 + k_3^-) c_2 - k_3^+ b_3 c_1 + k_3^+ b_2 b_3 = 0. \end{aligned} \tag{23}$$

The above equations can be simplified as follows

$$\begin{aligned} \left(1 + \frac{b_2}{b_1} + \frac{k_1^-}{k_1^+ b_1}\right) c_1 + c_2 = b_2 + O\left(\frac{c_1 c_2}{b_1}\right), \\ \left(\frac{b_2}{b_1} + \frac{b_3}{b_1} + \frac{k_3^-}{k_3^+ b_1}\right) c_2 = \frac{b_2 b_3}{b_1} + O\left(\frac{c_1 c_2}{b_1}\right). \end{aligned} \tag{24}$$

The analytical solutions for c_1 and c_2 are

$$\begin{aligned} c_1 &\approx \frac{b_1 b_2 (b_2 + k_3^*)}{(b_1 + b_2 + K_1^*)(b_2 + b_3 + k_3^*)}, \\ c_2 &\approx \frac{b_2 b_3}{b_2 + b_3 + k_3^*}, \end{aligned} \tag{25}$$

where $k_1^* = k_1^- / k_1^+$ and $k_3^* = k_3^- / k_3^+$. The analytical solution for s, e and i are obtained:

$$\begin{aligned} s(b_1, b_2, b_3) &= b_1 - \frac{b_1 b_2 (b_2 + k_3^*)}{(b_1 + b_2 + K_1^*)(b_2 + b_3 + k_3^*)}, \\ e(b_1, b_2, b_3) &= \frac{b_2 (b_2 + k_3^*)(b_2 + k_1^*)}{(b_1 + b_2 + K_1^*)(b_2 + b_3 + k_3^*)}, \\ i(b_1, b_2, b_3) &= \frac{b_3 (b_3 + k_3^*)}{b_2 + b_3 + k_3^*}. \end{aligned}$$

4 An Algorithm for Identifying Slow and Fast Reactions

Identifying slow and fast reactions become a difficult task analytically, and it may be impossible for complex biochemical reaction networks. Therefore, we need an algorithm that gives us a good step forward in identifying slow and fast reactions. In this study, we propose some steps for identifying slow and fast reactions.

Step One: Consider a chemical reaction network with m reversible reactions and n variables

$$\sum_{i=1}^{n} \alpha_{ij} A_i \underset{k_j^-}{\overset{k_j^+}{\rightleftharpoons}} \sum_{i=1}^{n} \beta_{ij} A_i, \quad j = 1, 2, ..., m.$$

Step Two: Calculate all forward and backward reaction rates

$$\begin{aligned} v_j^f(t) &= k_j^+ \prod_{i=1}^{n} [A_i]^{\alpha_{ij}}(t), \\ v_j^b(t) &= k_j^- \prod_{i=1}^{n} [A_i]^{\alpha_{ij}}(t), \quad j = 1, 2, ..., m, \end{aligned}$$

Step Three: Compute

$$|v_j(t)| = |v_j^f(t) - v_j^b(t)| \quad \texttt{for}\ t \in T \subset R.$$

Step Four: If $|v_j(t)| < \epsilon$, for $j = 1, 2, ..., p$ where $p \leqslant m$ and $0 < \epsilon \ll 1$, then the fast reactions are

$$R^f = \left\{ v_j : |v_j(t)| < \epsilon, j = 1, 2, ..., p.p \leqslant m \right\}$$

and the slow reactions are

$$R^s = \left\{ v_j : |v_j(t)| \not< \epsilon, j = 1, 2, ..., k.k \leqslant m \right\}$$

where $k + p = m$, and $R = R^f \cup R^s$; R is a set of all reactions, R^f is a set of all fast reactions and R^s is a set of all slow reactions.

4.1 Dihydrofolate Reductase (DHFR) Pathways

Dihydrofolate Reductase is one of the most important enzymes for DNA synthesis. It produces cofactor for DNA and other processes in cell signalling. In humans, the DHFR enzyme is encoded by the DHFR gene it is found in the region of chromosome 5. DHFR is a notable drug target for the design of anti–malarial, anti–bacterial, and anti–cancer drugs. DHFR inhibitors are in wide use as antibacterial and antiprotozoal agents [26]. The chemical reaction network for DHRF model is shown in Fig. 2. All reactions in the model are considered to be reversible. This includes thirteen reversible reactions (28 parameters) and thirteen state variables (concentrations). The stationary values for model initial species and parameters are given in Table 1. Using mass action law, the model equations are expressed as a system of ordinary differential equations:

$$\begin{aligned}
&\frac{d[E]}{dt} = -v_1 - v_2 + v_8 + v_9, && \frac{d[H2F]}{dt} = -v_1 + v_5 + v_{13},\\
&\frac{d[EH2F]}{dt} = v_1 - v_3 - v_{12}, && \frac{d[NH]}{dt} = -v_2 - v_3 + v_{11},\\
&\frac{d[ENH]}{dt} = v_2 + v_5 - v_{10}, && \frac{d[ENHH2F]}{dt} = v_3 - v_4 - v_5,\\
&\frac{d[ENH4F]}{dt} = v_4 + v_6 - v_7, && \frac{d[EH4F]}{dt} = -v_6 - v_9 + v_{11},\\
&\frac{d[N]}{dt} = -v_6 + v_8 - v_{12}, && \frac{d[EN]}{dt} = v_7 - v_8 + v_{13},\\
&\frac{d[H4F]}{dt} = v_7 + v_9 - v_{10}, && \frac{d[ENHH4F]}{dt} = v_{10} - v_{11},\\
&\frac{d[ENH2F]}{dt} = v_{12} - v_{13},
\end{aligned} \tag{26}$$

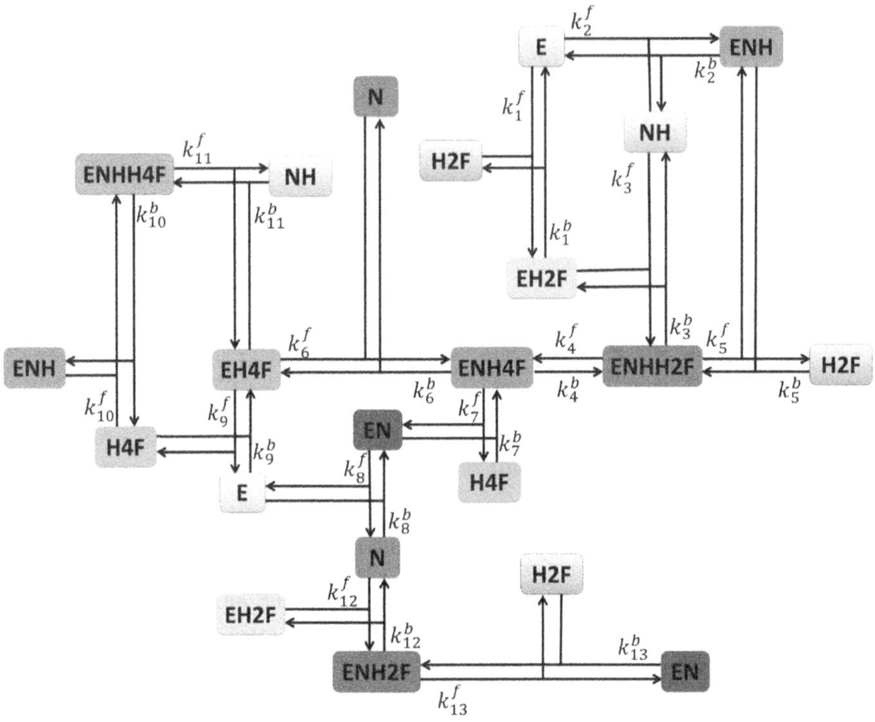

Fig. 2. The Dihydrofolate reductase DHFR pathways [26].

Table 1. The model initial concentrations and parameters used for computational simulations.

Initial concentrations	Parameters
E = 0.5	k1f = 264, k1b = 14
H2F = 0.3	k2f = 38, k2b = 1.7
EH2F = 0	k3f = 24, k3b = 19
NH = 0.08	k4f = 1360, k4b = 37
ENH = 0	k5f = 94, k5b = 98
ENHH2F = 0	k6f = 0.7, k6b = 84
ENH4F = 0	k7f = 46, k7b = 24
EH4F = 0.05	k8f = 32, k8b = 17
N = 0.06	k9f = 5.1, k9b = 117
EN = 0	k10f = 14, k10b = 225
H4F = 0	k11f = 100, k11b = 4.4
ENHH4F = 0	k12f = 20, k12b = 4.6
ENH2F = 0	k13f = 110, k13b = 1.3

where k_i^f and k_i^b for $i = 1, 2, ..., 13$ are the forward and backward reaction constants, respectively. The model reaction rates are given

$$\begin{aligned}
v_1 &= k_1^f[E](t)[H2F](t) - k_1^b[EH2F](t), \\
v_2 &= k_2^f[E](t)[NH](t) - k_2^b[ENH](t), \\
v_3 &= k_3^f[EH2F](t)[NH](t) - k_3^b[ENHH2F](t), \\
v_4 &= k_4^f[ENHH2F](t) - k_4^b[ENH4F](t), \\
v_5 &= k_5^f[ENHH2F](t) - k_5^b[ENH](t)[H2F](t), \\
v_6 &= k_6^f[EH4F](t)[N](t) - k_6^b[ENH4F](t), \\
v_7 &= k_7^f[ENH4F](t) - k_7^b[EN](t)[H4F](t), \\
v_8 &= k_8^f[EN](t) - k_8^b[E](t)[N](t), \\
v_9 &= k_9^f[EH4F](t) - k_9^b[E](t)[H4F](t), \\
v_{10} &= k_{10}^f[ENH](t)[H4F](t) - k_{10}^b[ENHH4F](t), \\
v_{11} &= k_{11}^f[ENHH4F](t) - k_{11}^b[EH4F](t)[NH](t), \\
v_{12} &= k_{12}^f[EH2F](t)[N](t) - k_{12}^b[ENH2F](t), \\
v_{13} &= k_{13}^f[ENH2F](t) - k_{13}^b[EN](t)[H2F](t).
\end{aligned} \tag{27}$$

We applied the suggested steps of the above algorithm on DHFR signalling pathways in order to identify slow and fast reactions in computational simulations for different time intervals; see Tables 2 and 3. Results show that fast reactions are more than slow reactions when $t \in [0, 92]$. On the other hand, the number of slow reactions is much greater than fast reactions when $t > 92$. Particularly, the model has no more fast reactions when $t \geq 110$. We can conclude that the fast reactions are mainly get equilibrium very quickly on a short interval of time whereas for a long range of time interval they are slowly reach equilibrium.

The results show that our proposed steps of identifying slow and fast reactions become a great step forward in model reduction. This is because identifying fast subsystems help us to analyse the model dynamics and to compare the reduced models with the original models. The reliability and accuracy of our proposed algorithm is mainly depended on the slow–fast parameter ϵ. This is normally a very small constant between zero and one $(0 < \epsilon \ll 1)$.

Applying the technique of QEA analytically is for lower dimensional models such as Eq. 7. This model needs some calculations and simplifications to get reduced models. On the other hand, applying the method for complex cell signaling pathways with higher dimensional elements, it will be a difficult task or it may be impossible to test all cases for fast and slow reactions. Therefore, we have proposed the algorithm firstly to identify slow and fast reactions in computational simulations. After that the QEA can be applied to obtain reduced models and simplifications. But again it is not easy to solve many equations analytically, this needs some numerical tools to have some numerical approximate solutions.

Table 2. Identifying slow and fast reactions for DHFR signalling pathways based on the suggested steps in computational simulations for different time intervals ($t \in [0, 92]$, $t = 95$ and $t = 100$).

When $t \in [0, 92]$		When $t = 95$		When $t = 100$	
Slow	Fast	Slow	Fast	Slow	Fast
reaction 1	reaction 3	reaction 1	reaction 4	reaction 1	reaction 7
reaction 2	reaction 4	reaction 2	reaction 7	reaction 2	reaction 10
reaction 6	reaction 5	reaction 3	reaction 10	reaction 3	reaction 13
reaction 8	reaction 7	reaction 5	reaction 13	reaction 4	
reaction 9	reaction 10	reaction 6		reaction 5	
reaction 11	reaction 12	reaction 8		reaction 6	
	reaction 13	reaction 9		reaction 8	
		reaction 11		reaction 9	
		reaction 12		reaction 11	
				reaction 12	

Table 3. Identifying slow and fast reactions for DHFR signalling pathways based on the suggested steps in computational simulations for different time intervals ($t = 105, 106$ and $t \geq 110$).

When $t = 105$		When $t = 106$		When $t \geq 110$	
Slow	Fast	Slow	Fast	Slow	Fast
reaction 1	reaction 7	reaction 1	reaction 7	reaction 1	
reaction 2	reaction 13	reaction 2		reaction 2	
reaction 3		reaction 3		reaction 3	
reaction 4		reaction 4		reaction 4	
reaction 5		reaction 5		reaction 5	
reaction 6		reaction 6		reaction 6	
reaction 8		reaction 8		reaction 7	
reaction 9		reaction 9		reaction 8	
reaction 10		reaction 10		reaction 9	
reaction 11		reaction 11		reaction 10	
reaction 12		reaction 12		reaction 11	
		reaction 13		reaction 12	
				reaction 13	

5 Conclusions

The complex models in biochemical reactions need some simplifications and reductions. Therefore, methods of model reduction play an important role to minimize the number of parameters and variables. We defined the mathematical modelling for non-competitive inhibition enzymatic reactions based on mass action law, with constant rates. Then, QEA method has been used in two different cases. This is for separating model equations into slow and fast reaction rates. All fast subsystems are analysed and slow manifold are calculated. Results give us some analytical approximate solutions for the reduced models. The simplified models in this study are great step forward because they are simpler, easier to understand and manipulate.

Applying QEA method is sometimes not easy especially when chemical reaction networks have a high number of species and reactions. In this situation, separating reaction rates into slow and fast subsystems become a difficult task. Therefore, we generalized the idea of QEA for complex biochemical reaction networks when the models have many mechanisms and pathways. Accordingly, we proposed the algorithm to identify slow and fast reactions. Furthermore, we applied the suggested steps on dihydrofolate reductase (DHFR) cell signalling pathways to divide the model reactions into fast-slow reactions. The computational simulations are calculated using MATLAB for given initial concentrations and parameters; see Tables 2 and 3.

The accuracy of our results is mainly related to the parameter ϵ $(0 < \epsilon \ll 1)$. This is because in reality if the difference between the forward and backward reaction rates becomes smaller and smaller then some of the reactions go to their equilibrium very quickly.

The suggested algorithm in this work helps us for further studying and understanding the complex biochemical reaction networks in many ways. Firstly, identifying slow and fast reactions become a great step forward to analyse fast subsystems and calculate slow manifolds. Secondly, calculating model solutions provides suggestions for its future development. Interestingly, the results provide us all reversible reactions go to their equilibrium very fast on the short interval of time while they are reached to equilibrium slowly on the long range of time. The proposed algorithm may help biologist and chemists for separating reactions into slow and fast particularly for high dimension cell signalling pathways.

References

1. Jones, C.K.R.T.: Geometric singular perturbation theory. In: Dynamical Systems. Lecture Notes in Mathematics, vol. 1609, pp. 44–118 (1995)
2. Briggs, G.E., Haldane, J.B.: A Note on the Kinetics of Enzyme Action. Biochem. J. **19**, 338–339 (1925). https://doi.org/10.1042/bj0190338
3. Vasiliev, V.M., Volpert, A.I., Hudiaev, S.I.: A method of quasi stationary concentrations for chemical kinetics equations. Zhurnal vychislitel noimatematiki matematicheskoi fiziki **13**, 683–697 (1973). https://doi.org/10.1016/0041-5553(73)90108-0
4. Schnell, S., Maini, P.K.: Enzyme kinetics far from the standard quasi steady state and equilibrium approximations. Math. Comput. Model. **35**, 137–144 (2002). https://doi.org/10.1016/S0895-7177(01)00156-X
5. Gorban, A.N., Radulescu, O., Zinovyev, A.Y.: Asymptotology of chemical reaction networks. Chem. Eng. **65**, 2310–2324 (2010). https://doi.org/10.1016/j.ces.2009.09.005

6. Prescott, T.P., Papachristodoulou, A.: Layered decomposition for the model order reduction of timescale separated biochemical reaction networks. J. Theor. Biol. **356**, 113–122 (2014). https://doi.org/10.1016/j.jtbi.2014.04.007
7. Khoshnaw, S.H.A.: Model Reductions in Biochemical Reaction Networks. Thesis. University of Leicester, UK (2015)
8. Huang, Y.J., Yong, W.A.: Partial equilibrium approximations in apoptosis I. The intracellular-signaling subsystem. Math. Biosci. **246**, 27–37 (2013). https://doi.org/10.1016/j.mbs.2013.09.003
9. Kijima, H., Kijima, S.K.: Steady/equilibrium approximation in relaxation and fluctuation: II. Mathematical theory of approximations in first–order reaction. Biophys. Chem. **17**, 261–283 (1983). https://doi.org/10.1016/0301-4622(83)80012-X
10. Volk, L., Richardson, W., Lau, K., Hall, M., Lin, S.: Steady state and equilibrium approximations in reaction kinetics. J. Chem. Educ. **54**, 95 (1977). https://doi.org/10.1021/ed054p95
11. Khoshnaw, S. H. A.: Reduction of a kinetic model of active export of importins. In: AIMS Conference on Dynamical Systems, Differential Equations and Applications, Madrid, pp. 7–11 (2015). https://doi.org/10.3934/proc.2015.0705
12. Khoshnaw, S.H., Mohammad, N.A., Salih, R.H.: Identifying critical parameters in SIR model for spread of disease. Open J. Model. Simul. **5**, 32 (2016). https://doi.org/10.4236/ojmsi.2017.51003
13. Gorban, A.N., Karlin, I.V.: Method of invariant manifold for chemical kinetics. Chem. Eng. Sci. **58**, 4751–4768 (2003). https://doi.org/10.1016/j.ces.2002.12.001
14. Khoshnaw, S.H.A.: Iterative approximate solutions of kinetic equations for reversible enzyme reactions. Nat. Sci. **5**, 740–755 (2013). https://doi.org/10.4236/ns.2013.56091
15. Ciliberto, A., Capuani, F., Tyson, J.: Modeling networks of coupled enzymatic reactions using the total quasi-steady state approximation. PLoS Comput. Biol. **3**, e45 (2007). https://doi.org/10.1371/journal.pcbi.0030045
16. Goeke, A., Schilli, C., Walcher, S., Zerz, E.: Computing quasi-steady state reductions. J. Math. Chem. **50**, 1495–1513 (2012). https://doi.org/10.1007/s10910-012-9985-x
17. Hannemann-tamas, R., Gabor, A., Szederkenyi, G., Hangos, K.M.: Model complexity reduction of chemical reaction networks using mixed-integer quadratic programming. Comput. Math. Appl. **65**, 1575–1595 (2013). https://doi.org/10.1016/j.camwa.2012.11.024
18. Klonowski, W.: Simplifying principles for chemical and enzyme reaction kinetics. Biophys. Chem. **18**, 73–87 (1983). https://doi.org/10.1016/0301-4622(83)85001-7
19. Khoshnaw, S.H.A.: Dynamic analysis of a predator and prey model with some computational simulations. J. Appl. Bioinf. Comput. Biol. **6**, 2329–9533 (2017). https://doi.org/10.4172/2329-9533.1000137
20. Okino, M.S., Mavrovouniotis, M.L.: Simplification of mathematical models of chemical reaction systems. Chem. Rev. **98**, 391–408 (1998). https://doi.org/10.1021/cr950223l
21. Petrov, V., Nikolova, E., Wolkenhauer, O.: Reduction of nonlinear dynamic systems with an application to signal transduction pathways. IET Syst. Biol. **1**, 2–9 (2007). https://doi.org/10.1049/iet-syb:20050030
22. Rao, S., Van der Schaft, A., Van Eunen, K., Bakker, B.M., Jayawardhana, B.: A model reduction method for biochemical reaction networks. BMC Syst. Biol. **8**, 52 (2014)

23. Schneider, K.R., Wilhelm, T.: Model reduction by extended quasi steady state approximation. J. Math. Biol. **40**, 443–450 (2000). https://doi.org/10.1007/s002850000026
24. Akgul, A., Khoshnaw, S.H.A., Mohammed, W.H.: Mathematical model for the ebola virus disease. J. Adv. Phys. **7**, 190–198 (2018). https://doi.org/10.1166/jap.2018.1407
25. Chow, M.L., Troussicot, L., Martin, M., Doumeche, B., Guillière, F., Lancelin, J.M.: Predicting and understanding the enzymatic inhibition of human peroxiredoxin 5 by 4-substituted pyrocatechols by combining funnel metadynamics. Solut. NMR Steady State Kinet. Biochem. **55**, 3469–3480 (2016). https://doi.org/10.1021/acs.biochem.6b00367
26. Lee, J., Yennawar, N.H., Gam, J., Benkovic, S.J.: Kinetic and structural characterization of dihydrofolate reductase from streptococcus pneumoniae. Biochemistry **49**, 195–206 (2009). https://doi.org/10.1021/bi901614m

Stress Analysis in Anisotropic Rock Massif by the 2Dimensional BIE Method

Sh. A. Dildabayev and G. K. Zakir'yanova(✉)

Institute of Mechanics and Mechanical Engineering, Almaty, Kazakhstan
shdilda@bk.ru, gulmzak@mail.ru

Abstract. Up to now remains open the question of constructing fundamental solutions of the two-dimensional statics of an elastic body with arbitrary anisotropy. Also in the scope of BEM method the question of calculating stresses in boundary points and points located close to the boundary of the region is still remain actual. In this work fundamental solutions of the static problem for elastic plane with arbitrary anisotropic properties are obtained as the sum of residues of complex variable function. The assessment of fundamental solution and theirs derivatives are presented in closed form. In the distribution space are obtained the regular representations for the Somigliana formulas and the stress calculation formulas. The numerical implementation of the BIE method in direct formulation has been realized in standard way. The test results performed for circular hole in anisotropic plane of rhombic system show a higher compliance for the boundary values of displacements and stresses and for nodes placed close to boundary. The results of analysis of the stress-strain state in the vicinity of rectangular mining chambers located in deep from day surface are presented in tables and pictures of isolines.

Keywords: Elastic · Anisotropy · Interior · Exterior · Boundary value problem · Distributions · Singular · Regular · Fundamental solution · Convolution · Stress · Mining chamber · Pillar

1 Introduction

The construction of fundamental solution and BEM implementation mainly for orthotropic plane domain were considered in [1–7]. Fundamental solutions for transversely isotropic magneto-electro-elastic media and boundary integral formulation where provided in [8]. The questions of justification of BEM for infinite domains were discussed in [10] for isotropic media and in [11] for orthotropic media. In [12] described the transformation technique of weakly singular and hyper singular integrals over arbitrary convex polygon into the regular contour integrals that can be easily calculated analytically or numerically.

This work is intended to show the technique to construct the fundamental solution for elastic plane with arbitrary anisotropic properties and demonstrate the ways to obtain the regular representation of singular integrals usually take place in practice of BEM method usage.

M. Zeki Sarıkaya et al. (Eds.): ICMRS 2019, LNNS 123, pp. 70–86, 2020.
https://doi.org/10.1007/978-3-030-43002-3_7

2 Statement of Problem

Let's consider the domain D in plane R^2 bounded by contour S, the contour may have corner nodes. For the interior problem we will consider simply connected domain inside the contour S (Fig. 1a). For exterior problem we will consider the doubly connected domain placed outside of S (Fig. 1b), where circle S_R with radius R and center at O is sufficiently large so that the region bounded by S_R covers S. In case of $R \to \infty$ we will have infinite domain with internal boundary S. Hereafter the unit outer normal is denoted by $\mathbf{n}$.

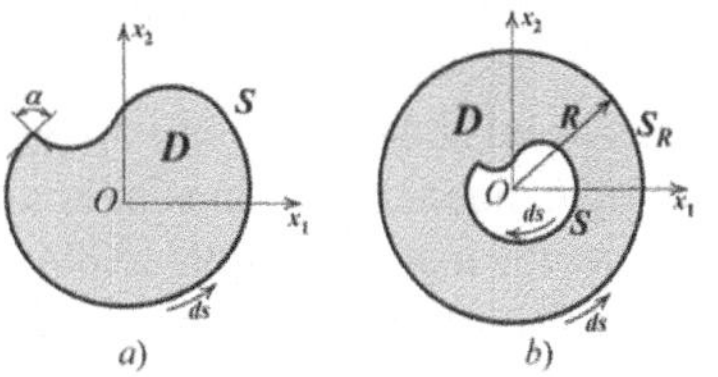

Fig. 1. Interior (a) and exterior (b) problem domains.

The differential equation of equilibrium state for homogenous anisotropic elastic body, occupied domain D, has the next form [13]

$$E_{ijkl}u_{k,lj}(\mathbf{x}) + F_i(\mathbf{x}) = 0, \quad \mathbf{x} \in D, \quad i, j, k, l = 1, 2, \tag{1}$$

where $u_i(\mathbf{x})$ are the components of displacement vector $\mathbf{u}$ in point $\mathbf{x} = (x_1, x_2)$, $F_i(\mathbf{x})$ are body force components, $\{E_{ijkl}\}$ are matrix of elasticity constants with next symmetry property $E_{ijkl} = E_{jikl} = E_{ijlk} = E_{klij}$. According to index notation the indices after comma denotes the second derivatives with respect to spatial variable x_l and x_j, and repeated index denotes summation.

The symmetry property lets present $\{E_{ijkl}\}$ matrix as 6×6 matrix $E_{\alpha\beta}$ (α, $\beta = 1$, .., 6). The relationship between pairs of (ij), (ml) indices and α, β indices established by next scheme $(11) \leftrightarrow 1$, $(22) \leftrightarrow 2$, $(33) \leftrightarrow 3$, $(23) = (32) \leftrightarrow 4$, $(31) = (13) \leftrightarrow 5$, $(21) = (12) \leftrightarrow 6$.

The first and the second boundary-value problem of plane-strain elasticity is posed as follows.

Find $\boldsymbol{u} \in C^2(D) \cap C^1(S)$ such that (1) is satisfied in the domain D with given values of boundary displacement $f_i(\mathbf{x})$ or boundary loading $g_i(\mathbf{x})$:

$$u_i(\mathbf{x}) = f_i(\mathbf{x}), \quad \mathbf{x} \in S, \quad i = 1, 2, \tag{2}$$

$$E_{ijkl}u_{k,l}(\mathbf{x})n_j(\mathbf{x}) = g_i(\mathbf{x}), \quad \mathbf{x} \in S, \quad i = 1, 2. \tag{3}$$

3 Somigliana Formulas in Distribution Space

3.1 Basic Equation in Distributions Space

Hereafter introduce abbreviations BVP1 and BVP2 for boundary value problem (1, 2) and (1, 3) consequently. For getting solution of BVP1 and BVP2 extend functions trough all plane R^2 by entering distributions in accordance with [14]

$$\begin{aligned} \hat{u}_i(\mathbf{x}) &= H_D(\mathbf{x})u_i(\mathbf{x}), \\ \hat{F}_i(\mathbf{x}) &= H_D(\mathbf{x})F_i(\mathbf{x}), \end{aligned} \tag{4}$$

where $H_D(\mathbf{x})$ is the characteristic function of domain D. Its value we determine according to [15]

$$H_D(\mathbf{x}) = \lim_{\varepsilon\to 0} \frac{\mu(S_\varepsilon(\mathbf{x}) \cap D)}{\mu(S_\varepsilon)} = \begin{cases} 1, & \mathbf{x} \in D, \\ \alpha/(2\pi) & \mathbf{x} \in S, \\ 0, & \mathbf{x} \notin D. \end{cases} \tag{5}$$

Here $S_\varepsilon(\mathbf{x})$ is a ring with radius ε and center in point $\mathbf{x}$, $\mu(\bullet)$- is square of region, α is an angle value between left and right side tangent lines in boundary point $\mathbf{x}$ (see Fig. 1a). The definition of $H_D(\mathbf{x})$ given by (5) makes possible to determine its value not only for interior and exterior points, but also for boundary points. For points on smooth part of contour we always have $H_D(\mathbf{x}) = 0.5$. So the characteristic function is always number.

The distributions $\hat{u}_i(\mathbf{x})$, $\hat{F}_i(\mathbf{x})$ given by (4) are equal to $u_i(\mathbf{x})$ and $F_i(\mathbf{x})$ inside of D end equal to zero outside of D. Calculation second derivatives of $\hat{u}(\mathbf{x})$ with respect to spatial variables gives us next:

$$\hat{u}_{i,kj}(\mathbf{x}) = -n_j u_{i,k}(\mathbf{x})\delta_S - \partial_j(n_k u_i(\mathbf{x})\delta_S) + u_{i,kj}(\mathbf{x})H_D(\mathbf{x}), \tag{6}$$

here $n_j u_{i,k}(\mathbf{x})\delta_S$ и $\partial_j(n_k u_i(\mathbf{x})\delta_S)$ are consequently well known single layered and double layered potential distributions on contour S [14]. By using (6) we can get next expression for generalized equilibrium equation

$$E_{ijkl}\hat{u}_{k,lj}(\mathbf{x}) + \hat{F}_i(\mathbf{x}) = -E_{ijkl}n_j u_{k,l}(\mathbf{x})\delta_S - E_{ijkl}\partial_j(n_l u_k(\mathbf{x})\delta_S). \tag{7}$$

So from solving of BVP1 or BVP2 we come to find the generalized solution in D of Eq. (7). Solution $\hat{u}_i(\mathbf{x})$ of Eq. (7) in view of its definition by (4) coincides in D with solution of BVP1 or BVP2.

3.2 Fundamental Solution, Somigliana and Stress Formulas

To obtain fundamental solution of differential Eq. (1) we let body force component to be $F_i(\mathbf{x}) = \delta_{i\beta}\delta(\mathbf{x})$, that represent concentrated force acting in x_β direction and applied in point $\mathbf{x}$, here $\delta_{i\beta}$ and $\delta(\mathbf{x})$ are Kronecker–delta and Dirac–delta. The solution for this force is entitled as Green tensor and be denoted as $U_{i\beta}(\mathbf{x})$. To construct the Green's tensor, it is convenient to use the Fourier transform, which brings the system (1) to the next system of linear equation in transform space

$$E_{ijkl}\omega_j\omega_l U^F_{k\beta}(\omega_1, \omega_2) = \delta_{i\beta}, \tag{8}$$

here ω_i is Fourier transform parameter. The solution of system (8) give us

$$U^F_{k\beta}(\omega_1, \omega_2) = \frac{Q_{k\beta}(\omega_1, \omega_2)}{Q(\omega_1, \omega_2)}, \tag{9}$$

Where $Q(\omega_1, \omega_2)$ is uniform polynomial of degree 4 and represent the determinant of system, $Q_{k\beta}(\omega_1, \omega_2)$ are uniform polynomials of degree 2 and

$$\begin{aligned}
Q_{11}(\omega_1, \omega_2) &= E_{66}\omega_1^2 + 2E_{26}\omega_1\omega_2 + E_{22}\omega_2^2,\\
Q_{12}(\omega_1, \omega_2) &= -E_{16}\omega_1^2 - (E_{12} + E_{66})\omega_1\omega_2 - E_{26}\omega_2^2,\\
Q_{22}(\omega_1, \omega_2) &= E_{11}\omega_1^2 + 2E_{16}\omega_1\omega_2 + E_{66}\omega_2^2,\\
Q(\omega_1, \omega_2) &= Q_{12}(\omega_1, \omega_2)Q_{12}(\omega_1, \omega_2) \ - \ Q_{12}^2(\omega_1, \omega_2).
\end{aligned} \tag{10}$$

By using the complex function theory the inversion of (9) is obtained in next form

$$U_{k\beta}(\mathbf{x}) = -\frac{1}{\pi^2}\sum_{p=1}^{8} \underset{\operatorname{Im} z_p>0}{res}\left\{\frac{(z_p^2+1)Q_{k\beta}(z_p^2-1,\ 2z_p)}{Q(z_p^2-1,\ 2z_p)}\ln\left[x_1(z_p^2-1)/2 + x_2 z_p\right]\right\}. \tag{11}$$

Here $z_p(p = 1, \ldots, 8)$ are the roots of polynomial $Q(z)$ of degree 8, and theirs value in view of (10) only depend on the elastic constants. In view of this matter, it is easy to obtain the derivatives of Green tensor with respect to spatial variables:

$$\begin{aligned}
U_{k\beta,l}(\mathbf{x}) &= -\frac{1}{\pi^2}\sum_{p=1}^{8} \underset{\operatorname{Im} z_p>0}{res}\left\{\frac{\delta_{1l}(z_p^2-1)/2 + \delta_{2l}z_p}{x_1(z^2-1)/2 + x_2 z} \times \frac{(z_p^2+1)\,Q_{k\beta}(z^2-1,\ 2z_p)}{Q(z^2-1, 2z_p)}\right\},\\
U_{k\beta,lj}(\mathbf{x}) &= -\frac{1}{\pi^2}\sum_{p=1}^{8} \underset{\operatorname{Im} z_p>0}{res}\left\{\frac{\left[\delta_{1l}(z_p^2-1)/2 + \delta_{2l}z_p\right]\left[\delta_{1j}(z_p^2-1)/2 + \delta_{2j}z_p\right]}{\left[x_1(z_p^2-1)/2 + x_2 z_p\right]^2}\right.\\
&\quad\left.\times\frac{(z^2+1)\,Q_{k\beta}(z_p^2-1,\ 2z_p)}{Q(z_p^2-1,\ 2z_p)}\right\}.
\end{aligned} \tag{12}$$

For assessment of Green tensor and its derivatives, it is useful to present them in next form:

$$\begin{aligned}
U_{k\beta}(\mathbf{x}) &= -\frac{\ln r}{\pi^2}\sum_{p=1}^{8} \underset{\operatorname{Im} z_p>0}{res}\left\{\frac{(z^2+1)\,Q_{k\beta}(z_p^2-1,\ 2z_p)}{Q(z_p^2-1,\ 2z_p)}\right\}\ln\left[\cos\theta(z_p^2-1)/2 + \sin\theta z_p\right],\\
U_{k\beta,l}(\mathbf{x}) &= -\ \frac{1}{\pi^2 r}\sum_{p=1}^{8} \underset{\operatorname{Im} z_p>0}{res}\left\{\frac{\delta_{1l}(z_p^2-1)/2 + \delta_{2l}z_p}{\cos\theta(z^2-1)/2 + \sin\theta z_p} \times \frac{(z^2+1)\,Q_{k\beta}(z^2-1,\ 2z_p)}{Q(z^2-1, 2z_p)}\right\},\\
U_{k\beta,lj}(\mathbf{x}) &= -\frac{1}{\pi^2 r^2}\sum_{p=1}^{8} \underset{\operatorname{Im} z_p>0}{res}\left\{\frac{\left[\delta_{1l}(z_p^2-1)/2 + \delta_{2l}z_p\right]\left[\delta_{1j}(z_p^2-1)/2 + \delta_{2j}z_p\right]}{\left[\cos\theta(z_p^2-1)/2 + \sin\theta z_p\right]^2}\right.,\\
&\quad\left.\times\frac{(z^2+1)\,Q_{k\beta}(z_p^2-1,\ 2z_p)}{Q(z_p^2-1,\ 2z_p)}\right\},\quad r = \sqrt{x_1^2\ +\ x_1^2},\quad \cos\theta = x_1/r,\quad \sin\theta = x_2/r.
\end{aligned} \tag{13}$$

According to (13) for any direction $\{\cos\theta, \sin\theta\}$ on plane R^2 we have next assessments for behavior of Green tensor and its derivatives when $r \to \infty$ or $r \to 0$

$$\left|U_{k\beta}(\mathbf{x})\right| \le A|\ln r|, \qquad \left|U_{k\beta,l}(\mathbf{x})\right| \le B/r\ , \qquad \left|U_{k\beta,lj}(\mathbf{x})\right| \le C/r^2\ . \tag{14}$$

Here A, B, C are some real positive constants.

As stated by distributions theory the convolution of Green tensor with left and right side of (7) gives its solution in next form

$$\hat{u}_\beta(\mathbf{x},t) = \int\limits_S \left[U_{k\beta}(\mathbf{x},\mathbf{y}) p_k(\mathbf{y}) - T_{k\beta}(\mathbf{x},\mathbf{y}) u_k(\mathbf{y}) \right] dS_y + \int\limits_D U_{k\beta}(\mathbf{x},\mathbf{y}) F_k(\mathbf{y})\, dV_y. \tag{15}$$

In (15) we introduce next designations

$$p_k(\mathbf{x}) = u_{k,j}(\mathbf{x}) n_j(\mathbf{x}), \tag{16}$$

$$\begin{aligned} T_{1\beta} &= \left[E_{11}U_{1\beta,1} + E_{12}U_{2\beta,3} + E_{16}(U_{1\beta,2} + U_{2\beta,1})\right] n_1 \\ &\quad + \left[E_{16}U_{1\beta,1} + E_{36}U_{2\beta,2} + E_{66}(U_{1\beta,2} + U_{2\beta,1})\right] n_2 \,, \\ T_{2\beta} &= \left[E_{16}U_{1\beta,1} + E_{26}U_{2\beta,2} + E_{66}(U_{1\beta,2} + U_{2\beta,1})\right] n_1 \\ &\quad + \left[E_{12}U_{1\beta,1} + E_{22}U_{2\beta,2} + E_{36}(U_{1\beta,2} + U_{2\beta,1})\right] n_2 \,, \end{aligned} \tag{17}$$

here p_k are component of boundary loading, and $T_{k\beta}$ are traction tensor components generated by Green tensor, n_k components of outer normal.

In (15) the generalized displacement in any point of D expressed by sum of integral of single layered potentials of boundary loading values p_k, double layered potential of boundary displacement values and Newton potential of body force.

Formula (15) is obtained for distributions. Note that in (15) on the right and on the left there are regular generalized functions. From du Bois-Reymond's Lemma [14], it is known that every locally integrable function f defines a regular generalized function by the formula (f, φ) (φ - from the space of basic functions) and contrary every regular distribution is defined by a unique locally integrable function. Due to this Eq. (15) are also valid in the usual sense. Similar formulas for elastic media were obtained using Betty's theorem of the elasticity theory. The approach based on the use of distributions and performed here is more correct, since the fundamental solutions are singular distributions and to work with them it is necessary to use the same space of functions in which they are defined.

By derivation displacement given by (15) with respect to spatial variable x_i and using elastic constants the stress formulae is obtained

$$\sigma_{km}(\mathbf{x}) = \int\limits_S \left[D_{ikm}(\mathbf{x},\mathbf{y}) p_i(\mathbf{y}) - V_{ikm}(\mathbf{x},\mathbf{y}) u_i(\mathbf{y})\right] dS_y + \int\limits_D W_{ikm}(\mathbf{x},\mathbf{y}) F_i(\mathbf{y})\, dV_y \tag{18}$$

where potentials kernel are

$$\begin{aligned} D_{i11} &= E_{11}U_{i1,1} + E_{12}U_{i2,2} + E_{16}\left(U_{i1,2} + U_{i2,1}\right), \\ D_{i22} &= E_{12}U_{i1,1} + E_{22}U_{i2,2} + E_{26}\left(U_{i1,2} + U_{i2,1}\right), \\ D_{i21} &= E_{16}U_{i1,1} + E_{26}U_{i2,2} + E_{66}\left(U_{i1,2} + U_{i2,1}\right). \end{aligned} \tag{19}$$

$$\begin{aligned} V_{i11} &= E_{11}T_{i1,1} + E_{12}T_{i2,2} + E_{16}\left(T_{i1,2} + T_{i2,1}\right), \\ V_{i22} &= E_{12}T_{i1,1} + E_{22}T_{i2,2} + E_{26}\left(T_{i1,2} + T_{i2,1}\right), \\ V_{i21} &= E_{16}T_{i1,1} + E_{26}T_{i2,2} + E_{66}\left(T_{i1,2} + T_{i2,1}\right). \end{aligned} \tag{20}$$

$$\begin{aligned} W_{i11} &= E_{11}U_{i1,1} + E_{12}U_{i2,2} + E_{16}\left(U_{i1,2} + U_{i2,1}\right), \\ W_{i22} &= E_{12}U_{i1,1} + E_{22}U_{i2,2} + E_{26}\left(U_{i1,2} + U_{i2,1}\right), \\ W_{i21} &= E_{16}U_{i1,1} + E_{26}U_{i2,2} + E_{66}\left(U_{i1,2} + U_{i2,1}\right). \end{aligned} \tag{21}$$

So the Somigliana formula give the solution of BVP1 or BVP2 thereat for BVP1 we have Fredholm equations of first kind, and second kind for BVP2 when $\mathbf{x} \in S$.

3.3 Gauss Formula and Regular Presentation of Somigliana and Stress Formulas

In view of assessments (14) for Green tensor and its derivatives the potential kernels $T_{i\beta}(\mathbf{x}, \mathbf{y})$ in Somigliana formulas and potential kernels $D_{ikj}(\mathbf{x}, \mathbf{y})$, $V_{ikj}(\mathbf{x}, \mathbf{y})$ has singularities on contour S when $\mathbf{x} = \mathbf{y}$. For performing, the regular presentation of (15) is used Gauss formula for double-layered potential. In this objection, lets take convolution of characteristic function $H_D(\mathbf{x})$ (5) of finite domain D with equilibrium equation for concentrated body force

$$E_{ijkl}U_{\beta k,lj}(\mathbf{x}) + \delta_{i\beta}\delta(\mathbf{x}) = 0, \quad \mathbf{x} \in R^2.$$

Performing convolution with first term gives

$$\begin{aligned} &\int\limits_{R^2} E_{ijkl}U_{\beta k,lj}(\mathbf{x}-\mathbf{y})H_D(\mathbf{x})dV_y = \int\limits_{D} E_{ijkl}U_{\beta k,lj}(\mathbf{x}-\mathbf{y})dV_y \\ &\quad = \int\limits_{S} E_{ijkl}U_{\beta k,l}(\mathbf{x}-\mathbf{y})n_j(\mathbf{y})dS_y = \int\limits_{S} T_{i\beta}(\mathbf{x}-\mathbf{y})dS_y \end{aligned}$$

In last expression on pass from body integral to contour integral the Gauss formula is used. Convolution with second terms gives

$$\int\limits_{R^2} H_D(\mathbf{x})\delta_{i\beta}\delta(\mathbf{x}-\mathbf{y})dV_y = \delta_{i\beta}\int\limits_{D} \delta(\mathbf{x}-\mathbf{y})dV_y = \delta_{i\beta}H_D(\mathbf{x})$$

Finally, for double potential kernel $T_{i\beta}$ we obtain the Gauss formulas for finite domain D with finite contour

$$\int\limits_{S} T_{i\beta}(\mathbf{x}, \mathbf{y})\, dS_y = -\delta_{i\beta}H_D(\mathbf{x}) \tag{22}$$

For infinite domain consider the doubly connected domain D placed outside of S (Fig. 1b), where circle S_R with finite radius R with center at O is sufficiently large so that the region bounded by S_R covers S. For this finite domain we have

$$\int\limits_{S \cup S_R} T_{i\beta}(\mathbf{x}, \mathbf{y})\, dS_y = -\delta_{i\beta}H_D(\mathbf{x})$$

or

$$\int_S T_{i\beta}(\mathbf{x},\mathbf{y})\,dS_y + \int_{S_R} T_{i\beta}(\mathbf{x},\mathbf{y})\,dS_y = -\delta_{i\beta}H_D(\mathbf{x})$$

For finite contour S_R with arbitrary large radius R integral of double potential with kernel $T_{i\beta}$ in view of (22) is equal to $-\delta_{i\beta}H_D(\mathbf{x})$ and we have

$$\int_S T_{i\beta}(\mathbf{x},\mathbf{y})\,dS_y - \delta_{i\beta}H_D(\mathbf{x}) = -\delta_{i\beta}H_D(\mathbf{x})$$

And as the last equality is valid for arbitrary contour of large diameter R so it also valid for $R \to \infty$ and finally we have for infinite domain outside of S

$$\int_S T_{i\beta}(\mathbf{x},\mathbf{y})\,dS_y = 0, \tag{23}$$

Using Gauss formulae we can obtain the regular representation of Somigliana formulas. By adding to (15) Gauss formulas (22) multiplied on $u_k(\mathbf{x})$ in case when $\mathbf{x} \in D$ we have next expression

$$(1 - H_D(\mathbf{x}))u_\beta(\mathbf{x}) = \int_S \left[U_{k\beta}(\mathbf{x},\mathbf{y})p_k(\mathbf{y}) - T_{k\beta}(\mathbf{x},\mathbf{y})(u_k(\mathbf{y}) - u_k(\mathbf{x}))\right]dS_y + \int_D U_{k\beta}(\mathbf{x},\mathbf{y})F_k(\mathbf{y})\,dV_y\,, \quad x \in D. \tag{24}$$

In view definition (5) of $H_D(\mathbf{x})$ the left side of (24) for interior BVP equal to zero and for exterior problems the left side is equal to $u_\beta(\mathbf{x})$. In this representation when integration point $\mathbf{y}$ is very close to $\mathbf{x}$ ($\mathbf{x}$ in domain or on contour) we have integrable singularity.

In case $\mathbf{x} \in S$ the expression (24) becomes boundary integral equation (BIE), and exclusion of singularity in this manner is very useful during numerical solution of BIE.

By derivation displacement given by (24) with respect to spatial variable x_i and using Hook's law the regular representation of stress formulae is obtained

$$\sigma_{km}(\mathbf{x}) = \int_S \left[D_{ikm}(\mathbf{x},\mathbf{y})p_i(\mathbf{y}) - V_{ikm}(\mathbf{x},\mathbf{y})(u_k(\mathbf{y}) - u_k(\mathbf{x}))\right]dS_y + \sigma_{im}(\mathbf{x})\int_S T_{i\beta}(\mathbf{x},\mathbf{y})\,dS_y + \int_D W_{ikm}(\mathbf{x},\mathbf{y})F_i(\mathbf{y})\,dV_y\,, \quad \mathbf{x} \in D. \tag{25}$$

In case when $\mathbf{x} \in S$ we have next regularized formula for boundary stress calculation

$$\sigma_{km}(\mathbf{x}) = \frac{1}{1 - H_D(\mathbf{x}) + \alpha/(2\pi)}\left\{\int_S \left[D_{ikm}(\mathbf{x},\mathbf{y})p_i(\mathbf{y}) - V_{ikm}(\mathbf{x},\mathbf{y})(u_k(\mathbf{y}) - u_k(\mathbf{x}))\right]dS_y + \int_D W_{ikm}(\mathbf{x},\mathbf{y})F_i(\mathbf{y})\,dV_y\right\}, \quad \mathbf{x} \in S. \tag{26}$$

Where $H_D(\mathbf{x}) = 1$ for interior BVP of finite regions, and $H_D(\mathbf{x}) = 0$ for exterior BVP of infinite domain.

4 Some Numerical Results

The numerical implementation of the BIE method in direct formulation has been realized in standard way. The boundary contours of region were approximated by using 3-nodal isoperimetric quadrilateral elements. The discrete analogues of regularized boundary integral Eq. (24) and stress formulae's (26) are obtained by performing Gauss quadrature on boundary elements.

For roots calculation of polynomials $Q(z^2+1, 2z)$ the Muller method was performed.

4.1 Compare with Analytical Solution for Exterior Problem of Infinite Domain

The numerical results were compared with analytical solution given in [13] for cylindrical hole with unit radius $R = 1$ under inner pressure. Anisotropic media of rhombic system (aragonite ($CaCO_3$)) [16] is considered with elastic constants

$$E_{11} = 1.60, E_{22} = 0.87, E_{33} = 0.85, E_{44} = 0.41, E_{55} = 0.26,$$
$$E_{66} = 0.43, E_{12} = 0.37, E_{13} = 0.02, E_{23} = 0.168 \tag{27}$$

The numerical results are computed by solving integral Eq. (24) and stress calculstion by (26) for approximation of hole with 16 of 3-nodal isoperimetric elements.

The epure of obtained circular σ_θ stress is symmetrical about axe Oy (see Fig. 2) and σ_θ values are equal to 1.4530 at $\theta = 0^\circ$, and 1.6206 at $\theta = 90^\circ$.

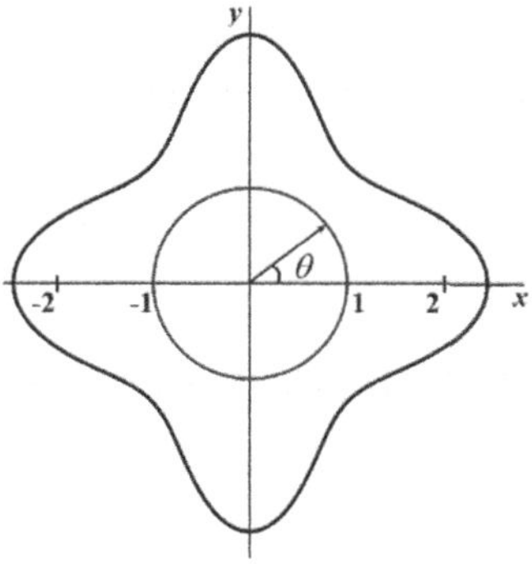

Fig. 2. Circular $\sigma_{\theta\theta}$ stress distribution on contour of cylindrical hole.

Below in Table 1 are presented numerical and analytical results for radial displacements u_R and circular stresses σ_θ in some boundary and regional points, placed close to hole contour.

The obtained numerical results are in good agreement with analytical values even for so rough approximation of boundary contour. This indicates the high efficiency of calculation using regular representation for Somigliana (24) and stress (26) formulas.

Table 1. Comparison of numerical and analytical radial displacement u_R and angular stress σ_θ values in boundary and regional points

R	x	y	u_{Rnum}	u_R	$\sigma_{\theta\ num}$	σ_θ
1.0	1.0	0.0	1.0160	1.0167	1.4530	1.4393
	0.924	−0.383	1.1437	1.1445	0.9982	0.9946
	0.707	−0.707	1.4050	1.4058	0.5982	0.6010
	0.0	−1.0	1.7077	1.7085	1.6206	1.6027
1.005	1.005	0.0	1.0267	1.0135	1.3901	1.4074
	0.928	−0.385	1.1946	1.1372	0.6869	0.9878
	0.711	−0.711	1.4023	1.3970	0.5682	0.6039
	0.0	−1.005	1.7451	1.7026	1.5027	1.5485
1.01	1.01	0.0	1.0244	1.0103	1.3764	1.3769
	0.933	−0.387	1.1207	1.1301	0.9621	0.9807
	0.714	−0.714	1.3836	1.3882	0.6213	0.6067
	0.0	−1.01	1.7023	1.6967	1.4973	1.4986
1.05	1.05	0.0	0.9857	0.9862	1.1727	1.1731
	0.97	−0.402	1.0772	1.0778	0.9188	0.9191
	0.742	−0.742	1.3194	1.3200	0.6246	0.6246
	0.0	−1.05	1.6514	1.6519	1.2019	1.2027
1.10	1.10	0.00	0.9581	0.9586	0.9886	0.9890
	1.0160	−0.4210	1.0210	1.0215	0.8387	0.8390
	0.7780	−0.7780	1.2396	1.2401	0.6363	0.6365
	0.0	−1.10	1.5998	1.6003	0.9768	0.9774

4.2 Stress Strain State of Rock Massif with Rectangular Chambers

Stresses in Untouched Rock Massive

In investigation of problems of mining geomechanics and determining the stress-strain state near underground structures, it is necessary to take into account the initial state of an untouched rock mass. The stress state of an untouched massif is determined on the hypothesis of A.N. Dinnik based on the possibility only vertical displacements in it

$$u_1^0 = 0, \qquad u_2^0 = w(y), \qquad u_3^0 = 0, \tag{28}$$

On the assumption of (28), the next components of strain tensor are equal to zero

$$\varepsilon_{xx}^0 = \varepsilon_{xy}^0 = \varepsilon_{xz}^0 = \varepsilon_{yz}^0 = \varepsilon_{zz}^0 = 0.$$

Suppose $\sigma_{yy}^{0} = -\gamma H$-where γ – the volume weight of massif, H– the depth of considered point, and using (28) is derived stress values

$$\begin{array}{lll} \sigma_{yy}^{0} = -\gamma H, & \sigma_{xx}^{0} = -\lambda_x \gamma H, & \sigma_{zz}^{0} = -\lambda_z \gamma H, \\ \sigma_{xy}^{0} = -\lambda_{xy} H, & \sigma_{xz}^{0} = \sigma_{yz}^{0} = 0, & \end{array} \quad (29)$$

where

$$\lambda_x = c_{12}/c_{22},\ \lambda_y = c_{23}/c_{22},\ \lambda_{xy} = c_{26}/c_{22}$$

Stresses in Vicinity of Two Rectangular Chambers in Rock Massive
Now consider two rectangular chambers of 8 m × 5 m of size placed on 70 m depth from the day surface with intechamber pillars width of 5 m. The elastic constant of massif are according to (27).

The numerical analysis was performed for each contour of chamber approximated by 80 quadrilateral 3–nodal elements with 320 surface nodes in total and 4726 regional nodes. The numerical results of stress component σ_{xx}, σ_{xy}, σ_{yy} and T_I– second invariant of stress tensor are presented for elements of chamber such as pillars, roofs and floors (Fig. 3).

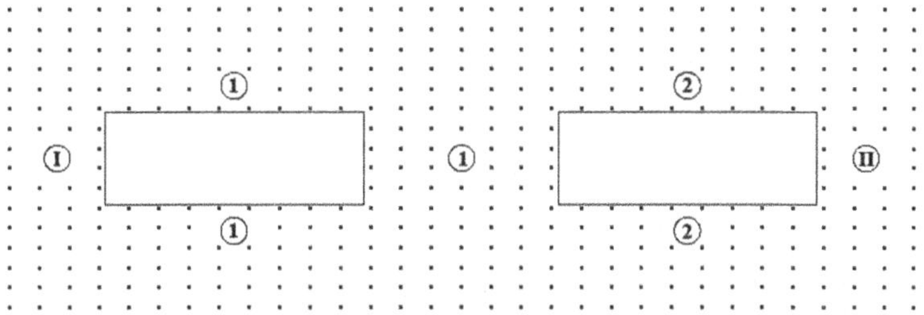

Fig. 3. Two rectangular mining chambers in rock massive and their elements: I, II– barrier pillars; 1 – interchamber pillar; 1, 2 – roof and floor of chambers.

The most loaded element of mining are the intechamber pillar 2 and then barrier pillars I, II (see Table 2), the roofs and floors of chambers are loaded less. The interchamber pillar is loaded more 14% than barriers pillars, more 192% than roofs and 112% than floors.

Table 2. Arithmetic mean values of stresses in pillars, Roof and Floor of chambers

	Pillars			Roof		Floor	
σ_{ik}\№	I	2	II	1	2	1	2
σ_{xx}	−11.4	−13.4	−11.4	1.99	1.99	1.33	1.33
σ_{xy}	0.61	0.00	−0.61	−0.85	0.85	1.30	−1.30
σ_{yy}	−122.9	−139.9	−122.9	−36.5	−36.5	−53.9	−53.9
T_I	57.1	64.9	57.1	22.2	22.2	30.6	30.6

The maximal stresses of mining elements are observed in the neighbor of corners of chambers and theirs values presented in Table 3.

Table 3. Maximal values of stresses in pillars, Roof and Floor of chambers

	Pillars			Roof		Floor	
σ_{ik}\№	I	2	II	1	2	1	2
σ_{xx}	−27.25	−31.60	−27.25	71.50	71.50	74.42	74.42
σ_{xy}	−24.80	28.16	−24.80	51.92	−51.92	−53.70	53.70
σ_{yy}	−203.9	−215.7	−203.9	−63.28	−63.28	−80.29	−80.29
T_I	103.1	109.1	103.1	58.30	58.30	60.69	60.69

The isolines of stresses σ_{yy} and T_I show that stress concentration are observed in points closest to chambers contour and the most concentration are around the corner of chambers (see Figs. 4 and 5). The stress values on isolines are presented in Tables 4 and 5.

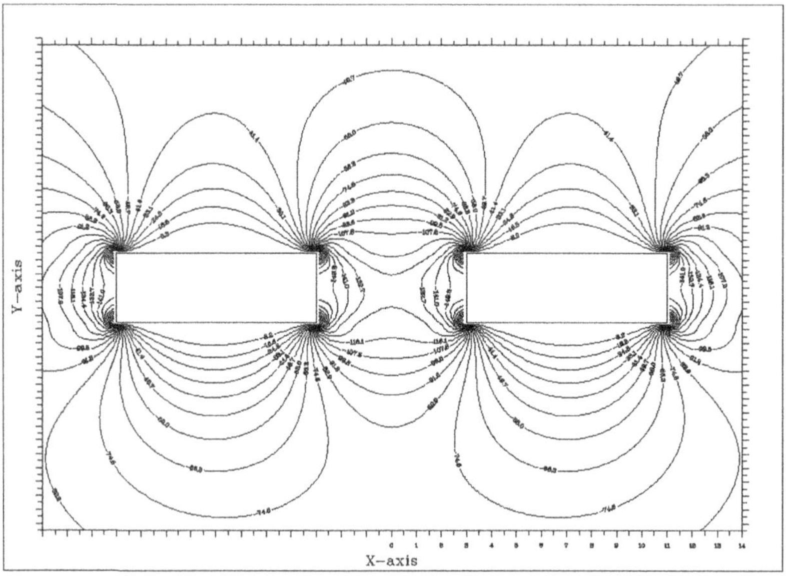

Fig. 4. Isolines of stress σ_{yy}.

Stresses in the Vicinity of Four Rectangular Chambers in Rock Massive

Now consider four rectangular chambers of 8 m × 5 m of size placed on 70 m depth from the day surface with intechamber pillars width of 5 m.

Table 4. Stress σ_{yy} value on isolines

№		№	σ_{yy}	№	σ_{yy}	№	σ_{yy}	№	σ_{yy}
1	−215.7	7	−171.1	13	−126.4	19	−81.8	25	−37.1
2	−208.3	8	−163.6	14	−118.9	20	−74.3	26	−29.7
3	−200.8	9	−156.2	15	−111.5	21	−66.9	27	−22.2
4	−193.4	10	−148.7	16	−104.1	22	−59.4	28	−14.8
5	−185.9	11	−141.3	17	−96.64	23	−52.0	29	−7.3
6	−178.5	12	−133.9	18	−89.2	24	−44.5	30	0.11

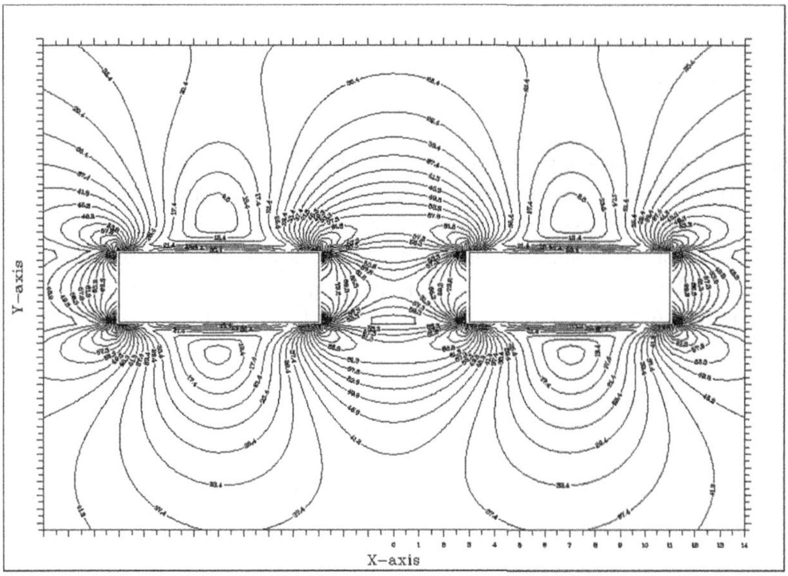

Fig. 5. Isolines of stress T_I.

Table 5. T_{I} stress value on isolines

№	T_{I}	№	T_{I}	№	T_{I}	№	T_{I}	№	T_{I}
1	5.47	7	26.9	13	48.36	19	69.80	25	91.3
2	9.04	8	30.5	14	51.93	20	73.38	26	94.8
3	12.6	9	34.1	15	55.51	21	76.9	27	98.4
4	16.2	10	37.6	16	59.08	22	80.5	28	101.9
5	19.9	11	41.21	17	62.65	23	84.1	29	105.5
6	23.3	12	44.78	18	66.23	24	87.7	30	109.1

The numerical analysis was performed for each contour of chamber approximated by 320 quadrilateral 3–nodal elements with 640 surface nodes in total and 9378 regional nodes. The numerical results of stress component σ_{xx}, σ_{xy}, σ_{yy} and T_I are presented for elements of chamber such as pillars, roofs and floors (Fig. 6).

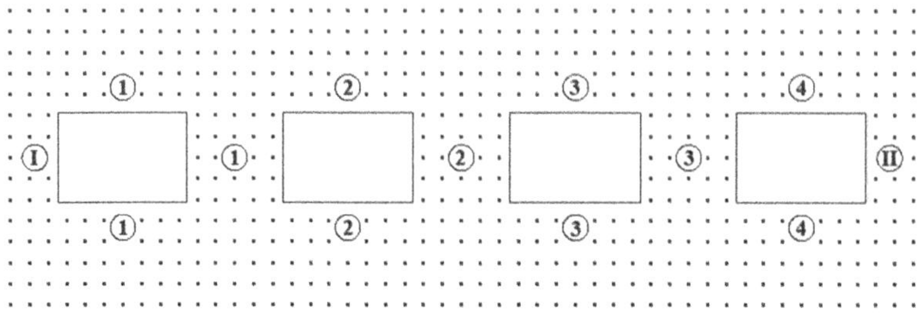

Fig. 6. Four rectangular mining chambers in rock massive and their elements: **I, II**– barrier pillars; **1–3**– interchamber pillars; **1–4** – roof and floor of chambers.

The most loaded element of mining is the middle intechamber pillar 2, then neighbor pillars (1, 3) and then the barrier pillars (I, II) (see Tables 6, 7 and 8), the roofs and floors of chambers are loaded less. The interchamber pillar 2 is loaded more 3% than neighbor pillars (1, 3), more 19% than barriers pillars (I, II), more 187% than neighbor roofs (2, 3), more 204% than roofs (1, 4). The maximal stresses of mining elements are observed in the neighbor of corners of chambers and theirs values presented in Tables 8 and 9.

Table 6. Arithmetic mean values of stresses in pillars of chamber

σ_{ik}\№	I	1	2	3	II
σ_{xx}	−12.14	−14.65	−15.21	−14.65	−12.14
σ_{xy}	0.77	0.26	0.00	−0.26	−0.77
σ_{yy}	−125.9	−147.4	−150.9	−147.4	−125.9
T_I	58.34	68.19	69.74	68.19	58.34

Table 7. Arithmetic mean values of stresses in Roof and Floor of chambers

	Roof				Floor			
σ_{ik}\№	1	2	3	4	1	2	3	4
σ_{xx}	1.41	2.39	2.39	1.41	0.72	1.95	1.95	0.72
σ_{xy}	−0.54	0.16	−0.16	0.54	1.35	0.04	−0.04	−1.35
σ_{yy}	−38.10	−39.79	−39.79	−38.10	−55.42	−56.80	−56.80	−55.42
T_I	22.93	24.33	24.33	22.93	31.23	32.56	32.56	31.23

Table 8. Maximal values of stresses in pillars of chamber

σ_{ik}\№	I	1	2	3	II
σ_{xx}	−29.17	−35.64	−36.32	−35.64	−29.17
σ_{xy}	26.29	31.13	31.59	−31.13	−26.29
σ_{yy}	−209.8	−231.7	−232.9	−231.7	−209.8
T_I	106.1	117.2	117.9	117.2	106.1

Table 9. Maximal values of stresses in Roof and Floor of chambers

	Roof				Floor			
σ_{ik}\№	1	2	3	4	1	2	3	4
σ_{xx}	70.36	69.50	69.50	70.36	72.87	72.37	72.37	72.87
σ_{xy}	54.80	56.60	−56.60	−54.80	−56.03	−58.43	58.43	56.03
σ_{yy}	−66.59	−68.93	−68.93	−66.59	−81.78	−80.61	−80.61	−81.78
T_I	60.59	62.24	62.24	60.59	62.42	64.63	64.63	62.42

The isolines of stresses σ_{yy} and T_I demonstrate that stress concentration are observed in the points closest to contour of chambers and the most concentration are around the corner of chambers (see Figs. 7 and 8). The stress values on isolines are presented in Tables 10 and 11.

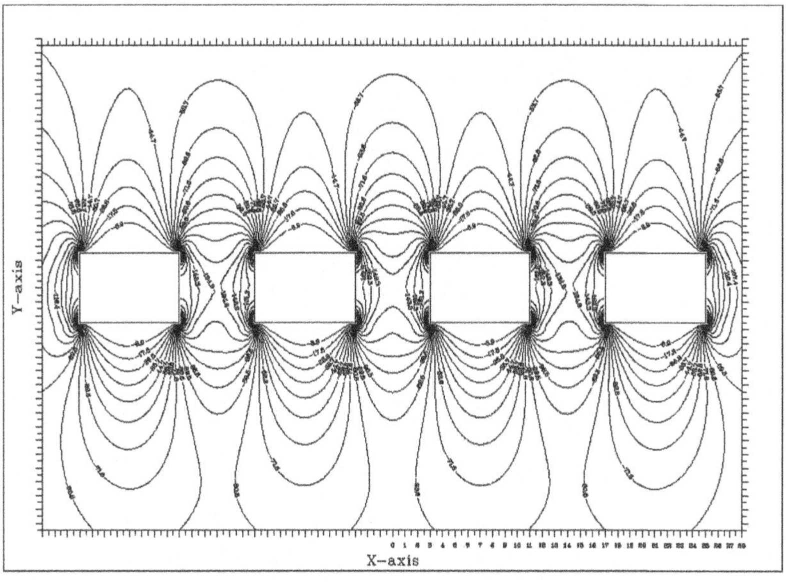

Fig. 7. Isolines of stress σ_{yy}.

Table 10. σ_{yy} stress value on isolines

№	σ_{yy}	№	σ_{yy}	№	σ_{yy}	№	σ_{yy}	№	σ_{yy}
1	−232.9	7	−184.7	13	−136.5	19	−88.29	25	−40.07
2	−224.9	8	−176.6	14	−128.4	20	−80.25	26	−32.03
3	−216.8	9	−168.6	15	−120.4	21	−72.21	27	−24.00
4	−208.8	10	−160.6	16	−112.3	22	−64.18	28	−15.96
5	−200.7	11	−152.5	17	−104.3	23	−56.14	29	−7.93
6	−192.7	12	−144.5	18	−96.32	24	−48.11	30	0.11

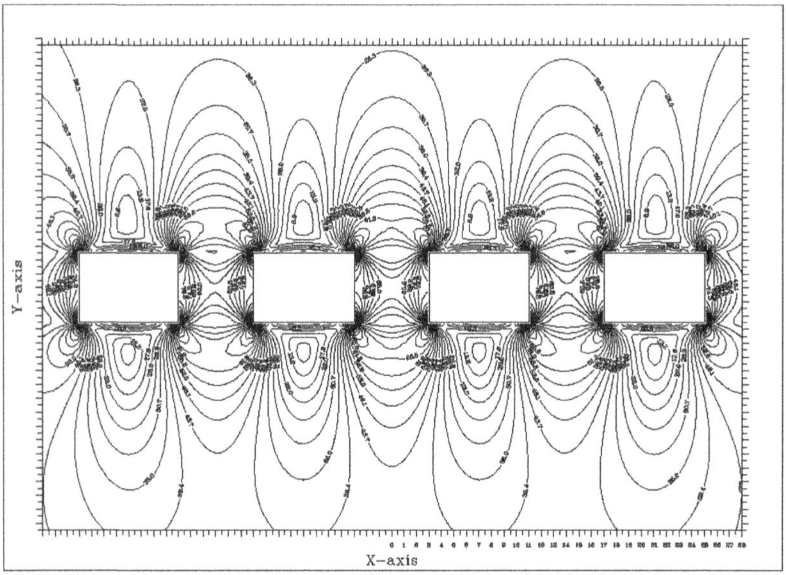

Fig. 8. Isolines of stress T_I.

Table 11. T_I Stress value on isolines

№	T_I	№	T_I	№	T_I	№	T_I	№	T_I
1	4.52	7	27.96	13	51.41	19	74.86	25	98.31
2	8.42	8	31.87	14	55.32	20	78.77	26	102.22
3	12.33	9	35.78	15	59.23	21	82.68	27	106.12
4	16.24	10	39.69	16	63,14	22	86.58	28	110.03
5	20.15	11	43.60	17	67.04	23	90.49	29	113.94
6	24.06	12	47.50	18	70.95	24	94.40	30	117.85

5 Conclusion

The fundamental solutions of the static problem for elastic plane with arbitrary anisotropic properties are obtained as the sum of residues of complex variable function. The assessment of fundamental solution and their derivatives are presented in closed form. In the distribution space are obtained the regular representations for the Somigliana formulas and the stress calculation formulas. These results are new and represent the subsequent development of the BEM method.

The test results performed for circular hole in anisotropic plane of rhombic system show a higher compliance for the boundary values of displacements and stresses calculated by proposed regular formulas. The results of analysis of the stress-strain state in the vicinity of rectangular mining chambers located in deep from day surface are presented in tables and pictures of isolines.

Acknowledgments. This research is financially supported by a grant from the Ministry of Science and Education of the Republic of Kazakhstan (Grant No. AP05135494).

References

1. Tomlin, G.: Numerical analysis of continuum problems of zoned anisotropic media. Ph.D. thesis, Southampton University (1972)
2. Brebbia, C., Telles, J., Wrobel, L.: Boundary Element Techniques: Theory and Applications in Engineering. Springer, Berlin (1984)
3. Vable, M., Sikarskie, D.: Stress analysis in plane orthotropic material by the boundary element method. Int. J. Solids Struct. **24**(1), 1–11 (1988)
4. Sun, X., Cen, Z.: Further improvement on fundamental solutions of plane problems for orthotropic materials. Acta Mech. Solida Sin. **15**(2), 171–181 (2002)
5. Liu, Y., Huang, L., Sun, X., Cen, Z.: Boundary element analysis for elastic and elastoplastic problems of 2D orthotropic media with stress concentration. Acta. Mech. Sin. **21**(5), 472–484 (2005)
6. Kolosova, E.: Fundamental solutions for anisotropic media and their applications. Ph.D. thesis, South Federal University, Rostov-na-Donu (2007). (in Russian)
7. Hasebe, N., Sato, M.: Mixed boundary value problem for quasi-orthotropic elastic plane. Acta Mech. **226**(2), 527–545 (2015)
8. Ding, H., Jiang, A., Chen, W.: Fundamental solutions for transversely isotropic magneto-electro-elastic media and boundary integral formulation. Sci. China Ser. E **46**(6), 607–619 (2003)
9. Berger, J., Martin, P., Mantič, V., Gray, L.: Fundamental solutions for steady-state heat transfer in an exponentially graded anisotropic material. Z. Angew. Math. Phys. **56**, 293–303 (2005)
10. Schiavone, P., Ru, C.: Integral equation methods in plane-strain elasticity with boundary reinforcement. Proc. R. Soc. Lond. A **454**, 2223–2242 (1998)
11. Szeidl, G., Dudra, J.: Boundary integral equations for plane orthotropic bodies and exterior regions. Electron. J. Boundary Elem. **8**(2), 10–23 (2010)
12. Zozulya, V.: Regularization of the divergent integrals I. General consideration. Electron. J. Boundary Elem. **4**(2), 49–57 (2006)

13. Lekhnitskii, S.: Theory of Elasticity of an Anisotropic Body. Mir Publishers, Moscow (1981)
14. Vladimirov V.: Equations of Mathematical Physics. Nauka, Moscow (1981). (in Russian)
15. Dildabayev, Sh., Zakir'yanova, G.: Fundamental solutions of the first and second boundary value problems of dynamics for an anisotropic elastic half-plane. Izvestya NAN Respuliki Kazakhstan ser. Fiz.-mat **5**, 65–70 (1993). (in Russian)
16. Clark, S.: Handbook of physical constants. Geological Society of America, New Haven (1966)

Class of Integral Operators for a Set of Boehmians Functions

Shrideh K. Q. Al-Omari(✉)

Department of Physics and Basic Sciences, Faculty of Engineering Technology, Al-Balqa Applied University, Amman 11134, Jordan
shridehalomari@bau.edu.jo, s.k.q.alomari@fet.edu.jo

Abstract. This paper investigates a class of Meijer type integrals on a quotient space of generalized functions known as Boehmians. Two spaces of Boehmians with certain topology are obtained. The Meijer type integral has been estimated and some properties are explored. Further results are also discussed in a generalized sense.

Keywords: Meijer function · Generalized integral · Convolution theorem · Boehmian

1991 Mathematics Subject Classification. 26A33 · 44A10 · 33C05 · 33C20 · 60E05 · 65R10

1 Preliminaries

Bessel functions and their modifications have received great attention due to their applications in many branches of mathematical physics and the solution of differential equations. Bessel functions have also an extensive use in problems appearing in physics, signal processing, engineering and mathematical physics. At the time where a wide range of phenomena in magnetism, electricity, optical transmission, heat conduction, and acoustical vibrations have their usual presentation by the Bessel functions j_v and k_v. The modified Bessel functions $\epsilon_v(x) = x^{\frac{-v}{2}} j_v(2\sqrt{x})$ and $l_v(x) = x^{\frac{-v}{2}} k_v(2\sqrt{x})$ are solutions of the equation $xy'' + (v+1)y' - y = 0$ and $l_v^{(n)} = (-1)^n l_{v+n}$, $\epsilon_v^{(n)} = \epsilon_{v+n}$, for every $n \in \mathbb{N}$. Further, we have [12,13,28]

$$l_{\alpha-1}(\vartheta\eta) = \frac{1}{2}\int_0^\infty \tau^{-\alpha_1} e^{-\left(\tau+\frac{\vartheta\eta}{\tau}\right)} d\tau.$$

In the recent past, various variants of Meijer's type integrals have been studied in [12,24,29,30] whereas various generalizations of Meijer's type integrals are given in [5–8,31]. In [28], a version of Meijer's type integral $M_{\alpha,\beta}$ - integral has been introduced as a generalization of those given in [4,6,27]. This article, considers the variant given by [1]

$$l_{\alpha_1,\alpha_2}(\phi(\eta))(\vartheta) = \vartheta^{-\alpha_2}\int_0^\infty \eta^{\alpha_1+\alpha_2-1} l_{\alpha_1-1}(\vartheta\eta)\,\phi(\eta)\,d\eta \qquad (1)$$

M. Zeki Sarıkaya et al. (Eds.): ICMRS 2019, LNNS 123, pp. 87–95, 2020.
https://doi.org/10.1007/978-3-030-43002-3_8

with the inversion formula

$$l_{\alpha_1,\alpha_2}^{-1}(\phi(\vartheta))(\eta)=\frac{\eta^{-\alpha_2}}{i\pi}\int_{C_T}\vartheta^{\alpha_1+\alpha_2-1}\epsilon_{\alpha_1-1}(\vartheta\eta)\,l_{\alpha_1,\alpha_2}(\phi(\eta))(\vartheta)\,d\vartheta, \tag{2}$$

where $C_T=\{w\in\ :Re\sqrt{2w}=T>0\}$.
The Boehmian's algebra is given as follows: let Y and X be sets such that $X\subseteq Y$. For $\kappa,h\in Y,\phi,\psi\in X$, there should be defined a product $*$ such that $\phi*\psi\in X$, $\phi*\psi=\psi*\phi$, $(\kappa*\phi)*\psi=\kappa*(\phi*\psi)$, $(\kappa+h)*\phi=\kappa*\phi+h*\phi$, $k(\kappa*\phi)=(k\kappa)*\phi=\kappa*(k\phi)$, $k\in\mathbb{R}$. A family of sequences Δ of X is said to be a set of delta sequences or approximating identities if: $\kappa*\epsilon_n=h*\epsilon_n\Rightarrow\kappa=h,n\in\mathbb{N}$, $(\epsilon_n)\in\Delta,(\epsilon_n),(\tau_n)\in\Delta\Rightarrow(\epsilon_n*\tau_n)\in\Delta$.
Let $Z=\{((\kappa_n),(\epsilon_n)):(\kappa_n)\subseteq Y,(\epsilon_n)\in\Delta\},\forall n\in\mathbb{N}$. Then $((\kappa_n),(\epsilon_n))\in Z$ defines a quotient of sequences whenever $\kappa_n*\epsilon_m=\kappa_m*\epsilon_n,\forall n,m\in\mathbb{N}$. $((\kappa_n),(\epsilon_n))\approx((h_n),(\tau_n))$ if $\kappa_n*\epsilon_m=h_m*\tau_n,\forall n,m\in\mathbb{N}$. The equivalence class including $((\kappa_n),(\epsilon_n))$ is introduced as $\frac{\kappa_n}{\epsilon_n}$. Boehmians are the sets of such equivalence classes. Such a space is written as $B(Y,X,\Delta,*)$. The usual scalar multiplications and additions are introduced as

$$\frac{\kappa_n}{\epsilon_n}+\frac{g_n}{\tau_n}=\frac{\kappa_n*\tau_n+g_n*\epsilon_n}{\epsilon_n*\tau_n}\text{ and }\zeta\frac{\kappa_n}{\epsilon_n}=\frac{\zeta\kappa_n}{\epsilon_n},$$

ζ being complex number. $*$ and D^α are given by

$$\frac{\kappa_n}{\epsilon_n}*\frac{g_n}{\tau_n}=\frac{\kappa_n*g_n}{\epsilon_n*\tau_n}\text{ and }D^\alpha\frac{\kappa_n}{\epsilon_n}=\frac{D^\alpha\kappa_n}{\epsilon_n}.$$

Further details are given in [9–11] and [13–18], [20–23].

2 Convolution Products and Necessary Theorems

Convolution products have found their applications in diverse fields of applied mathematics, physics and engineering problems. Convolution products have many applications in signal processing and in combining two signals to yield another signal. Despite utilities of convolution products, there has been found a lack of background information of this theory as all modern books usually discuss almost the theory and applications of integral transforms. Not going so far in this field of research, we state and prove a convolution theorem for the Meijer type integral transform that will be well used in establishing the Boehmian spaces. However, the convolution theorem we obtain here doesn't have a simplicity elegance comparable to convolution theorem of the Fourier transform, which states that the Fourier convolution of two classical functions is a product of their corresponding Fourier transforms.

Definition 1. Let ϕ and ψ be functions defined on $\mathbb{R}_+$. We define a convolution product between ϕ and ψ as

$$(\phi\bullet_{\alpha_1,\alpha_2}\psi)(\vartheta)=\int_0^\infty x^{\alpha_1+\alpha_2-1}\phi(\vartheta x)\,\psi(x)\,dx. \tag{3}$$

Definition 2. Let ϕ and ψ be two functions defined on $\mathbb{R}_+$. The convolution product $*$ is defined by [9] as

$$(\phi * \psi)(\eta) = \int_0^\infty x^{-1} \phi\left(\eta x^{-1}\right) \psi(x)\, dx \tag{4}$$

when the integral exists. The general properties of $*$ may be presented as:
$\phi * \psi = \psi * \phi$; $(\phi + \psi) * \varphi = \phi * \varphi + \psi * \varphi$; $\alpha\phi * \psi = \alpha(\psi * \phi)$; $((\phi * \psi) * \varphi)(t) = (\phi * (\psi * \varphi))(t)$, α is complex number.
Now we examine the convolution theorem of the l_{α_1,α_2} transform as follows.

Theorem 3. For two functions ϕ, ψ defined on $\mathbb{R}_+$ and $\vartheta \in \mathbb{R}_+$ we have

$$l_{\alpha_1,\alpha_2}(\phi * \psi)(\vartheta) = \left(l_{\alpha_1,\alpha_2}\phi \bullet_{\alpha_1,\alpha_2} \psi\right)(\vartheta).$$

Proof. By (1) and (4) we write

$$\begin{aligned} l_{\alpha_1,\alpha_2}(\phi * \psi)(\vartheta) &= \vartheta^{-\alpha_2} \int_0^\infty \eta^{\alpha_1+\alpha_2-1} l_{\alpha_1-1}(\vartheta\eta) \int_0^\infty x^{-1}\phi\left(\eta x^{-1}\right)\psi(x)\, dx d\eta \\ &= \vartheta^{-\alpha_2} \int_0^\infty x^{-1}\psi(x) \int_0^\infty \eta^{\alpha_1+\alpha_2-1} l_{\alpha_1-1}(\vartheta\eta)\, \phi\left(\eta x^{-1}\right) dx d\eta. \end{aligned}$$

By aid of the Fubini's theorem and change of variables, we reach to

$$l_{\alpha_1,\alpha_2}(\phi * \psi)(\vartheta) = \vartheta^{-\alpha_2} \int_0^\infty \psi(x)\, x^{\alpha_1+\alpha_2-1} y^{\alpha_1+\alpha_2-1} l_{\alpha_1-1}(\vartheta(xy))\, \phi(y)\, dy dx.$$

Hence, simplifications yield

$$\begin{aligned} l_{\alpha_1,\alpha_2}(\phi * \psi)(\vartheta) &= \int_0^\infty \psi(x)\, x^{\alpha_1+\alpha_2-1} \\ &\quad \times \left((\vartheta x)^{-\alpha_2} \int_0^\infty y^{\alpha_1+\alpha_2-1} l_{\alpha_1-1}((\vartheta x)\, y)\, \phi(y)\, dy\right) dx \\ &= \int_0^\infty x^{\alpha_1+\alpha_2-1} l_{\alpha_1,\alpha_2}\phi(\vartheta x)\, \psi(x)\, dx. \end{aligned}$$

This finishes the proof of the theorem.

Theorem 4. Let ϕ, ψ be defined on $\mathbb{R}_+$ and $y \in \mathbb{R}_+$. Then, we have

$$\left(\phi \bullet_{\alpha_1,\alpha_2} (\varphi * \psi)\right)(y) = \left(\left(\phi \bullet_{\alpha_1,\alpha_2} \varphi\right) \bullet_{\alpha_1,\alpha_2} \psi\right)(y).$$

Proof. By using (4), (3) and the Fubini's theorem we get

$$\begin{aligned} \left(\phi \bullet_{\alpha_1,\alpha_2} (\varphi * \psi)\right)(y) &= \int_0^\infty \zeta^{\alpha_1+2\alpha_2-1} \phi(y\zeta) \int_0^\infty x^{-1}\varphi\left(\zeta x^{-1}\right)\psi(x)\, dx d\zeta \\ &= \int_0^\infty x^{\alpha_1+2\alpha_2-1}\psi(x) \int_0^\infty \phi(yx\tau)\, \varphi(z)\, \tau^{\alpha_1+2\alpha_2-1} d\tau dx. \end{aligned}$$

Change of variables gives

$$(\phi \bullet (\varphi * \psi))(y) = \int_0^\infty x^{\alpha_1+2\alpha_2-1} (\phi \bullet_{\alpha_1,\alpha_2} \varphi)(yx)\, \psi(x)\, dx.$$

The proof is therefore completed.

Let C_0^∞ be the space of test functions of compact supports defined on $\mathbb{R}_+$ and $\tilde{S}$ be the subset of those $\phi \in C_0^\infty$ such that

$$\int_0^\infty |x^\alpha \phi(x)|\, dx < \infty \qquad (5)$$

for every real number α.

Theorem 5. Let $\phi \in C_0^\infty$. Then, we have $l_{\alpha_1,\alpha_2}(\phi) \in C_0^\infty$.

Proof. Let $\phi \in C_0^\infty$. Then, by aid of (1) and [[1], p. 5] we get

$$\begin{aligned} |D^m l_{\alpha_1,\alpha_2}(\phi)(\vartheta)| &\le \int_0^\infty \eta^{\alpha_1+2\alpha_2-1} \left|D^m \left(\vartheta^{-\alpha_2} l_{\alpha_1-1}(\vartheta\eta)\right)\right| |\phi(\eta)|\, d\eta \\ &\le \int_0^\infty \eta^{\alpha_1+2\alpha_2-1} \left|D^m \left(K\left(\eta^{\frac{3}{4}-\frac{\alpha_1}{2}} + \eta^{\frac{\alpha_1}{2}-\frac{1}{4}}\right)\right)\right| |\phi(\eta)|\, d\eta \\ &\le KB\frac{1}{2}\eta^n \int_0^\infty \eta^{\alpha_1+2\alpha_2-1} |\phi(\eta)|\, d\eta \end{aligned}$$

where K and B are certain positive constants.
This finishes the proof.

3 Spaces of Boehmians

Now, we establish the Boehmian space $\Omega^1 \left(\Omega^1 \cong B\left(C_0^\infty, \tilde{S}, *, \bullet_{\alpha_1,\alpha_2}\right)\right)$ with the set C_0^∞, the subset $\tilde{S}$, the products $*$ and $\bullet_{\alpha_1,\alpha_2}$ and, the collection Δ of delta sequence (δ_n) such that

$$\Delta_1 : \int_0^\infty \delta_n = 1;$$

$$\Delta_2 : \int_0^\infty |\delta_n| < M \ \ (M \in \mathbb{R}, M > 0)$$

$$\Delta_3 : supp\delta_n \to 0 \text{ as } n \to \infty,$$

and

$$\Delta_4 : \delta_n \in \tilde{S} \ \ (\forall n \in \mathbb{N}).$$

We omit the details of establishing the space $\Omega^2 \left(\text{or} B\left(C_0^\infty, \tilde{S}, *, *\right)\right)$ as the construction is obvious when the properties of $*$ are being considered.
We prove the axioms which are sufficient for this construction.
We state without proof the following theorem as the proof follows from Theorem 4.

Theorem 6. Let $\phi \in C_0^\infty$ and $\varphi, \psi \in \tilde{S}$. Then, we have $\phi \bullet_{\alpha_1,\alpha_2} (\varphi * \psi) = (\phi \bullet_{\alpha_1,\alpha_2} \varphi) \bullet_{\alpha_1,\alpha_2} \psi$.

Theorem 7. Let $\phi \in C_0^\infty$ and $\varphi \in \tilde{S}$. Then, we have $\phi \bullet_{\alpha_1,\alpha_2} \varphi \in C_0^\infty$.

Proof. By the Fubini's theorem we write

$$(\phi \bullet_{\alpha_1,\alpha_2} \varphi)(\vartheta) \leq \int_k |\varphi(x)| \left| x^{\alpha_1+2\alpha_2-1} \phi(\vartheta x) \right| dx,$$

where $k \supseteq supp\varphi$. On aid of Δ_1, we obtain

$$|\phi \bullet_{\alpha_1,\alpha_2} \varphi| \leq M M_1,$$

where $M_1 = \left| x^{\alpha_1+2\alpha_2-1} \phi(\vartheta x) \right|$, $M = \int_k |\varphi(x)| \, dx$.
Hence, the proof is finished.
The following theorem follows from [25]. Hence, we omit the details.

Theorem 8. For given sequences (δ_n), $(\psi_n) \in \Delta$, the sequence $(\delta_n * \psi_n) \in \Delta$.

Following theorems are straightforward. Hence, the details are deleted.

Theorem 9. Let $\phi_1, \phi_2 \in C_0^\infty, \psi_1, \psi_2 \in \tilde{S}$. Then, the following hold:
$(\phi_1 + \phi_2) \bullet_{\alpha_1,\alpha_2} \psi_1 = \phi_1 \bullet_{\alpha_1,\alpha_2} \psi_1 + \phi_2 \bullet_{\alpha_1,\alpha_2} \psi_1$.

Theorem 10. If $\phi_n \to \phi$ in C_0^∞ as $n \to \infty$ and $\psi \in \tilde{S}$, then $\phi_n \bullet_{\alpha_1,\alpha_2} \psi \to \phi \bullet_{\alpha_1,\alpha_2} \psi$ as $n \to \infty$.
As final, we derive the following theorem.

Theorem 11. For $\phi \in C_0^\infty$ and $(\delta_n) \in \Delta$, $\phi \bullet_{\alpha_1,\alpha_2} \delta_n \to \phi$ as $n \to \infty$.

Proof. Let $k \subseteq \mathbb{R}_+$ be compact. Then, Δ_1 implies

$$\begin{aligned} |(\phi \bullet_{\alpha_1,\alpha_2} \delta_n)(\vartheta) - \phi(\vartheta)| &= \left| \int_0^\infty x^{\alpha_1+2\alpha_2-1} \phi(\vartheta x) \delta_n(x) \, dx - \int_0^\infty \phi(\vartheta) \delta_n(x) \, dx \right| d\vartheta \\ &\leq \int_0^\infty \left| x^{\alpha_1+2\alpha_2-1} \phi(\vartheta x) - \phi(\vartheta) \right| |\delta_n(x)| \, dx d\vartheta. \end{aligned}$$

Hence $|(\phi \bullet_{\alpha_1,\alpha_2} \delta_n)(\vartheta) - \phi(\vartheta)| \to 0$ as $n \to \infty$.
Therefore $\phi \bullet_{\alpha_1,\alpha_2} \delta_n \to \phi$ as $n \to \infty$.
This finishes the proof.

Ω^1 is therefore can be accepted as a Boehmian space. Construction of the space Ω^2 is analogous.

The operation of addition and scalar multiplication in Ω^1 has the natural expression of Boehmian spaces. Between C_0^∞ and Ω^1 there is the embedding

$$x \to \frac{x \bullet_{\alpha_1,\alpha_2} \delta_n}{\delta_n}.$$

The operation $\bullet_{\alpha_1,\alpha_2}$ can be extended to $\Omega^1 \times C_0^\infty$ by

$$\frac{x_n}{\delta_n} \bullet_{\alpha_1,\alpha_2} t = \frac{x_n \bullet_{\alpha_1,\alpha_2} t}{\delta_n}.$$

In Ω^1, two types of convergence which can be followed from the author citations.

4 The Meijer Type Integral for Boehmians

In this section we define the generalized Meijer's type integral and obtain some results in the quotient space of Boehmians. Theorem 5 ensures that following definition is well-defined.

Definition 12. Let $\frac{(\phi_n)}{(\delta_n)} \in \Omega^2$. Then, we define the extended Meijer's type integral transform of $\frac{(\phi_n)}{(\delta_n)}$ as

$$E_{\alpha_1,\alpha_2}\frac{(\phi_n)}{(\delta_n)} = \frac{l_{\alpha_1,\alpha_2}(\phi_n)}{(\delta_n)}$$

in Ω^1.
We now discuss some properties of E_{α_1,α_2}.

Theorem 13. Let $\beta_1, \beta_2 \in \Omega^2$. Then $E_{\alpha_1,\alpha_2}(\beta_1 * \beta_1) = E_{\alpha_1,\alpha_2}\beta_1 \bullet_{\alpha_1,\alpha_2} \beta_2$ in Ω^1.

Proof. If $\beta_1, \beta_2 \in \Omega^2$, then for some $(\phi_n), (\kappa_n) \in C_0^\infty$, $(\varphi_n), (\delta_n) \in \Delta, \beta_1 = \frac{(\phi_n)}{(\varphi_n)}$ and $\beta_2 = \frac{(\kappa_n)}{(\delta_n)}$. Therefore, we write $E_{\alpha_1,\alpha_2}(\beta_1 * \beta_2) = E_{\alpha_1,\alpha_2}\left(\frac{(\phi_n)*(\kappa_n)}{(\varphi_n)*(\delta_n)}\right)$. By the assistance of Definition 12 we get

$$E_{\alpha_1,\alpha_2}(\beta_1 * \beta_2) = \frac{l_{\alpha_1,\alpha_2}((\phi_n) * (\kappa_n))}{(\varphi_n) * (\delta_n)}.$$

With the aid of Theorem 3 we obtain $E_{\alpha_1,\alpha_2}(\beta_1 * \beta_2) = \frac{(l_{\alpha_1,\alpha_2}\phi_n)\bullet_{\alpha_1,\alpha_2}(\kappa_n)}{(\varphi_n)\bullet_{\alpha_1,\alpha_2}(\delta_n)}$. Separating yields

$$E_{\alpha_1,\alpha_2}(\beta_1 * \beta_2) = \frac{(l_{\alpha_1,\alpha_2}\phi_n)}{(\varphi_n)} \bullet_{\alpha_1,\alpha_2} \frac{(\kappa_n)}{(\delta_n)}.$$

Hence, we have obtained $E_{\alpha_1,\alpha_2}(\beta_1 * \beta_2) = E_{\alpha_1,\alpha_2}(\beta_1) \bullet_{\alpha_1,\alpha_2} \beta_2$.
This finishes the proof.
Proof of the following result is straightforward. We delete the details.

Theorem 14. E_{α_1,α_2} defines a linear mapping from Ω^2 into Ω^1.

Theorem 15. For $\frac{(\phi_n)}{(\delta_n)} \in \Omega^2$ and $\delta \in \tilde{S}$ we have

$$E_{\alpha_1,\alpha_2}\left(\frac{(\phi_n)}{(\delta_n)} * \delta\right) = \frac{(l_{\alpha_1,\alpha_2}\phi_n)}{(\delta_n)} \bullet_{\alpha_1,\alpha_2} \delta.$$

Proof. Let $\frac{(\phi_n)}{(\delta_n)} \in \Omega^2$ and $\delta \in \tilde{S}$. By aid of Definition 12 we obtain

$$E_{\alpha_1,\alpha_2}\left(\frac{(\phi_n)}{(\delta_n)} * \delta\right) = \frac{l_{\alpha_1,\alpha_2}((\phi_n) * \delta)}{(\delta_n)}.$$

Once again by Definition 12 and Theorem 3 we get

$$E_{\alpha_1,\alpha_2}\left(\frac{(\phi_n)}{(\delta_n)} * \delta\right) = \frac{(l_{\alpha_1,\alpha_2}\phi_n) \bullet_{\alpha_1,\alpha_2} \delta}{(\delta_n)} = \frac{(l_{\alpha_1,\alpha_2}\phi_n)}{(\delta_n)} \bullet_{\alpha_1,\alpha_2} \delta.$$

This finishes the proof.

Theorem 16. E_{α_1,α_2} and l_{α_1,α_2} are consistent.

Proof. For every $\phi \in C_0^\infty$, let $\beta \in \Omega^2$ be its representative in the space Ω^2 of Boehmians, then, $\forall n \in \mathbb{N}$, $\delta_n \in \Delta, \beta = \frac{\phi*(\delta_n)}{(\delta_n)}$. $\forall n \in \mathbb{N}$ it can be easily noted that (δ_n) is independent from the representative.
We have

$$E_{\alpha_1,\alpha_2}(\beta) = E_{\alpha_1,\alpha_2}\left(\frac{\phi * (\delta_n)}{(\delta_n)}\right) = \frac{l_{\alpha_1,\alpha_2}(\phi * (\delta_n))}{(\delta_n)} = \frac{l_{\alpha_1,\alpha_2}\phi * (\delta_n)}{(\delta_n)}$$

which is the representative of $E_{\alpha_1,\alpha_2}\phi$ in the space C_0^∞.
The proof is finished.

Theorem 17. $\frac{(g_n)}{(\psi_n)} \in \Omega^1$ is in the range space of $E_{\alpha_1,\alpha_2} \Leftrightarrow g_n$ belongs to range space of l_{α_1,α_2} for every $n \in \mathbb{N}$.

Proof. Let $\frac{(g_n)}{(\psi_n)}$ be a member in the range space of E_{α_1,α_2}. Then ofcourse g_n is a member of the range space of l_{α_1,α_2}, $\forall n \in \mathbb{N}$. To show the converse is true, let g_n be in the range space of $l_{\alpha_1,\alpha_2}, \forall n \in \mathbb{N}$. Then, there is $\phi_n \in C_0^\infty$ such that $l_{\alpha_1,\alpha_2}\phi_n = g_n$, $n \in \mathbb{N}$. Since $\frac{(g_n)}{(\psi_n)} \in \Omega^2$ we get

$$g_n \bullet_{\alpha_1,\alpha_2} \psi_m = g_m \bullet_{\alpha_1,\alpha_2} \psi_n, \forall m, n \in \mathbb{N}.$$

Therefore, Theorem 3 yields

$$l_{\alpha_1,\alpha_2}(\phi_n * \delta_m) = l_{\alpha_1,\alpha_2}(\phi_m * \delta_n), \forall m, n \in \mathbb{N},$$

where $\phi_n \in C_0^\infty$ and $(\delta_n) \in \Delta$, $\forall n \in \mathbb{N}$. Thus $\phi_n * \delta_m = \phi_m * \delta_n, m, \forall n \in \mathbb{N}$.
Hence, we have

$$\frac{(\phi_n)}{(\delta_n)} \in \Omega^1 \text{ and } E_{\alpha_1,\alpha_2}\left(\frac{(\phi_n)}{(\delta_n)}\right) = \frac{(g_n)}{(\psi_n)}.$$

The proof is finished.

References

1. Betancor, J.J.: On a varient of the Meijer integral transformation. Portugaliae Mathematica **45**(3), 251–264 (1988)
2. Boehme, T.K.: The support of Mikusinski operators. Transit. Am. Math. Soc. **176**, 319–334 (1973)

3. Brown, D., Dernek, N., Yurekli, O.: Identities for the $E_{2,1}$-transform and their applications. Appl. Math. Comput. **187**, 1557–1566 (2007)
4. Conlan, C., Koh, S.L.: On the Meijer transformation. Int. J. Math. Math. Sci. **1**, 145–159 (1978)
5. Kratzel, E.: Integral transformation of Bessel-type. In: Proceeding of International Conference on Generalized Functions and Operational Calculus, Varna, pp. 148–155 (1975)
6. Kratzel, E.: Bemerkunger zur Meljer-transformatlon und anwendungen. Math. Nachr. **30**, 327–334 (1965)
7. Kratzel, E.: Eine Verallgemeinerung der Laplace und Meljer transformation. Wiss. Z. Univ. Jena. Math. Naturw. Reihe. **5**, 369–381 (1965)
8. Kratzel, E.: Die faltung der L-transformation. Wiss. Univ. Jena. Math. Naturw. Reihe. **5**, 383–390 (1965)
9. Al-Omari, S.K.Q., Baleanu, D.: Quaternion fourier integral operators for spaces of generalized quaternions. Math. Methods Appl. Sci. **41**, 9477–9484 (2018)
10. Karunakaran, V., Vembu, R.: Hilbert transform on periodic Boehmians. Houst. J. Math. **29**, 439–454 (2005)
11. Al-Omari, S.K.Q.: Some remarks on short-time Fourier integral operators and classes of rapidly decaying functions. Math. Methods Appl. Sci. **41**, 1–8 (2018)
12. Meijer, C.S.: Eine nene erweiterung der Laplace transformation. Nederl. Akad. Wetench. Proc. Ser. A **44**, 727–739 (1941)
13. Al-Omari, S.K.Q., Kilicman, A.: An estimate of Sumudu transform for Boehmians. Adv. Differ. Equ. **77**, 1–12 (2013)
14. Al-Omari, S.K.Q., Agarwal, P., Choi, J.: Real covering of the generalized Hankel-Clifford transform of Fox kernel type of a class of Boehmians. Bull. Korean Math. Soc. **52**(5), 1607–1619 (2015)
15. Al-Omari, S.K.Q.: On a widder potential transform and its extension to a space of locally integrable Boehmians. J. Assoc. Arab Univ. Basic Appl. Sci. **85**(18), 94–98 (2015)
16. Al-Omari, S.K.Q.: Some estimate of a generalized Bessel-Struve transform on certain space of generalized functions. Ukrainian Math. J. **69**(9), 1155–1165 (2017)
17. Al-Omari, S.K.Q., Baleanu, D.: Convolution theorems associated with some integral operators and convolutions. Math. Methods Appl. Sci. **42**, 541–552 (2019)
18. Mikusinski, P.: Boehmians and pseudoquotients. Appl. Math. Inf. Sci. **5**, 192–204 (2011)
19. Mikusinski, P., Zayed, A.: The Radon transform of Boehmians. Am. Math. Soc. **118**(2), 561–570 (1993)
20. Al-Omari, S.K.Q., Al-Omari, J.F.: Some extensions of a certain integral transform to a quotient space of generalized functions. Open Math. **13**, 816–825 (2015)
21. Al-Omari, S.K.Q.: Hartley transforms on certain space of generalized functions. Georgian Math. J. **20**(3), 415–426 (2013)
22. Nemzer, D.: Periodic Boehmians II. Bull. Aust. Math. Soc. **44**, 271–278 (1991)
23. Al-Omari, S.K.Q., Agarwal, P.: Some general properties of a fractional Sumudu transform in the class of Boehmians. Kuwait J. Sci. **43**(2), 206–220 (2016)
24. Pandey, R.N.: A generalization of Meijer transform. Proc. Camb. Philos. Soc. **67**, 339–345 (1970)
25. Al-Omari, S.K.Q., Baleanu, D.: A Lebesgue integrable space of Boehmians fo a class of D_k transformations. J. Comput. Anal. Appl. **25**(1), 85–95 (2018)
26. Al-Omari, S.K.Q.: Some characteristics of S transforms in a class of rapidly decreasing Boehmians. J. Pseudo-Diff. Oper. Appl. **5**(4), 527–537 (2014)

27. Rodriguez, J.: Sobre una variante de la K-transformation. Actas V Congreso de Ecuaciones Diferenciales y Aplicaciones, pp. 503-53, Tenerife (1982)
28. Rodriguez, J.: A new variant for the Meijer's integral transform. Commentationes Math. Univ. Carol. **31**(3), 543–555 (1990)
29. Srivastava, H.M., Vyas, O.D.: A relation between Meijer and generalized Hankel transforms. Indagationes Math. **72**(2), 140–144 (1969)
30. Srivastava, H.M., Vyas, O.D.: A theorem relating generalized Hankel and Whittaker transforms. Nederl. Akad. Wetensch. Proc. Ser. A **72**(2), 1–13 (1969)
31. Waphare, B.B.: Meijer type transformation and related results. Int. J. Phys. Chem. Math. Fundam. **4**(1), 1846–2278 (2013)
32. Zemanian, A.H.: Generalized Integral Transformation. Dover Publications Inc., New York (1968). First Published by Interscience Publishers

Inequalities for Curve and Surface Integrals

Zlatko Pavić(✉)

Mechanical Engineering Faculty in Slavonski Brod,
University of Osijek, Slavonski Brod, Croatia
Zlatko.Pavic@sfsb.hr

Abstract. We deal with double inequalities that contain convex combinations and integral arithmetic means. This approach involves the connection between the Jensen and Hermite-Hadamard inequalities. As a result, we get very general inequalities that can be applied to curve and surface integrals.

Keywords: Convex function · Curve integral · Surface integral

1 Introduction

Let a and b be distinct points in the line $\mathbb{R}$. Then each point $x \in \mathbb{R}$ can be represented by the affine combination of points a and b as

$$x = \alpha(x)a + \beta(x)b, \tag{1}$$

where

$$\alpha(x) = \frac{x-b}{a-b} = \frac{\begin{vmatrix} x & 1 \\ b & 1 \end{vmatrix}}{\begin{vmatrix} a & 1 \\ b & 1 \end{vmatrix}}, \quad \beta(x) = \frac{a-x}{a-b} = \frac{\begin{vmatrix} a & 1 \\ x & 1 \end{vmatrix}}{\begin{vmatrix} a & 1 \\ b & 1 \end{vmatrix}}. \tag{2}$$

The above coefficients indicate affine functions $\alpha, \beta : \mathbb{R} \to \mathbb{R}$. In a concise form $\alpha(x) = \alpha_0 + \alpha_1 x$, where α_0 and α_1 are real constants. The same goes for $\beta(x)$. Since the points a and b are distinct, the affine combination in formula (1) is unique. The convex combinations in formula (1), accentuating points x with $\alpha(x), \beta(x) \geq 0$, delineate the closed interval with endpoints a and b as the set

$$\Delta_{ab} = \left\{ \overline{\alpha} a + \overline{\beta} b : \overline{\alpha}, \overline{\beta} \geq 0,\ \overline{\alpha} + \overline{\beta} = 1 \right\}.$$

Let $X \subset \mathbb{R}^n$ be a bounded closed set having a positive n-volume measured by the multiple Riemann integral $\int_X dx$ (where $dx = dx_1 \dots dx_n$), let $p : X \to \mathbb{R}$ be a nonnegative integrable function with a positive integral, let $g : X \to \mathbb{R}$ be an integrable function, let Δ_{ab} be an interval containing the image of g, and let

$$\overline{g} = \frac{\int_X g(x)p(x)\,dx}{\int_X p(x)\,dx} \tag{3}$$

M. Zeki Sarıkaya et al. (Eds.): ICMRS 2019, LNNS 123, pp. 96–107, 2020.
https://doi.org/10.1007/978-3-030-43002-3_9

be the integral arithmetic mean of g respecting p. By applying formula (1) to the point $\overline{g}$, and utilizing the affinity of functions α and β, we get the representation

$$\overline{g} = \alpha(\overline{g})a + \beta(\overline{g})b = \overline{\alpha}a + \overline{\beta}b, \tag{4}$$

where

$$\overline{\alpha} = \frac{\int_X \alpha(g(x))p(x)\,dx}{\int_X p(x)\,dx}, \quad \overline{\beta} = \frac{\int_X \beta(g(x))p(x)\,dx}{\int_X p(x)\,dx}. \tag{5}$$

Since $g(x) \in \Delta_{ab}$ for all $x \in X$, it follows that $\alpha(g(x)), \beta(g(x)) \geq 0$, and therefore $\overline{\alpha}, \overline{\beta} \geq 0$. Thus the combination $\overline{g} = \overline{\alpha}a + \overline{\beta}b$ is convex, which means that $\overline{g} \in \Delta_{ab}$. By including a convex function $f : \Delta_{ab} \to \mathbb{R}$, we achieve the double inequality

$$f\left(\frac{\int_X g(x)p(x)\,dx}{\int_X p(x)\,dx}\right) \leq \frac{\int_X f(g(x))p(x)\,dx}{\int_X p(x)\,dx} \leq \overline{\alpha}f(a) + \overline{\beta}f(b). \tag{6}$$

The inequality of first and second members is the expanded integral form of Jensen's inequality, see [6]. The inequality of second and third members can be obtained by applying the convexity of f to convex combinations $g(x) = \alpha(g(x))a + \beta(g(x))b$, and using the coefficients in formula (5).

We will generalize the inequality in formula (6) to convex functions defined on triangles and tetrahedrons.

2 Affine Combinations in the Plane and Space

Let $a = (a_1, a_2)$, $b = (b_1, b_2)$ and $c = (c_1, c_2)$ be non-collinear points in the plane $\mathbb{R}^2$. Then each point $x = (x_1, x_2) \in \mathbb{R}^2$ can be represented by the affine combination of points a, b and c as

$$x = \alpha(x)a + \beta(x)b + \gamma(x)c, \tag{7}$$

where

$$\alpha(x) = \frac{\begin{vmatrix} x_1 & x_2 & 1 \\ b_1 & b_2 & 1 \\ c_1 & c_2 & 1 \end{vmatrix}}{\begin{vmatrix} a_1 & a_2 & 1 \\ b_1 & b_2 & 1 \\ c_1 & c_2 & 1 \end{vmatrix}}, \quad \beta(x) = \frac{\begin{vmatrix} a_1 & a_2 & 1 \\ x_1 & x_2 & 1 \\ c_2 & c_2 & 1 \end{vmatrix}}{\begin{vmatrix} a_1 & a_2 & 1 \\ b_1 & b_2 & 1 \\ c_1 & c_2 & 1 \end{vmatrix}}, \quad \gamma(x) = \frac{\begin{vmatrix} a_1 & a_2 & 1 \\ b_1 & b_2 & 1 \\ x_1 & x_2 & 1 \end{vmatrix}}{\begin{vmatrix} a_1 & a_2 & 1 \\ b_1 & b_2 & 1 \\ c_1 & c_2 & 1 \end{vmatrix}}. \tag{8}$$

The functions $\alpha, \beta, \gamma : \mathbb{R}^2 \to \mathbb{R}$ are affine. For example, $\alpha(x_1, x_2) = \alpha_0 + \alpha_1 x_1 + \alpha_2 x_2$ with α_0, α_1 and α_2 as real constants. Since the points a, b and c are non-collinear, the affine combination in formula (7) is unique. The convex combinations in formula (7), emphasizing points x with $\alpha(x), \beta(x), \gamma(x) \geq 0$, designate the triangle with vertices a, b and c as the set

$$\Delta_{abc} = \left\{\overline{\alpha}a + \overline{\beta}b + \overline{\gamma}c : \overline{\alpha}, \overline{\beta}, \overline{\gamma} \geq 0,\ \overline{\alpha} + \overline{\beta} + \overline{\gamma} = 1\right\}.$$

Let $a = (a_1, a_2, a_3)$, $b = (b_1, b_2, b_3)$, $c = (c_1, c_2, c_3)$ and $d = (d_1, d_2, d_3)$ be points in the space $\mathbb{R}^3$ such that differences $a - b$, $a - c$ and $a - d$ are linearly independent. Then each point $x = (x_1, x_2, x_3) \in \mathbb{R}^3$ can be represented by the affine combination of points a, b, c and d as

$$x = \alpha(x)a + \beta(x)b + \gamma(x)c + \delta(x)d, \tag{9}$$

where

$$\alpha(x) = \frac{\begin{vmatrix} x_1 & x_2 & x_3 & 1 \\ b_1 & b_2 & b_3 & 1 \\ c_1 & c_2 & c_3 & 1 \\ d_1 & d_2 & d_3 & 1 \end{vmatrix}}{\begin{vmatrix} a_1 & a_2 & a_3 & 1 \\ b_1 & b_2 & b_3 & 1 \\ c_1 & c_2 & c_3 & 1 \\ d_1 & d_2 & d_3 & 1 \end{vmatrix}},\ \beta(x) = \frac{\begin{vmatrix} a_1 & a_2 & a_3 & 1 \\ x_1 & x_2 & x_3 & 1 \\ c_1 & c_2 & c_3 & 1 \\ d_1 & d_2 & d_3 & 1 \end{vmatrix}}{\begin{vmatrix} a_1 & a_2 & a_3 & 1 \\ b_1 & b_2 & b_3 & 1 \\ c_1 & c_2 & c_3 & 1 \\ d_1 & d_2 & d_3 & 1 \end{vmatrix}},\ \gamma(x) = \frac{\begin{vmatrix} a_1 & a_2 & a_3 & 1 \\ b_1 & b_2 & b_3 & 1 \\ x_1 & x_2 & x_3 & 1 \\ d_1 & d_2 & d_3 & 1 \end{vmatrix}}{\begin{vmatrix} a_1 & a_2 & a_3 & 1 \\ b_1 & b_2 & b_3 & 1 \\ c_1 & c_2 & c_3 & 1 \\ d_1 & d_2 & d_3 & 1 \end{vmatrix}},\ \delta(x) = \frac{\begin{vmatrix} a_1 & a_2 & a_3 & 1 \\ b_1 & b_2 & b_3 & 1 \\ c_1 & c_2 & c_3 & 1 \\ x_1 & x_2 & x_3 & 1 \end{vmatrix}}{\begin{vmatrix} a_1 & a_2 & a_3 & 1 \\ b_1 & b_2 & b_3 & 1 \\ c_1 & c_2 & c_3 & 1 \\ d_1 & d_2 & d_3 & 1 \end{vmatrix}}. \tag{10}$$

The functions $\alpha, \beta, \gamma, \delta : \mathbb{R}^3 \to \mathbb{R}$ are affine. Since the differences $a - b$, $a - c$ and $a - d$ are linearly independent, the affine combination in formula (9) is unique. The convex combinations in formula (9), highlighting points x with $\alpha(x), \beta(x), \gamma(x), \delta(x) \geq 0$, appoint the tetrahedron with vertices a, b, c and d as the set

$$\Delta_{abcd} = \left\{\overline{\alpha}a + \overline{\beta}b + \overline{\gamma}c + \overline{\delta}d : \overline{\alpha}, \overline{\beta}, \overline{\gamma}, \overline{\delta} \geq 0,\ \overline{\alpha} + \overline{\beta} + \overline{\gamma} + \overline{\delta} = 1\right\}.$$

3 Main Results

We continue to use a bounded closed set $X \subset \mathbb{R}^n$ having a positive n-volume measured by the multiple integral $\int_X dx$, where dx abbreviates $dx_1 \ldots dx_n$.

Lemma 1. *Let $p : X \to \mathbb{R}$ be a nonnegative integrable function with a positive integral, let $g_1, g_2 : X \to \mathbb{R}$ be integrable functions, let Δ_{abc} be a triangle containing the image of the mapping $g = (g_1, g_2) : X \to \mathbb{R}^2$, and let $\overline{\alpha}a + \overline{\beta}b + \overline{\gamma}c$ be the convex combination such that*

$$\left(\frac{\int_X g_1(x)p(x)\,dx}{\int_X p(x)\,dx}, \frac{\int_X g_2(x)p(x)\,dx}{\int_X p(x)\,dx}\right) = \overline{\alpha}a + \overline{\beta}b + \overline{\gamma}c. \tag{11}$$

Then each convex function $f : \Delta_{abc} \to \mathbb{R}$ satisfies the double inequality

$$\begin{aligned} &f\left(\frac{\int_X g_1(x)p(x)\,dx}{\int_X p(x)\,dx}, \frac{\int_X g_2(x)p(x)\,dx}{\int_X p(x)\,dx}\right) \\ &\leq \frac{\int_X f(g_1(x), g_2(x))p(x)\,dx}{\int_X p(x)\,dx} \leq \overline{\alpha}f(a) + \overline{\beta}f(b) + \overline{\gamma}f(c). \end{aligned} \tag{12}$$

Proof. As regards the equality in formula (11), we have to demonstrate the convex combination $\overline{\alpha}a + \overline{\beta}b + \overline{\gamma}c$ representing the point

$$\overline{g} = \left(\frac{\int_X g_1(x)p(x)\,dx}{\int_X p(x)\,dx}, \frac{\int_X g_2(x)p(x)\,dx}{\int_X p(x)\,dx}\right).$$

It will be backed up by the fact that each affine function $h : \mathbb{R}^2 \to \mathbb{R}$ meets the integral equality

$$h\left(\frac{\int_X g_1(x)p(x)\,dx}{\int_X p(x)\,dx}, \frac{\int_X g_2(x)p(x)\,dx}{\int_X p(x)\,dx}\right) = \frac{\int_X h(g_1(x), g_2(x))p(x)\,dx}{\int_X p(x)\,dx}.$$

By applying formula (7) to the point $\overline{g}$, utilizing the affinity of functions α, β and γ, and writing points $(g_1(x), g_2(x))$ as $g(x)$, we obtain the affine combination

$$\begin{aligned}\overline{g} &= \alpha(\overline{g})a + \beta(\overline{g})b + \gamma(\overline{g})c \\ &= \frac{\int_X \alpha(g(x))p(x)\,dx}{\int_X p(x)\,dx}a + \frac{\int_X \beta(g(x))p(x)\,dx}{\int_X p(x)\,dx}b + \frac{\int_X \gamma(g(x))p(x)\,dx}{\int_X p(x)\,dx}c \\ &= \overline{\alpha}a + \overline{\beta}b + \overline{\gamma}c,\end{aligned}$$

where apparently

$$\overline{\alpha} = \frac{\int_X \alpha(g(x))p(x)\,dx}{\int_X p(x)\,dx}, \quad \overline{\beta} = \frac{\int_X \beta(g(x))p(x)\,dx}{\int_X p(x)\,dx}, \quad \overline{\gamma} = \frac{\int_X \gamma(g(x))p(x)\,dx}{\int_X p(x)\,dx}. \tag{13}$$

These coefficients are nonnegative because $g(x) \in \Delta_{abc}$ for all $x \in X$, and so the above affine combination is convex. The point $\overline{g}$ lies inside the triangle Δ_{abc}.

To prove the inequality in formula (12), we consider three cases of the convex combination representation $\overline{g} = \overline{\alpha}a + \overline{\beta}b + \overline{\gamma}c$, referring to the number of positive coefficients.

If $\overline{g} = \overline{\alpha}a + \overline{\beta}b + \overline{\gamma}c$ with $\overline{\alpha}, \overline{\beta}, \overline{\gamma} > 0$, then $\overline{g}$ lies inside the interior of Δ_{abc}. By including a support plane f_* of f at $\overline{g}$, and the secant plane f^* of f over Δ_{abc}, we have the coincidences

$$f_*(\overline{g}) = f(\overline{g}), \quad f^*(a) = f(a), \quad f^*(b) = f(b), \quad f^*(c) = f(c)$$

and the inequality

$$f_*(g(x)) \le f(g(x)) \le f^*(g(x))$$

which holds for all $x \in X$. By relying on the above coincidences and inequality, and applying the affinity of f_* and f^*, we get the multiple inequality

$$\begin{aligned} f(\overline{g}) = f_*(\overline{g}) &= \frac{\int_X f_*(g(x))p(x)\,dx}{\int_X p(x)\,dx} \le \frac{\int_X f(g(x))p(x)\,dx}{\int_X p(x)\,dx} \\ &\le \frac{\int_X f^*(g(x))p(x)\,dx}{\int_X p(x)\,dx} = f^*(\overline{g}) = \overline{\alpha}f^*(a) + \overline{\beta}f^*(b) + \overline{\gamma}f^*(c) \\ &= \overline{\alpha}f(a) + \overline{\beta}f(b) + \overline{\gamma}f(c). \end{aligned} \tag{14}$$

The first, fourth and eight members of the above inequality represents formula (12). The composition $f(g)$ is integrable because it is bounded on X, and continuous almost everywhere on X. So, the product $f(g)p$ is integrable.

If $\overline{g} = \overline{\alpha}a + \overline{\beta}b$ with $\overline{\alpha}, \overline{\beta} > 0$, then the equation $\overline{\gamma} = 0$ via formula (13) implies

$$\int_X \gamma(g(x))p(x)\,dx = 0.$$

Since $\gamma(g(x))p(x) \geq 0$ for all $x \in X$, and $\int_X p(x)\,dx > 0$, it follows that $\gamma(g(x)) = 0$ for almost all $x \in \operatorname{supp}(p)$. Thus $g(x) \in \Delta_{ab}$ for almost all $x \in \operatorname{supp}(p)$.

Since $\overline{g}$ lies inside the relative interior of the side Δ_{ab}, we can use the restriction f/Δ_{ab} as follows. Let h_* be a plane such that the restriction h_*/Δ_{ab} is the support line of the restriction f/Δ_{ab} at $\overline{g}$, and let h^* be a plane such that h^*/Δ_{ab} is the secant line of f/Δ_{ab} over Δ_{ab} (f^* can serve as h^*). The inequality

$$h_*(g(x)) \leq f(g(x)) \leq h^*(g(x))$$

holds for almost all $x \in \operatorname{supp}(p)$. The multiple inequality in formula (14) remains valid if we put in $\overline{\gamma} = 0$, h_* instead of f_*, and h^* instead of f^*. Namely, all integrals remains unchanged if we put in $\operatorname{supp}(p)$ instead of X. Thus we reach formula (12) with $\overline{\gamma} = 0$. Similar arguments apply for $\overline{\beta} = 0$ or $\overline{\alpha} = 0$.

If $\overline{g} = a$, then the equation $\overline{\alpha} = 1$ via formula (13) implies that $g(x) = a$ for almost all $x \in \operatorname{supp}(p)$ (the same conclusion arises from the system of equations $\overline{\beta} = 0$ and $\overline{\gamma} = 0$). It turns out that

$$\frac{\int_X f(g(x))p(x)\,dx}{\int_X p(x)\,dx} = \frac{\int_X f(a)p(x)\,dx}{\int_X p(x)\,dx} = f(a),$$

and so the trivial double inequality $f(a) \leq f(a) \leq f(a)$ represents formula (12). A similar reasoning applies for $\overline{g} = b$ or $\overline{g} = c$. □

The functional approach to the Jensen and Hermite-Hadamard inequality on the triangle can be found in [11].

The generalization of Lemma 1 to tetrahedrons is as follows.

Theorem 1. *Let $p : X \to \mathbb{R}$ be a nonnegative integrable function with a positive integral, let $g_1, g_2, g_3 : X \to \mathbb{R}$ be integrable functions, let Δ_{abcd} be a tetrahedron containing the image of the mapping $g = (g_1, g_2, g_3) : X \to \mathbb{R}^3$, and let $\overline{\alpha}a + \overline{\beta}b + \overline{\gamma}c + \overline{\delta}d$ be the convex combination such that*

$$\left(\frac{\int_X g_1(x)p(x)\,dx}{\int_X p(x)\,dx}, \frac{\int_X g_2(x)p(x)\,dx}{\int_X p(x)\,dx}, \frac{\int_X g_3(x)p(x)\,dx}{\int_X p(x)\,dx}\right) = \overline{\alpha}a + \overline{\beta}b + \overline{\gamma}c + \overline{\delta}d. \tag{15}$$

Then each convex function $f : \Delta_{abcd} \to \mathbb{R}$ satisfies the double inequality

$$\begin{aligned} & f\left(\frac{\int_X g_1(x)p(x)\,dx}{\int_X p(x)\,dx}, \frac{\int_X g_2(x)p(x)\,dx}{\int_X p(x)\,dx}, \frac{\int_X g_3(x)p(x)\,dx}{\int_X p(x)\,dx}\right) \\ & \leq \frac{\int_X f(g_1(x), g_2(x), g_3(x))p(x)\,dx}{\int_X p(x)\,dx} \leq \overline{\alpha}f(a) + \overline{\beta}f(b) + \overline{\gamma}f(c) + \overline{\delta}f(d). \end{aligned} \tag{16}$$

Proof. The representation in formula (15) can be obtained by applying formula (9) to the point

$$\overline{g}=\left(\frac{\int_X g_1(x)p(x)\,dx}{\int_X p(x)\,dx},\frac{\int_X g_2(x)p(x)\,dx}{\int_X p(x)\,dx},\frac{\int_X g_3(x)p(x)\,dx}{\int_X p(x)\,dx}\right).$$

So, relying on the affinity of functions α, β, γ and δ, we find that the required nonnegative coefficients are represented by

$$\overline{\alpha}=\frac{\int_X \alpha(g(x))p(x)\,dx}{\int_X p(x)\,dx},\tag{17}$$

and just so for $\overline{\beta}$, $\overline{\gamma}$ and $\overline{\delta}$.

To prove the inequality in formula (16), we consider four cases of the point $\overline{g}=\overline{\alpha}a+\overline{\beta}b+\overline{\gamma}c+\overline{\delta}d$, referring to the number of positive coefficients.

If $\overline{g}=\overline{\alpha}a+\overline{\beta}b+\overline{\gamma}c+\overline{\delta}d$ with $\overline{\alpha},\overline{\beta},\overline{\gamma},\overline{\delta}>0$, then $\overline{g}$ is in the interior of Δ_{abcd}. By using a support hyperplane f_* of f at $\overline{g}$, and the secant hyperplane f^* of f over Δ_{abcd}, we get the multiple inequality in formula (14) for the tetrahedron Δ_{abcd}.

If $\overline{g}=\overline{\alpha}a+\overline{\beta}b+\overline{\gamma}c$ with $\overline{\alpha},\overline{\beta},\overline{\gamma}>0$, then the equation $\overline{\delta}=0$ (via formula (17) for $\overline{\delta}$) implies that

$$g(x)\in\Delta_{abc}$$

for almost all $x\in\operatorname{supp}(p)$.
Since $\overline{g}$ is in the relative interior of the triangle Δ_{abc}, we can use a hyperplane h_* such that the restriction h_*/Δ_{abc} is the support plane of the restriction f/Δ_{abc} at $\overline{g}$, and a hyperplane h^* such that h^*/Δ_{abc} is the secant plane of f/Δ_{abc} over Δ_{abc}. Then $h_*(g(x))\le f(g(x))\le h^*(g(x))$ for almost all $x\in\operatorname{supp}(p)$. By continuing as in the proof of Lemma 1, we reach the multiple inequality in formula (14) for the triangle Δ_{abc}. The same applies to other triangles as 2-faces of Δ_{abcd}.

If $\overline{g}=\overline{\alpha}a+\overline{\beta}b$ with $\overline{\alpha},\overline{\beta}>0$, then the system of equations $\overline{\gamma}=0$ and $\overline{\delta}=0$ (via formula (17) for $\overline{\gamma}$ and $\overline{\delta}$) implies that

$$g(x)\in\Delta_{abd}\cap\Delta_{abc}=\Delta_{ab}$$

for almost all $x\in\operatorname{supp}(p)$. By proceeding as in the previous case, we reach the multiple inequality in formula (14) for the interval Δ_{ab}. The same applies to other intervals as 1-faces of Δ_{abcd}.

If $\overline{g}=a$, then the equation $\overline{\alpha}=1$ (via formula (17) for $\overline{\alpha}$) implies that $g(x)=a$ for almost all $x\in\operatorname{supp}(p)$. It further produces the trivial double inequality $f(a)\le f(a)\le f(a)$, representing formula (16). The same applies to other vertices. □

The inequality in formula (16) taken in the form

$$f\left(\overline{\alpha}a+\overline{\beta}b+\overline{\gamma}c+\overline{\delta}d\right)\le\frac{\int_X f(g_1(x),g_2(x),g_3(x))p(x)\,dx}{\int_X p(x)\,dx}\le\overline{\alpha}f(a)+\overline{\beta}f(b)+\overline{\gamma}f(c)+\overline{\delta}f(d).$$

presents the generalization of the Hermite-Hadamard inequality for simplices. The above inequality is reduced to the basic form (referring to the tetrahedron barycenter)

$$f\left(\frac{a+b+c+d}{4}\right) \leq \frac{\int_{\Delta_{abcd}} f(x)\,dx}{\mathrm{vol}(\Delta_{abcd})} \leq \frac{f(a)+f(b)+f(c)+f(d)}{4}.$$

if $p(x) = 1$ and $g_i(x) = x_i$ for $x = (x_1, x_2, x_3) \in X = \Delta_{abcd}$. In this reduction, the correlations $\overline{\alpha} = \overline{\beta} = \overline{\gamma} = \overline{\delta} = 1/4$ and $\int_{\Delta_{abcd}} dx = \mathrm{vol}(\Delta_{abcd})$ arise.

The classic form of the Hermite-Hadamard inequality (referring to the interval midpoint) springs from articles [5] and [4]. The development of the Hermite-Hadamard inequality for simplices can be tracked through articles [2,8–10], and many others published in the last ten years.

Theorem 1 can be applied to curves and surfaces in the space $\mathbb{R}^3$, as well as their barycenters. Curves and surfaces are fundamental objects of profound and complicated modern theory of manifolds. Comprehensive introduction to manifolds can be found in [12] and [7]. Differential geometry of curves and surfaces was discussed in [1]. Surface integrals were considered in [3].

We are interested only in the integration over parametric (parameterized) curves and surfaces, or more specifically, over curve arcs and surface arches.

4 Inequalities for Curve Integrals

In this section, the set X will be bounded closed interval $I \subset \mathbb{R}$ with distinct endpoints. Theorem 1 will be applied to curve arcs in the space $\mathbb{R}^3$.

Let $J \subseteq \mathbb{R}$ be an open interval, let $r_1, r_2, r_3 : J \to \mathbb{R}$ be continuously differentiable functions (having continuous derivatives) such that the mapping $r = (r_1, r_2, r_3) : J \to \mathbb{R}^3$ is injective. Then the image of r represents the parametric curve in $\mathbb{R}^3$ over J. We are interested in the curve arc over the bounded closed interval $I \subset J$, depicted as the set

$$\mathcal{C} = \{r(x) = (r_1(x), r_2(x), r_3(x)) : x \in I\},$$

and graphically presented in Fig. 1. To facilitate the application, we employ the unitary coordinate vectors $\mathbf{i} = (1,0,0)$, $\mathbf{j} = (0,1,0)$ and $\mathbf{k} = (0,0,1)$, and use the representation

$$\mathbf{r}(x) = r_1(x)\mathbf{i} + r_2(x)\mathbf{j} + r_3(x)\mathbf{k},$$

as well as its derivative

$$\mathbf{r}'(x) = r_1'(x)\mathbf{i} + r_2'(x)\mathbf{j} + r_3'(x)\mathbf{k}.$$

Since the curve integrals are expressed and calculated via definite integrals, if we say a function $g : \mathcal{C} \to \mathbb{R}$ is integrable, then it assumes the existence of the integral

$$\int_{\mathcal{C}} g(r)\,ds = \int_I g(r(x))\|\mathbf{r}'(x)\|\,dx. \tag{18}$$

The barycenter of the curve arc $\mathcal{C}$ is the point $\overline{x} \in \mathbb{R}^3$ whose coordinates $\overline{x}_j$ are specified by the curve integral equations

$$\int_{\mathcal{C}} (r_j - \overline{x}_j)\, ds = 0,$$

from which it comes out

$$\overline{x} = \left(\frac{\int_{\mathcal{C}} r_1\, ds}{\int_{\mathcal{C}} ds}, \frac{\int_{\mathcal{C}} r_2\, ds}{\int_{\mathcal{C}} ds}, \frac{\int_{\mathcal{C}} r_3\, ds}{\int_{\mathcal{C}} ds} \right).$$

So, the barycenter coordinates of the arc $\mathcal{C}$ are integral arithmetic means of projections $g_i : \mathcal{C} \to \mathbb{R}$ defined by $g_i(r_1(x), r_2(x), r_3(x)) = r_i(x)$. The injectivity of the mapping r ensures that the length $\int_{\mathcal{C}} ds$ of the curve $\mathcal{C}$ is positive. It can be proved that the barycenter $\overline{x}$ belongs to the convex hull of the set of points of the curve $\mathcal{C}$, that is, $\overline{x} \in \operatorname{conv}\mathcal{C}$.

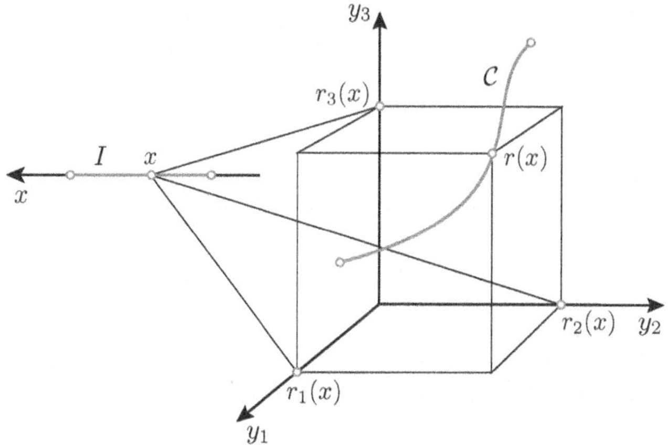

Fig. 1. Curve in the three-dimensional space

The influence of Theorem 1 to curve arcs in the space $\mathbb{R}^3$ is as follows.

Corollary 1. *Let $\mathcal{C} = \{r(x) = (r_1(x), r_2(x), r_3(x)) : x \in I\}$ be a parametric curve arc in the space $\mathbb{R}^3$, let $p : \mathcal{C} \to \mathbb{R}$ be a nonnegative integrable function with a positive integral, let $g_1, g_2, g_3 : \mathcal{C} \to \mathbb{R}$ be integrable functions, let Δ_{abcd} be a tetrahedron containing the image of the mapping $g = (g_1, g_2, g_3) : \mathcal{C} \to \mathbb{R}^3$, and let $\overline{\alpha}a + \overline{\beta}b + \overline{\gamma}c + \overline{\delta}d$ be the convex combination such that*

$$\left(\frac{\int_{\mathcal{C}} g_1(r)p(r)\, ds}{\int_{\mathcal{C}} p(r)\, ds}, \frac{\int_{\mathcal{C}} g_2(r)p(r)\, ds}{\int_{\mathcal{C}} p(r)\, ds}, \frac{\int_{\mathcal{C}} g_3(r)p(r)\, ds}{\int_{\mathcal{C}} p(r)\, ds} \right) = \overline{\alpha}a + \overline{\beta}b + \overline{\gamma}c + \overline{\delta}d. \quad (19)$$

Then each convex function $f : \Delta_{abcd} \to \mathbb{R}$ satisfies the double inequality

$$\begin{aligned} & f\left(\frac{\int_C g_1(r)p(r)\,ds}{\int_C p(r)\,ds}, \frac{\int_C g_2(r)p(r)\,ds}{\int_C p(r)\,ds}, \frac{\int_C g_3(r)p(r)\,ds}{\int_C p(r)\,ds}\right) \\ & \leq \frac{\int_C f(g_1(r), g_2(r), g_3(r))p(r)\,ds}{\int_C p(r)\,ds} \leq \overline{\alpha} f(a) + \overline{\beta} f(b) + \overline{\gamma} f(c) + \overline{\delta} f(d). \end{aligned} \tag{20}$$

Proof. We consider the norm of $\mathbf{r}'(x)$ as the function $q : I \to \mathbb{R}$ determined by

$$q(x) = \|\mathbf{r}'(x)\| = \sqrt{r_1'(x)^2 + r_2'(x)^2 + r_3'(x)^2},$$

and include the functions $\widetilde{p}, \widetilde{g}_1, \widetilde{g}_2, \widetilde{g}_3 : I \to \mathbb{R}$ defined by

$$\widetilde{p}(x) = p(r(x))q(x) \text{ and } \widetilde{g}_j(x) = g_j(r(x)).$$

Then the application of Theorem 1 to the functions $\widetilde{p}$ and $\widetilde{g}_i$ produces the representation

$$\left(\frac{\int_I \widetilde{g}_1(x)\widetilde{p}(x)\,dx}{\int_I \widetilde{p}(x)\,dx}, \frac{\int_I \widetilde{g}_2(x)\widetilde{p}(x)\,dx}{\int_I \widetilde{p}(x)\,dx}, \frac{\int_I \widetilde{g}_3(x)\widetilde{p}(x)\,dx}{\int_I \widetilde{p}(x)\,dx}\right) = \overline{\alpha} a + \overline{\beta} b + \overline{\gamma} c + \overline{\delta} d$$

and the inequality

$$\begin{aligned} & f\left(\frac{\int_I \widetilde{g}_1(x)\widetilde{p}(x)\,dx}{\int_I \widetilde{p}(x)\,dx}, \frac{\int_I \widetilde{g}_2(x)\widetilde{p}(x)\,dx}{\int_I \widetilde{p}(x)\,dx}, \frac{\int_I \widetilde{g}_3(x)\widetilde{p}(x)\,dx}{\int_I \widetilde{p}(x)\,dx}\right) \\ & \leq \frac{\int_I f(\widetilde{g}_1(x), \widetilde{g}_2(x), \widetilde{g}_3(x))\widetilde{p}(x)\,dx}{\int_I \widetilde{p}(x)\,dx} \leq \overline{\alpha} f(a) + \overline{\beta} f(b) + \overline{\gamma} f(c) + \overline{\delta} f(d). \end{aligned}$$

The above relations are just definite integral forms of the curve integrals contained in formula (19) and formula (20). □

The above corollary contains the inequality including the curve arc barycenter.

Corollary 2. *Let $\mathcal{C} = \{r(x) = (r_1(x), r_2(x), r_3(x)) : x \in I\}$ be a parametric curve arc in the space $\mathbb{R}^3$, let Δ_{abcd} be a tetrahedron containing $\mathcal{C}$, and let $\overline{\alpha} a + \overline{\beta} b + \overline{\gamma} c + \overline{\delta} d$ be the convex combination representing the barycenter of $\mathcal{C}$.*

Then each convex function $f : \Delta_{abcd} \to \mathbb{R}$ satisfies the double inequality

$$\begin{aligned} & f\left(\frac{\int_C r_1\,ds}{\int_C ds}, \frac{\int_C r_2\,ds}{\int_C ds}, \frac{\int_C r_3\,ds}{\int_C ds}\right) \\ & \leq \frac{\int_C f(r_1, r_2, r_3)\,ds}{\int_C ds} \leq \overline{\alpha} f(a) + \overline{\beta} f(b) + \overline{\gamma} f(c) + \overline{\delta} f(d). \end{aligned} \tag{21}$$

Proof. Corollary 2 follows from Corollary 1 with functions $p, g_1, g_2, g_3 : \mathcal{C} \to \mathbb{R}$ determined by $p(r(x)) = 1$ and $g_i(r(x)) = r_i(x)$ for $x \in I$. □

5 Inequalities for Surface Integrals

In this section, the set X will be bounded closed connected set $D \subset \mathbb{R}^2$ having a positive area $\int_D dx$, where dx abbreviates $dx_1 dx_2$. Theorem 1 will be applied to surface arches in the space $\mathbb{R}^3$.

Let $U \subseteq \mathbb{R}^2$ be an open set, let $r_1, r_2, r_3 : U \to \mathbb{R}$ be continuously differentiable functions (having continuous partial derivatives) such that the mapping $r = (r_1, r_2, r_3) : U \to \mathbb{R}^3$ is injective. Then the image of r represents the parametric surface in $\mathbb{R}^3$ over U. We are interested in the surface arch over the two-dimensional bounded closed connected set $D \subset U$, typified as the set

$$\mathcal{S} = \{r(x) = (r_1(x), r_2(x), r_3(x)) : x \in D\},$$

and graphically presented in Fig. 2. If we say a function $g : \mathcal{S} \to \mathbb{R}$ is integrable, then it assumes the existence of the integral

$$\int_{\mathcal{S}} g(r)\, dS = \int_D g(r(x)) \left\| \frac{\partial \mathbf{r}}{\partial x_1} \times \frac{\partial \mathbf{r}}{\partial x_2} \right\| dx, \tag{22}$$

where the vector product

$$\frac{\partial \mathbf{r}}{\partial x_1} \times \frac{\partial \mathbf{r}}{\partial x_2} = \begin{vmatrix} \mathbf{i} & \mathbf{j} & \mathbf{k} \\ \frac{\partial r_1}{\partial x_1} & \frac{\partial r_2}{\partial x_1} & \frac{\partial r_3}{\partial x_1} \\ \frac{\partial r_1}{\partial x_2} & \frac{\partial r_2}{\partial x_2} & \frac{\partial r_3}{\partial x_2} \end{vmatrix}. \tag{23}$$

As regards the barycenter of the surface arch $\mathcal{S}$, it is determined by the point

$$\overline{x} = \left(\frac{\int_{\mathcal{S}} r_1\, dS}{\int_{\mathcal{S}} dS}, \frac{\int_{\mathcal{S}} r_2\, dS}{\int_{\mathcal{S}} dS}, \frac{\int_{\mathcal{S}} r_3\, dS}{\int_{\mathcal{S}} dS} \right).$$

The following is the effect of Theorem 1 to surface arches in $\mathbb{R}^3$.

Corollary 3. *Let $\mathcal{S} = \{r(x) = (r_1(x), r_2(x), r_3(x)) : x \in D\}$ be a parametric surface arch in the space $\mathbb{R}^3$, let $p : \mathcal{S} \to \mathbb{R}$ be a nonnegative integrable function with a positive integral, let $g_1, g_2, g_3 : \mathcal{S} \to \mathbb{R}$ be integrable functions, let Δ_{abcd} be a tetrahedron containing the image of the mapping $g = (g_1, g_2, g_3) : \mathcal{S} \to \mathbb{R}^3$, and let $\overline{\alpha} a + \overline{\beta} b + \overline{\gamma} c + \overline{\delta} d$ be the convex combination such that*

$$\left(\frac{\int_{\mathcal{S}} g_1(r) p(r)\, dS}{\int_{\mathcal{S}} p(r)\, dS}, \frac{\int_{\mathcal{S}} g_2(r) p(r)\, dS}{\int_{\mathcal{S}} p(r)\, dS}, \frac{\int_{\mathcal{S}} g_3(r) p(r)\, dS}{\int_{\mathcal{S}} p(r)\, dS} \right) = \overline{\alpha} a + \overline{\beta} b + \overline{\gamma} c + \overline{\delta} d. \tag{24}$$

Then each convex function $f : \Delta_{abcd} \to \mathbb{R}$ satisfies the double inequality

$$\begin{aligned} & f\left(\frac{\int_{\mathcal{S}} g_1(r) p(r)\, dS}{\int_{\mathcal{S}} p(r)\, dS}, \frac{\int_{\mathcal{S}} g_2(r) p(r)\, dS}{\int_{\mathcal{S}} p(r)\, dS}, \frac{\int_{\mathcal{S}} g_3(r) p(r)\, dS}{\int_{\mathcal{S}} p(r)\, dS} \right) \\ \leq\ & \frac{\int_{\mathcal{S}} f(g_1(r), (g_2(r), g_3(r)) p(r)\, dS}{\int_{\mathcal{S}} p(r)\, dS} \leq \overline{\alpha} f(a) + \overline{\beta} f(b) + \overline{\gamma} f(c) + \overline{\delta} f(d). \end{aligned} \tag{25}$$

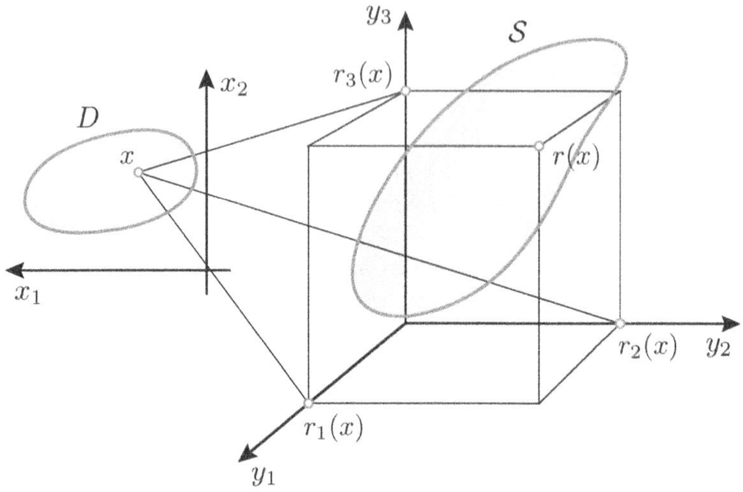

Fig. 2. Surface in the three-dimensional space

Proof. It fits in the proof of Corollary 1 if we employ the set D instead of the interval I, and if we use the function $q : D \to \mathbb{R}$ designated by

$$q(x) = q(x_1, x_2) = \left\| \frac{\partial \mathbf{r}}{\partial x_1} \times \frac{\partial \mathbf{r}}{\partial x_2} \right\|,$$

representing the norm of the vector product of partial derivatives of $\mathbf{r}(x)$. □

As for the surface arch barycenter, we have the following consequence.

Corollary 4. *Let $\mathcal{S} = \{r(x) = (r_1(x), r_2(x), r_3(x)) : x \in D\}$ be a parametric surface arch in the space $\mathbb{R}^3$, let Δ_{abcd} be a tetrahedron containing $\mathcal{S}$, and let $\overline{\alpha}a + \overline{\beta}b + \overline{\gamma}c + \overline{\delta}d$ be the convex combination representing the barycenter of $\mathcal{S}$.*

Then each convex function $f : \Delta_{abcd} \to \mathbb{R}$ satisfies the double inequality

$$\begin{aligned} & f\left(\frac{\int_{\mathcal{S}} r_1\, dS}{\int_{\mathcal{S}} dS}, \frac{\int_{\mathcal{S}} r_2\, dS}{\int_{\mathcal{S}} dS}, \frac{\int_{\mathcal{S}} r_3\, dS}{\int_{\mathcal{S}} dS}\right) \\ & \leq \frac{\int_{\mathcal{S}} f(r_1, r_2, r_3)\, dS}{\int_{\mathcal{S}} dS} \leq \overline{\alpha} f(a) + \overline{\beta} f(b) + \overline{\gamma} f(c) + \overline{\delta} f(d). \end{aligned} \tag{26}$$

References

1. Abate, M., Tovena, F.: Curves and Surfaces. Springer, Milan (2012). https://doi.org/10.1007/978-88-470-1941-6
2. Bessenyei, M.: The Hermite-Hadamard inequality on simplices. Am. Math. Mon. **115**, 339–345 (2008)
3. Driver, B.K.: Surfaces, Surface Integrals and Integration by Parts. http://www.math.ucsd.edu/~bdriver/231-02-03/Lecture_Notes/pde8.pdf

4. Hadamard, J.: Étude sur les propriétés des fonctions entières et en particulier d'une fonction considerée par Riemann. J. Math. Pures Appl. **58**, 171–215 (1893)
5. Hermite, C.: Sur deux limites d'une intégrale définie. Mathesis **3**, 82 (1883)
6. Jensen, J.L.W.V.: Sur les fonctions convexes et les inégalités entre les valeurs moyennes. Acta Math. **30**, 175–193 (1906)
7. Lee, J.M.: Introduction to Smooth Manifolds. Springer, New York (2013). https://doi.org/10.1007/978-0-387-21752-9
8. Mitroi, F.-C., Spiridon, C.I.: Refinements of Hermite-Hadamard inequality on simplices. Math. Rep. (Bucur.) **15**, (2013). arXiv:1105.5043
9. Neuman, E.: Inequalities involving multivariate convex functions II. Proc. Am. Math. Soc. **109**, 965–974 (1990)
10. Pavić, Z.: Improvements of the Hermite-Hadamard inequality for the simplex. J. Inequal. Appl. **2017**, 3 (2017). Article 3
11. Pavić, Z.: The Jensen and Hermite-Hadamard inequality on the triangle. J. Math. Inequal. **11**, 1099–1112 (2017)
12. Tu, L.W.: An Introduction to Manifolds. Springer, New York (2011). https://doi.org/10.1007/978-1-4419-7400-6

Numerical Simulation of Conformable Fuzzy Differential Equations

Mohammed Al-Smadi(✉)

Department of Applied Science, Ajloun College,
Al-Balqa Applied University, Ajloun 26816, Jordan
mhm.smadi@bau.edu.jo

Abstract. The analytical behavior of differential equations under uncertainty often seems confusing and cannot be fully understood or predicted. Therefore, finding suitable, comprehensive and highly efficient tools to address these issues is of great importance. In this study, an efficient analytical tool is developed to investigate the approximate solution of a class of fuzzy differential equations subject to uncertain initial conditions in the sense of conformable fuzzy fractional derivatives. The proposed technique relies on Taylor series expansion as well as minimizing residual-error function. The methodology is based on constructing a fractional power series in a rapid-convergent form under strongly generalized differentiability without any restrictive assumptions. Parametric characterizing of the solutions is obtained by converting the conformable fuzzy fractional differential equation to an equivalent crisp system of corresponding conformable fractional differential equations. This adaptive can be used as an alternative technique for solving many uncertain problems arising in diverse fields of engineering, chemistry, and biology. The effectiveness, validity, and potentiality of the proposed method are illustrated by verifying a numerical experiment. Numerical and graphical consequences indicate the accuracy and appropriateness of the suggested algorithm in dealing with fuzzy fractional models.

Keywords: Conformable fractional derivative · Fuzzy differential equation · Residual power series approach · Approximate solutions

1 Introduction

Fuzzy differential equations are an influential part of the theory of uncertainty analysis, which have become essential vital techniques to model and simulate various natural phenomena as well as to describe uncertain parameters of process levels. Uncertainty comes from different sources including measurement errors, missing data, data collection, cumulative errors, initial estimations, etc. In many cases, it is difficult to provide an explicit solution to these equations, which requires a reliable approximation to handle the complexities of uncertain issues and to achieve an accurate arithmetical structure that appropriately addresses the fuzzy initial value problems [1–4]. In the sense of fractional derivatives, the analytic-numeric approaches to fuzzy fractional calculus are significantly based mostly on the versions of Riemann-Liuoville, Caputo-Liuoville,

M. Zeki Sarıkaya et al. (Eds.): ICMRS 2019, LNNS 123, pp. 108–122, 2020.
https://doi.org/10.1007/978-3-030-43002-3_10

Graunwald-Letnikov or Riesz that have been discussed in scientific researches in the past few years [5–7]. A common likeness among these concepts is that the integral-differential operator with a singular kernel appears side by side with framing differentiability assumption from integrability approaches. But no one denies that it is better to utilize the fractional derivation in a more natural and productive way, similar to the classical approach to derivation, in the affirmative meaning of the limit [8]. In light of this, a novel construction was introduced for fuzzy fractional derivative, called the fuzzy conformable. Such fuzzy conformable fractional derivative seems to be a natural extension of $\mathbb{H}$-derivative with own mathematical details associated with defining of its two-sides limits. Using parameterization of the fuzzy concept, the conformable fuzzy differential equation can be switched to an equivalent system of crisp conformable fractional differential equations that are solved approximately [9–12].

The primary motivation in this work was devoted to the study of the approximate solution of a class of fuzzy fractional problems of fractional-order using a suitable concept of fuzzy conformable fractional derivatives based on an analytic approach. More specifically, let us consider the conformable fuzzy fractional differential equation (FFDE) in the following form:

$$\big(T^{\alpha}u\big)(x) = f(x, u(x)), a \le x \le b, \tag{1}$$

along with the following fuzzy initial condition

$$u(a) = \gamma, \tag{2}$$

where $\alpha \in (0, 1]$, T^{α} is the conformable fuzzy fractional derivative of order α, $\gamma \in \mathbb{R}_{\mathcal{F}}$, $f : [a, b] \times \mathbb{R}_{\mathcal{F}} \to \mathbb{R}_{\mathcal{F}}$ is continuous fuzzy function, and $a \in \mathbb{R}$ with $a > 0$. While a fuzzy function $u(x) : [a, b] \to \mathbb{R}_{\mathcal{F}}$ must be determined. $\mathbb{R}_{\mathcal{F}}$ stands for the set of all fuzzy numbers on $\mathbb{R}$. Also, we assume that FFDE (1) with condition (2) has a unique analytic solution on $[a, b]$.

The FPRSM is an effective analytical technique for identifying and finding the FPS solutions for different types of FDEs, partial FDEs, FFDEs, integral equations and so forth. This approach is easily applicable to create FPS solutions for both linear and nonlinear equations without any hypothetical conditions on the suggested problems [13–15]. Unlike the classical PS method, the FPS approach neither requires comparing the corresponding coefficients nor is a recursion relation needed as well. In addition, the coefficients of power series are calculated by an algebraic system consisting of one or more variables, which gives quick convergence for the series solution, especially when the exact solution is polynomial [16–19].

The aim of this work is to propose the residual power series method (RPSM) for solving FFDEs by utilizing conformable fractional concept under strongly generalized differentiability. The proposed conformable FFDE can be converted to a crisp system of conformable fractional equations based on the type of differentiability that can be solved directly via the RPSM. The structure of this analysis is as follows. In Sect. 2, necessary definitions and preliminaries relating to conformable fractional calculus and fuzzy differentiations are presented. In Sect. 3, description of conformable power series method for handling FFDEs is introduced under strongly generalized differentiability. In Sect. 4, a numerical example is tested to show conformable RPSM performance. The summary and future work are outlined in Sect. 5.

2 Conformable Fuzzy Fractional Derivative

In this section, some essential results are presented for conformable fractional calculus. Indeed, there are many definitions to fractional integration and differentiation, including Grunwald-Letnikov's, Riemann-Liouville's and Caputo's definitions [20–26]. In 2014, a novel definition of fractional derivative, called conformable fractional derivative, has been introduced by Khalil et al. [8]. So, researchers have begun to study this concept, for more details, we refer to [27–32].

Definition 1 [8] **.** Given a function $u : [0, \infty) \to \mathbb{R}$. Then, the conformable fractional derivative of u of order $\alpha > 0$ is defined by

$$\left(T^{\alpha}u\right)(x) = \lim_{\varepsilon \to 0} \frac{u\left(x + \varepsilon x^{1-\alpha}\right) - u(x)}{\varepsilon} \tag{3}$$

for all $x > 0$, and $\alpha \in (0, 1]$.

If the conformable derivative of u of order α exists, then we simply say that u is α-differentiable [27]. Further, if u is α-differentiable in some interval $(0, s)$, $s > 0$, provided that $\lim_{x \to 0^+} T^{\alpha}u(x)$ exists, then define $T^{\alpha}u(0) = \lim_{x \to 0^+} T^{\alpha}u(x)$. Consequently, if u and v are α-differentiable at a point $x > 0$, then T^{α} has the following nice properties:

1. $T^{\alpha}(au + bv) = aT^{\alpha}(u) + bT^{\alpha}(v)$, for all $a, b \in \mathbb{R}$.
2. $T^{\alpha}(x^{p}) = px^{p-\alpha}$, for all $p \in \mathbb{R}$.
3. $T^{\alpha}(c) = 0$, for all constant functions $u(x) = c$.

Definition 2. Given a fuzzy function $u{:}[a, b] \to \mathbb{R}_{\mathcal{F}}$ with $a > 0$. Then, the conformable fuzzy fractional derivative of u of order α is defined by

$$\left(T^{\alpha}u\right)(x) = \lim_{h \to 0} \frac{u\left(x + hx^{1-\alpha}\right) \ominus u(x)}{h} \tag{4}$$

for all $x > 0$ and $\alpha \in (0, 1]$.

If $\lim_{x \to 0^+} T^{\alpha}u(x)$ exists, then define $T^{\alpha}u(0) = \lim_{x \to 0^+} T^{\alpha}u(x)$. Further, if the conformable fuzzy fractional derivative of u of order α exists, then we say that u is α-fuzzy differentiable.

Definition 3. Let $u{:}[a, b] \to \mathbb{R}_{\mathcal{F}}$, $x_0 \in [a, b]$, $a > 0$, and $\alpha \in (0, 1]$. We say that u is strongly generalized α-fuzzy differentiable at x_0, if there exists an element $T^{\alpha}u(x_0) \in \mathbb{R}_{\mathcal{F}}$ such that either:

(I) For all $h > 0$ sufficiently close to 0, the $\mathbb{H}$-differences $u\left(x_0 + hx_0^{1-\alpha}\right) \ominus u(x_0)$, $u(x_0) \ominus u\left(x_0 - hx_0^{1-\alpha}\right)$ exist and

$$\lim_{h \to 0^+} \frac{u\left(x_0 + hx_0^{1-\alpha}\right) \ominus u(x_0)}{h} = \lim_{h \to 0^+} \frac{u(x_0) \ominus u\left(x_0 - hx_0^{1-\alpha}\right)}{h} = T^{\alpha}u(x_0); \tag{5}$$

(II) For all $h > 0$ sufficiently close to 0, the $\mathbb{H}$-differences $u(x_0) \ominus u\left(x_0 + hx_0^{1-\alpha}\right)$, $u\left(x_0 - hx_0^{1-\alpha}\right) \ominus u(x_0)$ exist and

$$\lim_{h \to 0^+} \frac{u(x_0) \ominus u\left(x_0 + hx_0^{1-\alpha}\right)}{-h} = \lim_{h \to 0^+} \frac{u\left(x_0 - hx_0^{1-\alpha}\right) \ominus u(x_0)}{-h} = T^{\alpha} u(x_0), \quad (6)$$

where h and $-h$ at the denominators mean $\frac{1}{h} \odot$ and $\frac{1}{h} \odot$, respectively. The limit is taken in the metric space $(\mathbb{R}_{\mathcal{F}}, d_{\infty})$. Moreover, let u be differentiable at each point $x \in [a, b]$, then u is said to be α-fuzzy differentiable on $[a, b]$.

Definition 4. Let $u: [a, b] \to \mathbb{R}_{\mathcal{F}}$, $a > 0$, and $\alpha \in (0, 1]$. We call u a (1)-α-fuzzy differentiable on $[a, b]$ if u is differentiable in the sense (I) of Definition 3 and its derivative is denoted by $T_1^{\alpha} u(x)$. Similarly, u is a (2)-α-fuzzy differentiable on $[0, \infty)$ if u is differentiable in the sense (II) of Definition 3 and its derivative is denoted by $T_2^{\alpha} u(x)$.

Theorem 5. Let $u: [a, b] \to \mathbb{R}_{\mathcal{F}}, \alpha \in (0, 1]$, and $a > 0$. Put $[u(x)]^r = [u_{1r}(x), u_{2r}(x)]$ for each $r \in [0, 1]$. Then

(I) If u is (1)-α-fuzzy differentiable, then u_{1r} and u_{2r} are α-differentiable functions on $[a, b]$ and

$$\left[T_1^{\alpha} u(x)\right]^r = \left[T^{\alpha} u_{1r}(x), T^{\alpha} u_{2r}(x)\right]; \quad (7)$$

(II) If u is (2)-α-fuzzy differentiable, then u_{1r} and u_{2r} are α-differentiable functions on $[a, b]$ and

$$\left[T_2^{\alpha} u(x)\right]^r = \left[T^{\alpha} u_{2r}(x), T^{\alpha} u_{1r}(x)\right]. \quad (8)$$

Corollary 6. If a fuzzy function $u: [a, b] \to \mathbb{R}_{\mathcal{F}}$ with $a > 0$ is (1)-α-fuzzy differentiable or (2)-α-fuzzy differentiable at $x_0 \in [a, b]$ and $\alpha \in (0, 1]$, then f is continuous at x_0.

Definition 7 [19]**.** A power series expansion of the following form, where $0 \le n - 1 < \alpha \le n$, and $x \ge a$:

$$\sum_{m=0}^{\infty} c_m (x - a)^{m\alpha} = c_0 + c_1 (x - a)^{\alpha} + c_2 (x - a)^{2\alpha} + \ldots, \quad (9)$$

is called fractional PS about $x = a$, where c_m denotes the coefficients of the series, $m \in \mathbb{N}$.

Theorem 8. Let u has the following fractional PS about $x = x_0$

$$u(x) = \sum_{m=0}^{\infty} c_m (x - x_0)^{m\alpha}, \; x_0 \le x < x_0 + R. \quad (10)$$

If $u(x) \in C[x_0, x_0 + R)$, and $T^{m\alpha} u(x) \in C(t_0, t_0 + R)$, for $m = 1, 2, \ldots$, then coefficients c_m in Eq. (10) can be given by $c_m = \frac{T^{m\alpha} u(x_0)}{(m\alpha)!}$, where $T^{m\alpha} = T^{\alpha} \cdot T^{\alpha} \cdots T^{\alpha}$ (m-times) and R is the radius of convergence.

3 Algorithm of Conformable FFDEs

In this section, description of conformable power series method for handling conformable FFDEs is explained utilizing the strongly generalized differentiability, whereas the considered conformable FFDE is switched to crisp system of conformable differential equations based on the type of differentiability. To solve conformable FFDEs (1) and (2), reformulate $u(x)$ in term of the r-cut representation as $[u(x)]^r = [u_{1r}(x), u_{2r}(x)]$, the fuzzy initial condition as $[\gamma]^r = [\gamma_{1r}, \gamma_{2r}]$, and the $f(x, u(x))$ such as $[f(x, u(x))]^r = [f_{1r}(x, u_{1r}(x), u_{2r}(x)), f_{2r}(x, u_{1r}(t), u_{2r}(x))]$. Subsequently, the conformable FFDEs (1) and (2) can be given equivalently by

$$\begin{aligned} [T^\alpha u(x)]^r &= [f(x, u(x))]^r, a \le x \le b, \\ u(a) &= [\gamma]^r. \end{aligned} \tag{11}$$

Definition 9. Let $u{:}[a, b] \to \mathbb{R}_{\mathcal{F}}$ with $a > 0$ such that $T_1^\alpha u(x)$ or $T_2^\alpha u(x)$ exists. If $u(x)$ and $T_1^\alpha u(x)$ satisfy Eq. (11), we say that $u(x)$ is a (1)-conformable fuzzy solution of FFDEs (1) and (2) in which $\alpha \in (0, 1]$. If $u(x)$ and $T_2^\alpha u(x)$ satisfy Eq. (11), we say that $u(x)$ is a (2)-conformable fuzzy solution of FFDEs (1) and (2) in which $\alpha \in (0, 1]$.

To obtain the solution of FFDEs (11), there are two cases relating to each type of differentiability:

Case A: Let $u(x)$ be (1)-α-fuzzy differentiable, that is, $[T_1^\alpha u(x)]^r = [T^\alpha u_{1r}(x), T^\alpha u_{2r}(x)]$. Then do the following: solve the resulted crisp system

$$\begin{aligned} T^\alpha u_{1r}(x) &= f_{1r}(x, u_{1r}(x), u_{2r}(x)), \\ T^\alpha u_{2r}(x) &= f_{2r}(x, u_{1r}(x), u_{2r}(x)), \end{aligned} \tag{12}$$

subject to crisp initial conditions $u_{1r}(a) = \gamma_{1r}$ and $u_{2r}(a) = \gamma_{2r}$. Ensure that $[u_{1r}(x), u_{2r}(x)]$, and $[T^\alpha u_{1r}(x), T^\alpha u_{2r}(x)]$ are valid for all r in $[0, 1]$. Hence, the (1)-conformable fuzzy solution $u(x) = [u_{1r}(x), u_{2r}(x)]$ is obtained for all $r \in [0, 1]$.

Case B: Let $u(x)$ be (2)-α-fuzzy differentiable, that is, $[T_2^\alpha u(x)]^r = [T^\alpha u_{2r}(x), T^\alpha u_{1r}(x)]$. Then do the following: solve the resulted crisp system

$$\begin{aligned} T^\alpha u_{1r}(x) &= f_{2r}(x, u_{1r}(x), u_{2r}(x)), \\ T^\alpha u_{2r}(x) &= f_{1r}(x, u_{1r}(x), u_{2r}(x)), \end{aligned} \tag{13}$$

subject to crisp initial conditions $u_{1r}(a) = \gamma_{1r}$ and $u_{2r}(a) = \gamma_{2r}$. Ensure that $[u_{1r}(x), u_{2r}(x)]$, and $[T^\alpha u_{2r}(x), T^\alpha u_{1r}(x)]$ are valid for all r in $[0, 1]$. Hence, the (2)-conformable fuzzy solution $u(x) = [u_{1r}(x), u_{2r}(x)]$ is obtained for each $r \in [0, 1]$.

To apply the fractional RPSM, let us consider the crisp system (12) by assuming the solutions as follows

$$u_{1r}(x) = \sum_{m=0}^{\infty} c_m (x-a)^{m\alpha},$$

$$u_{2r}(x) = \sum_{m=0}^{\infty} d_m (x-a)^{m\alpha}. \tag{14}$$

Since $u_{1r}(x)$ and $u_{2r}(x)$ satisfy the initial conditions $u_{1r}(a) = \gamma_{1r}$ and $u_{2r}(a) = \gamma_{2r}$, then it yields that

$$\begin{aligned} u_{1r}(x) &= \gamma_{1r} + \sum_{m=1}^{\infty} c_m (x-a)^{m\alpha}, \\ u_{2r}(x) &= \gamma_{2r} + \sum_{m=1}^{\infty} d_m (x-a)^{m\alpha}. \end{aligned} \tag{15}$$

Now, defined the kth-truncated series solution of $u_{1r}(x)$ and $u_{2r}(x)$ as follows

$$\begin{aligned} u_{k,1r}(x) &= \gamma_{1r} + \sum_{m=1}^{k} c_m (x-a)^{m\alpha}, \\ u_{k,2r}(x) &= \gamma_{2r} + \sum_{m=1}^{k} d_m (x-a)^{m\alpha}, \end{aligned} \tag{16}$$

and the kth residual error functions as follows

$$\begin{aligned} Res_{k,1r}(x) &= T^{\alpha} u_{k,1r}(x) - f_{1r}\big(x, u_{k,1r}(x), u_{k,2r}(x)\big), \\ Res_{k,2r}(x) &= T^{\alpha} u_{k,2r}(x) - f_{2r}\big(x, u_{k,1r}(x), u_{k,2r}(x)\big), \end{aligned} \tag{17}$$

where the residual functions $Res_{ir}(x) = \lim_{k\to\infty} Res_{k,ir}(x), i = 1, 2.$

Obviously, $Res_{ir}(x) = 0, i = 1, 2,$ for each $x > a$ that leads to $T^{\alpha} Res_{ir}(x) = 0, i = 1, 2,$ for each $x > a$. Also, $T^{m\alpha} Res_{ir}(x) = 0$ and $T^{m\alpha} Res_{k,ir}(x) = 0$ at $x = a$. Anyhow, to obtain the unknown coefficients c_m and $d_m, m = 1, 2, \ldots, k,$ of Eq. (16), we have to minimize the k th residual error functions, utilize the fact $T^{(k-1)\alpha} Res_{k,ir}(x)\big|_{x=a} = 0$ and then solve the resulted algebraic system. In a similar manner, one can solve the crisp system (13).

4 Numerical Simulation and Discussion

To confirm the high degree of accurateness and efficiency of the proposed FRPS method for solving the fuzzy differential equation of fractional-order, a numerical example is applied in this section. Meanwhile, a numerical comparison between the FRPSM and implicit Runge-Kutta method (IRKM) is discussed. The results obtained interpret that the FRPS method is highly effective and simple to solve different types of FFDEs. Numeric calculations in this work were performed by using the Mathematica 10.0 software package.

Example 1. Consider the fuzzy fractional differential equation

$$T^{\alpha}u(x) + u(x) = 0, 0 \le t \le 1, \tag{18}$$

with respect to the fuzzy initial condition

$$u(0) = \gamma, \tag{19}$$

where $x \in [0, 1]$, $\alpha \in (0.1]$, and $\gamma(\xi) = \max(0, 1 - |\xi|)$, $\xi \in \mathbb{R}$.

Here, it is worth noting that for each $r \in [0, 1]$; if $r = 1 - |\xi|$, then $\xi = r - 1$ or $\xi = 1 - r$. Therefore, $[\gamma]^r = [r - 1, 1 - r]$.

According to the type of differentiability, there are two cases as follows:

Case A: The crisp system corresponding to (1)-α-fuzzy differentiability, which is given by

$$\begin{aligned} T^{\alpha}u_{1r}(x) + u_{1r}(x) &= 0, \\ T^{\alpha}u_{2r}(x) + u_{2r}(x) &= 0, \end{aligned} \tag{20}$$

with respect to crisp initial conditions

$$\begin{aligned} u_{1r}(0) &= r - 1, \\ u_{2r}(0) &= 1 - r. \end{aligned}$$

The exact solutions at $\alpha = 1$ are given by

$$\begin{aligned} u_{1r}(x) &= (r - 1)e^{-x}, \\ u_{2r}(x) &= (1 - r)e^{-x}. \end{aligned}$$

Case B: The crisp system corresponding to (2)-α-fuzzy differentiability, which is given by

$$\begin{aligned} T^{\alpha}u_{1r}(x) + u_{2r}(x) &= 0, \\ T^{\alpha}u_{2r}(x) + u_{1r}(x) &= 0, \end{aligned} \tag{21}$$

with respect to crisp initial conditions

$$\begin{aligned} u_{1r}(0) &= r - 1, \\ u_{2r}(0) &= 1 - r. \end{aligned}$$

The exact solutions at $\alpha = 1$ are given by

$$\begin{aligned} u_{1r}(x) &= (r - 1)e^{x}, \\ u_{2r}(x) &= (1 - r)e^{x}. \end{aligned}$$

According the RPSM, let the solutions of the crisp system have the following form

$$u_{1r}(x) = \sum_{m=0}^{\infty} c_m x^{m\alpha},$$

$$u_{2r}(x) = \sum_{m=0}^{\infty} d_m x^{m\alpha}.$$

Now, to find out the (1)-conformable fuzzy solution $u(x) = [u_{1r}(x), u_{2r}(x)]$ of Case A, we have that

$$u_{1r}(x) = (r-1) + \sum_{m=1}^{\infty} c_m x^{m\alpha},$$
$$u_{2r}(x) = (1-r) + \sum_{m=1}^{\infty} d_m x^{m\alpha}. \tag{22}$$

Consequently, defined the kth residual error functions as follows

$$Res_{k,1r}(x) = T^{\alpha} u_{k,1r}(x) + u_{k,1r}(x),$$
$$Res_{k,2r}(x) = T^{\alpha} u_{k,2r}(x) + u_{k,2r}(x), \tag{23}$$

and then by substituting the k th-truncated series solutions of Eq. (22) into Eq. (23), one can get

$$Res_{k,1r}(x) = T^{\alpha}\left((r-1) + \sum_{m=1}^{k} c_m x^{m\alpha}\right) + \left((r-1) + \sum_{m=1}^{k} c_m x^{m\alpha}\right), \tag{24}$$
$$Res_{k,2r}(x) = T^{\alpha}\left((1-r) + \sum_{m=1}^{k} d_m x^{m\alpha}\right) + \left((1-r) + \sum_{m=1}^{k} d_m x^{m\alpha}\right).$$

Following the procedure of RPS algorithm in finding the coefficients $c_m, d_m, m = 1, 2, \ldots, k$, of Eq. (24), let $k = 1$ in Eq. (24) to get

$$\begin{aligned} Res_{1,1r}(x) &= T^{\alpha}\left((r-1) + c_1 x^{\alpha}\right) + \left((r-1) + c_1 x^{\alpha}\right) \\ &= \alpha c_1 + (r-1) + c_1 x^{\alpha}, \\ Res_{1,2r}(x) &= T^{\alpha}\left((1-r) + d_1 x^{\alpha}\right) + \left((1-r) + d_1 x^{\alpha}\right) \\ &= \alpha d_1 + (1-r) + d_1 x^{\alpha}, \end{aligned}$$

and depending on the $Res_{1,1r}(0) = Res_{1,2r}(0) = 0$, we have $c_1 = \frac{-(r-1)}{\alpha}$ and $d_1 = \frac{-(1-r)}{\alpha}$. Thus, the 1st RPS solutions are $u_{1,1r}(x) = (r-1) - (r-1)\frac{x^{\alpha}}{\alpha}$ and $u_{1,2r}(x) = (1-r) - (1-r)\frac{x^{\alpha}}{\alpha}$.

For $k = 2$, the second residual functions are

$$\begin{aligned} Res_{2,1r}(x) &= T^{\alpha}\left((r-1) - (r-1)\frac{x^{\alpha}}{\alpha} + c_2 x^{2\alpha}\right) \\ &\quad + \left((r-1) - (r-1)\frac{x^{\alpha}}{\alpha} + c_2 x^{2\alpha}\right) \\ &= 2\alpha c_2 x^{\alpha} - (r-1)\frac{x^{\alpha}}{\alpha} + c_2 x^{2\alpha}, \end{aligned}$$

$$\begin{aligned} Res_{2,2r}(x) &= T^{\alpha}\left((1-r)-(1-r)\frac{x^{\alpha}}{\alpha}+d_2x^{2\alpha}\right) \\ &\quad +\left((1-r)-(1-r)\frac{x^{\alpha}}{\alpha}+d_2x^{2\alpha}\right) \\ &= 2\alpha d_2x^{\alpha}-(1-r)\frac{x^{\alpha}}{\alpha}+d_2x^{2\alpha}. \end{aligned} \tag{25}$$

By operating T^{α} on both sides of Eq. (25), one can get

$$\begin{aligned} T^{\alpha}Res_{2,1r}(x) &= T^{\alpha}\left(2\alpha c_2x^{\alpha}-(r-1)\frac{x^{\alpha}}{\alpha}+c_2x^{2\alpha}\right) \\ &= (2\alpha)\alpha c_2-(r-1)+2\alpha c_2x^{\alpha}, \\ T^{\alpha}Res_{2,2r}(x) &= T^{\alpha}\left(2\alpha d_2x^{\alpha}-(1-r)\frac{x^{\alpha}}{\alpha}+d_2x^{2\alpha}\right) \\ &= (2\alpha)\alpha d_2-(1-r)+2\alpha d_2x^{\alpha}, \end{aligned}$$

and then using the facts $T^{\alpha}Res_{2,1r}(0)=T^{\alpha}Res_{2,2r}(0)=0$, it yields $c_2=\frac{(r-1)}{2\alpha!}$ and $d_2=\frac{(1-r)}{2\alpha!}$. Thus, the 2nd RPS solutions are $u_{2,1r}(x)=(r-1)-(r-1)\frac{x^{\alpha}}{\alpha}+(r-1)\frac{x^{2\alpha}}{2\alpha!}$ and $u_{2,2r}(x)=(1-r)-(1-r)\frac{x^{\alpha}}{\alpha}+(1-r)\frac{x^{2\alpha}}{2\alpha!}$.

For $k=3$, the third residual functions are

$$\begin{aligned} Res_{3,1r}(x) &= T^{\alpha}\left((r-1)-(r-1)\frac{x^{\alpha}}{\alpha}+(r-1)\frac{x^{2\alpha}}{2\alpha!}+c_3x^{3\alpha}\right) \\ &\quad +\left((r-1)-(r-1)\frac{x^{\alpha}}{\alpha}+(r-1)\frac{x^{2\alpha}}{2\alpha!}+c_3x^{3\alpha}\right) \\ &= 3\alpha c_3x^{2\alpha}+(r-1)\frac{x^{2\alpha}}{2\alpha!}+c_3x^{3\alpha}, \\ Res_{3,2r}(x) &= T^{\alpha}\left((1-r)-(1-r)\frac{x^{\alpha}}{\alpha}+(1-r)\frac{x^{2\alpha}}{2\alpha!}+d_3x^{3\alpha}\right) \\ &\quad +\left((1-r)-(1-r)\frac{x^{\alpha}}{\alpha}+(1-r)\frac{x^{2\alpha}}{2\alpha!}+d_3x^{3\alpha}\right) \\ &= 3\alpha d_3x^{2\alpha}+(1-r)\frac{x^{2\alpha}}{2\alpha!}+d_3x^{3\alpha}. \end{aligned} \tag{26}$$

By operating T^{α} on both sides of Eq. (26), one can get

$$\begin{aligned} T^{2\alpha}Res_{3,1r}(x) &= T^{\alpha}\left(T^{\alpha}\left(3\alpha c_3x^{2\alpha}+(r-1)\frac{x^{2\alpha}}{2\alpha!}+c_3x^{3\alpha}\right)\right) \\ &= (3\alpha)!c_3+(r-1)+(3\alpha)(2\alpha)c_3x^{\alpha}, \\ T^{2\alpha}Res_{3,2r}(x) &= T^{\alpha}\left(T^{\alpha}\left(3\alpha d_3x^{2\alpha}+(1-r)\frac{x^{2\alpha}}{2\alpha!}+d_3x^{3\alpha}\right)\right) \\ &= (3\alpha)!d_3+(1-r)+(3\alpha)(2\alpha)d_3x^{\alpha}, \end{aligned}$$

and then using the facts $T^{2\alpha}Res_{3,1r}(0) = T^{2\alpha}Res_{3,2r}(0) = 0$, it yields $c_3 = \frac{-(r-1)}{3\alpha!}$ and $d_2 = \frac{-(1-r)}{3\alpha!}$. Thus, the 3rd RPS solutions are $u_{3,1r}(x) = (r-1) - (r-1)\frac{x^\alpha}{\alpha} + (r-1)\frac{x^{2\alpha}}{2\alpha!} - (r-1)\frac{x^{3\alpha}}{3\alpha!}$ and $u_{3,2r}(x) = (1-r) - (1-r)\frac{x^\alpha}{\alpha} + (1-r)\frac{x^{2\alpha}}{2\alpha!} - (1-r)\frac{x^{3\alpha}}{3\alpha!}$.

Anyhow, by continuing with such process, one can get

$$u_{k,1r}(x) = (r-1)\left(1 - \frac{x^\alpha}{\alpha} + \frac{x^{2\alpha}}{2\alpha!} - \ldots + (-1)^k \frac{x^{k\alpha}}{k\alpha!}\right),$$

$$u_{k,2r}(x) = (1-r)\left(1 - \frac{x^\alpha}{\alpha} + \frac{x^{2\alpha}}{2\alpha!} - \ldots + (-1)^k \frac{x^{k\alpha}}{k\alpha!}\right).$$

Hence, the RPS solutions of Eq. (20) are

$$u_{1r}(x) = \lim_{k\to\infty} u_{k,1r}(x) = (r-1)\sum_{n=0}^{\infty}(-1)^n \frac{x^{n\alpha}}{n\alpha!},$$

$$u_{2r}(x) = \lim_{k\to\infty} u_{k,2r}(x) = (1-r)\sum_{n=0}^{\infty}(-1)^n \frac{x^{n\alpha}}{n\alpha!},$$

which coincide with the exact solutions at $\alpha = 1$. Furthermore, following the same fashion, one can get the RPS solutions of Eq. (21), Case B.

To show the accuracy of the method, some numerical results of the 8th RPS solutions are shown in Tables 1 and 2 at different truth number r in the interval [0, 1] at $\alpha = 1$. The comparison of FRPS results with the implicit Runge-Kutta method (IRKM) is listed in Table 3 at $\alpha = 1$ over the interval [0, 1] with step-size 0.16, $r = 0$ and $n = 8$. From this table, it is observed that the accuracy by the proposed method is compatible with the IRKM. While Tables 4 and 5 show the approximate fuzzy solutions of the two Cases A and B at different number of x in [0, 1], $r = 0$ and different values of α. From these results, it can be observed that the behavior of the approximated solutions is in good agreement with each other. Figures 1 and 2 show the 3D plots of the (i)-α-fuzzy approximated solution, $i = 1, 2$, for $x \in [0, 1]$ and $r \in [0, 1]$ with different values of α such that $\alpha \in \{0.9, 0.7, 0.5, 0.3\}$

Obviously, each plot is triangle fuzzy number at every type of differentiability, which depends continuously on the conformable fuzzy derivative used.

Table 1. Absolute errors of Example 1, Case A, at $\alpha = 1$ and $n = 8$

x	$r = 0.0$	$r = 0.25$	$r = 0.5$	$r = 0.75$
0.16	1.86406×10^{-13}	1.39777×10^{-13}	9.32032×10^{-14}	4.66016×10^{-14}
0.32	9.39439×10^{-11}	7.0458×10^{-11}	4.69719×10^{-11}	2.34860×10^{-11}
0.48	3.55601×10^{-9}	2.66701×10^{-9}	1.77800×10^{-9}	8.89002×10^{-10}
0.64	4.66412×10^{-8}	3.49809×10^{-8}	2.33206×10^{-8}	1.16603×10^{-8}
0.80	3.42295×10^{-7}	2.56722×10^{-7}	1.71148×10^{-7}	8.55739×10^{-8}
0.96	1.74003×10^{-6}	1.30502×10^{-6}	8.70013×10^{-7}	4.35007×10^{-7}

Table 2. Absolute errors of Example 1, Case B, at $\alpha = 1$ and $n = 8$

x	$r = 0.0$	$r = 0.25$	$r = 0.5$	$r = 0.75$
0.16	1.92735×10^{-13}	1.44551×10^{-13}	9.63674×10^{-14}	4.81837×10^{-14}
0.32	1.00154×10^{-10}	7.51157×10^{-11}	5.00771×10^{-11}	2.50385×10^{-11}
0.48	3.91447×10^{-9}	2.93585×10^{-9}	1.95723×10^{-9}	9.78616×10^{-10}
0.64	5.30152×10^{-8}	3.97614×10^{-8}	2.65076×10^{-8}	1.32538×10^{-8}
0.80	4.01762×10^{-7}	3.01322×10^{-7}	2.00881×10^{-7}	1.00441×10^{-8}
0.96	2.10902×10^{-6}	1.58176×10^{-6}	1.05451×10^{-7}	5.27254×10^{-7}

Table 3. Numerical results of Example 1 at $n = 8$, and $\alpha = 1$

$r = 0$	x	$u_{1r}(x)$		$u_{2r}(x)$	
		FRPS	IRKM	FRPS	IRKM
Case A	0.16	−0.8521438	−0.8521438	0.8521438	0.8521438
	0.32	−0.7261490	−0.7261489	0.7261490	0.7261488
	0.48	−0.6187834	−0.6187833	0.6187834	0.6187832
	0.64	−0.5272925	−0.5272923	0.5272925	0.5272920
	0.80	−0.4493293	−0.4493288	0.4493293	0.4493286
	0.96	−0.3828946	−0.3828941	0.3828946	0.3828939
Case B	0.16	−1.1735109	−1.1735108	1.1735109	1.1735108
	0.32	−1.3771278	−1.3771278	1.3771278	1.3771278
	0.48	−1.6160744	−1.6160739	1.6160744	1.6160737
	0.64	−1.8964808	−1.8964806	1.8964808	1.8964805
	0.80	−2.2255405	−2.2255410	2.2255405	2.2255400
	0.96	−2.6116944	−2.6116938	2.6116944	2.6116939

Table 4. The (1)-α-fuzzy approximate solution of Example 1 for $r = 0$

$n = 8$	x	$\alpha = 1$	$\alpha = 0.9$	$\alpha = 0.8$	$\alpha = 0.7$
$u_{1r}(x)$	0.2	−0.818731	−0.785762	−0.750625	−0.714437
	0.4	−0.670320	−0.641099	−0.614348	−0.590734
	0.6	−0.548812	−0.531114	−0.517571	−0.507946
	0.8	−0.449329	−0.444884	−0.444366	−0.447074
$u_{2r}(x)$	0.2	−0.818731	−0.785762	−0.750625	−0.714437
	0.4	−0.670320	−0.641099	−0.614348	−0.590734
	0.6	−0.548812	−0.531114	−0.517571	−0.507946
	0.8	−0.449329	−0.444884	−0.444366	−0.447074

Table 5. The (2)-α-fuzzy approximate solution of Example 1 for $r = 0$

$n = 8$	x	$\alpha = 1$	$\alpha = 0.9$	$\alpha = 0.8$	$\alpha = 0.7$
$u_{1r}(x)$	0.2	−1.221402	−1.280533	−1.357406	−1.459495
	0.4	−1.491825	−1.593745	−1.722608	−1.889771
	0.6	−1.822119	−1.968845	−2.152578	−2.389017
	0.8	−2.225541	−2.422938	−2.669133	−2.984817
$u_{2r}(x)$	0.2	−1.221402	−1.280533	−1.357406	−1.459495
	0.4	−1.491825	−1.593745	−1.722608	−1.889771
	0.6	−1.822119	−1.968845	−2.152578	−2.389017
	0.8	−2.225541	−2.422938	−2.669133	−2.984817

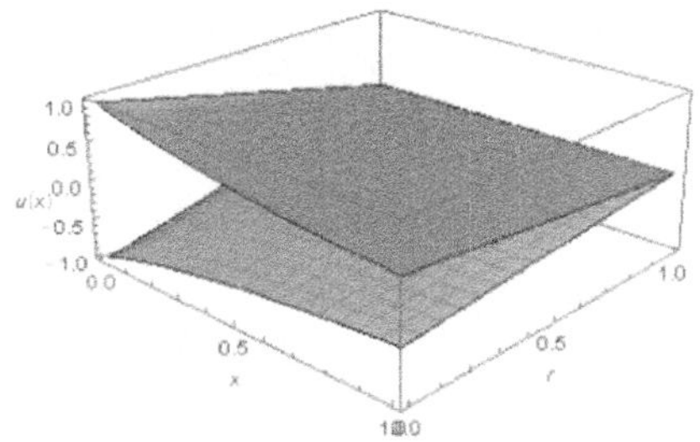

(a) $\alpha = 0.9$

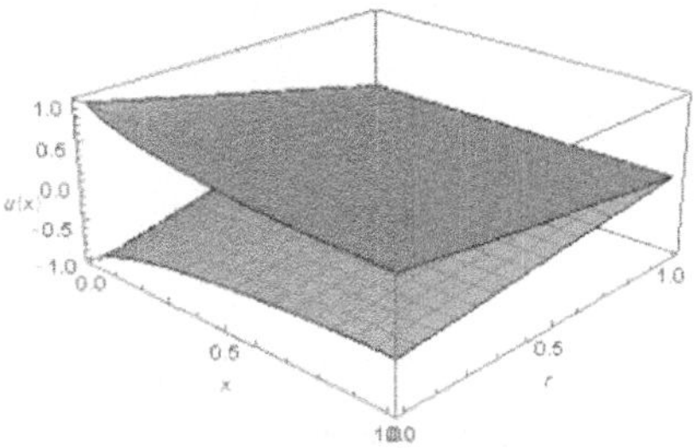

(b) $\alpha = 0.7$

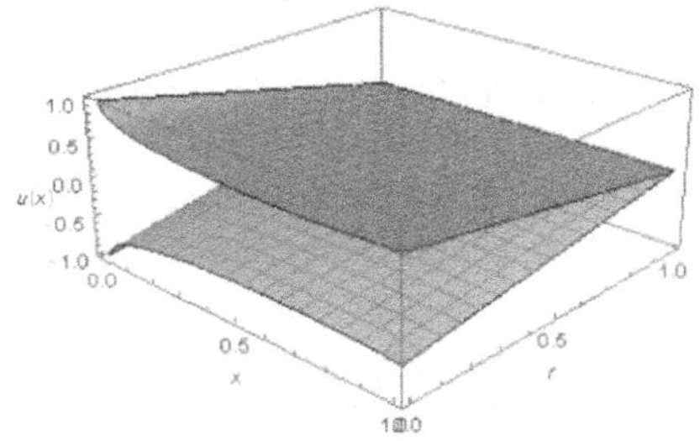

(c) $\alpha = 0.5$

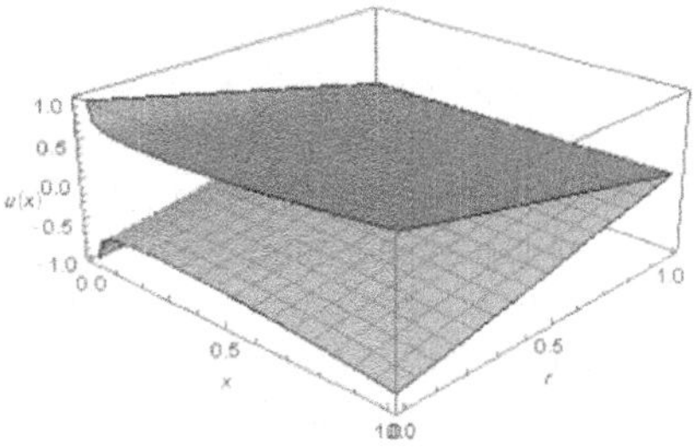

(d) $\alpha = 0.3$

Fig. 1. The 3D plots of (1)-α-fuzzy approximated solution for $x, r \in [0, 1]$.

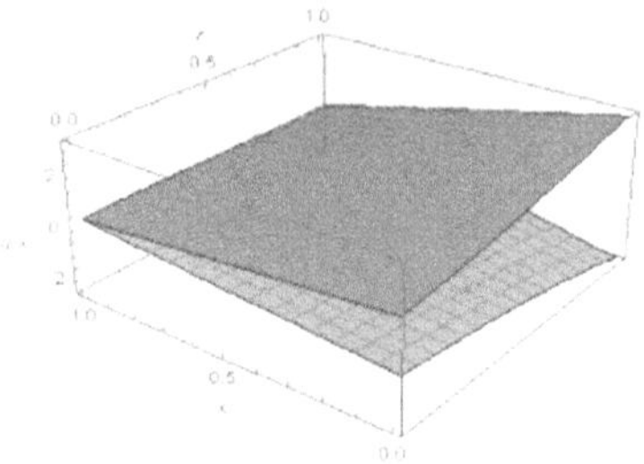

(a) $\alpha = 0.9$

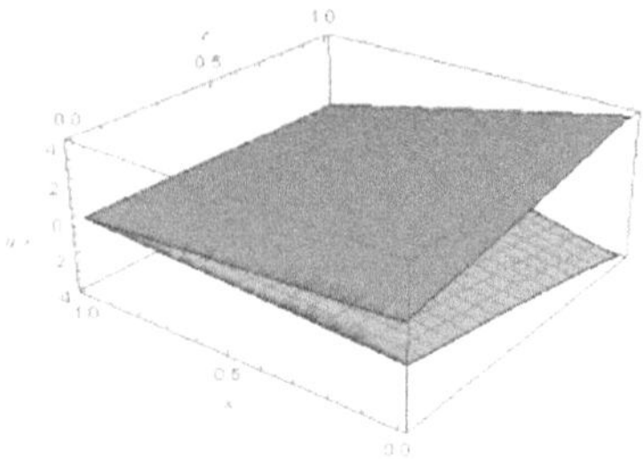

(b) $\alpha = 0.7$

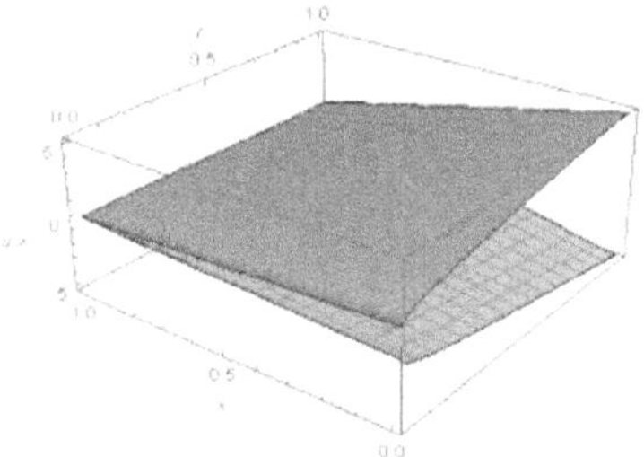

(c) $\alpha = 0.5$

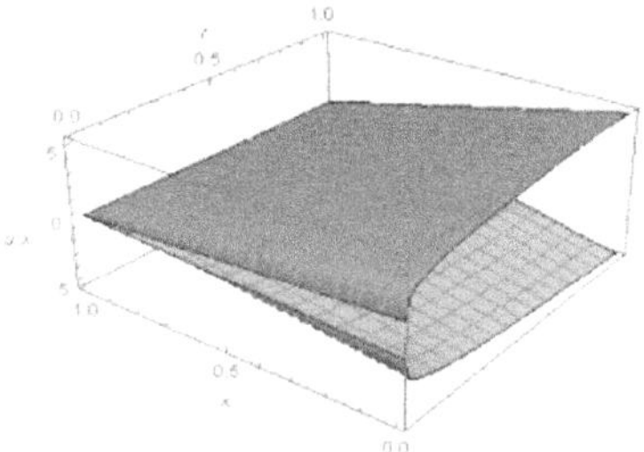

(d) $\alpha = 0.3$

Fig. 2. The 3D plots of (2)-α-fuzzy approximated solution for $x, r \in [0, 1]$.

5 Summary and Future Work

In summary, this research aims to propose the RPS method for investigating the approximate solution of fuzzy fractional differential equations subject to fuzzy initial condition based on conformable fuzzy fractional derivative. The aim has been accomplished successfully through extending the fractional RPS method to solve those FFDEs. The present method provides a solution in rapidly convergent FPS without linearization, or any limitations. A numerical example is performed to illustrate the efficiency and the reliability of the proposed method using Mathematica 10 software package. The obtained results show that the FRPS method is efficient and powerful algorithm to provide an approximate solution of such FFDEs. As future works, we plan to investigate the approximate solutions to the systems of FFDEs, FFIEs, and FFIDEs, to better understand the impact of the effectiveness of conformable fractional derivatives.

Acknowledgments. This research was financially supported by Al-Balqa Applied University, Jordan.

References

1. Bede, B., Gal, S.: Generalizations of the differentiability of fuzzy-number-valued functions with applications to fuzzy differential equations. Fuzzy Sets Syst. **151**, 581–599 (2005)
2. Alaroud, M., Al-Smadi, M., Ahmad, R.R., Salma Din, U.K.: An analytical numerical method for solving fuzzy fractional volterra integro-differential equations. Symmetry **11**(2), 205 (2019)
3. Gumah, G.N., Naser, M.F.M., Al-Smadi, M., Al-Omari, S.K.: Application of reproducing kernel Hilbert space method for solving second-order fuzzy Volterra integro-differential equations. Adv. Differ. Eqn. **2018**, 475 (2018)
4. Al-Smadi, M.: Reliable numerical algorithm for handling fuzzy integral equations of second kind in Hilbert spaces. Filomat **33**(2), 583–597 (2019)
5. Al-Smadi, M., Freihat, A., Khalil, H., Momani, S., Khan, R.A.: Numerical multistep approach for solving fractional partial differential equations. Int. J. Comput. Meth. **14**, 1750029 (2017)
6. Momani, S., Abu Arqub, O., Freihat, A., Al-Smadi, M.: Analytical approximations for Fokker-Planck equations of fractional order in multistep schemes. Appl. Comput. Math. **15**(3), 319–330 (2016)
7. Al-Smadi, M.: Simplified iterative reproducing kernel method for handling time-fractional BVPs with error estimation. Ain Shams Eng. J. **9**(4), 2517–2525 (2018)
8. Khalil, R., Horani, M., Yousef, A., Sababheh, M.: A new definition of fractional derivative. Comput. Appl. Math. **246**, 65–70 (2014)
9. Chalco-Cano, Y., Román-Flores, H.: On new solutions of fuzzy differential equations. Chaos, Solitons Fractals **38**, 112–119 (2008)
10. Kaleva, O.: Fuzzy differential equations. Fuzzy Sets Syst. **24**, 301–317 (1987)
11. Abu Arqub, O., Al-Smadi, M., Momani, S., Hayat, T.: Numerical solutions of fuzzy differential equations using reproducing kernel Hilbert space method. Soft. Comput. **20**(8), 3283–3302 (2016)
12. Abu Arqub, O., Al-Smadi, M., Momani, S., Hayat, T.: Application of reproducing kernel algorithm for solving second-order, two-point fuzzy boundary value problems. Soft. Comput. **21**(23), 7191–7206 (2017)
13. Freihet, A., Hasan, S., Al-Smadi, M., Gaith, M., Momani, S.: Construction of fractional power series solutions to fractional stiff system using residual functions algorithm. Adv. Differ. Eqn. **2019**, 95 (2019)
14. Komashynska, I., Al-Smadi, M., Ateiwi, A., Al-Obaidy, S.: Approximate analytical solution by residual power series method for system of Fredholm integral equations. Appl. Math. Inf. Sci. **10**(3), 975–985 (2016)
15. Hasan, S., Al-Smadi, M., Freihet, A., Momani, S.: Two computational approaches for solving a fractional obstacle system in Hilbert space. Adv. Differ. Eqn. **2019**, 55 (2019)
16. Moaddy, K., Al-Smadi, M., Hashim, I.: A novel representation of the exact solution for differential algebraic equations system using residual power-series method. Discr. Dyn. Nat. Soc. **2015** (2015). Art. ID 205207, 12 pages
17. Komashynska, I., Al-Smadi, M., Abu Arqub, O., Momani, S.: An efficient analytical method for solving singular initial value problems of nonlinear systems. Appl. Math. Inf. Sci. **10**(2), 647–656 (2016)
18. Alaroud, M., Al-Smadi, M., Ahmad, R.R., Salma Din, U.K.: Computational optimization of residual power series algorithm for certain classes of fuzzy fractional differential equations. Int. J. Differ. Eqn. **2018** (2018). Art. ID 8686502, 11 pages
19. El-Ajou, A., Abu Arqub, O., Al-Smadi, M.: A general form of the generalized Taylor's formula with some applications. Appl. Math. Comput. **256**, 851–859 (2015)

20. Abu Arqub, O., Al-Smadi, M.: Atangana-Baleanu fractional approach to the solutions of Bagley-Torvik and Painlevé equations in Hilbert space. Chaos Soliton Fractal **117**, 161–167 (2018)
21. Podlubny, I.: Fractional Differential Equations. Academic Press, New York (1999)
22. Al-Smadi, M., Abu Arqub, O.: Computational algorithm for solving Fredholm time-fractional partial integrodifferential equations of Dirichlet functions type with error estimates. Appl. Math. Comput. **342**, 280–294 (2019)
23. Abu Arqub, O., Odibat, Z., Al-Smadi, M.: Numerical solutions of time-fractional partial integrodifferential equations of Robin functions types in Hilbert space with error bounds and error estimates. Nonlinear Dyn. **94**(3), 1819–1834 (2018)
24. Allahviranloo, T., Salahshour, S., Abbasbandy, S.: Explicit solutions of fractional differential equations with uncertainty. Soft. Comput. **16**, 297–302 (2012)
25. Abu Arqub, O., Al-Smadi, M.: Numerical algorithm for solving time-fractional partial integrodifferential equations subject to initial and Dirichlet boundary conditions. Num. Meth. Partial Differ. Eqn. **34**(5), 1577–1597 (2017)
26. Altawallbeh, Z., Al-Smadi, M., Komashynska, I., Ateiwi, A.: Numerical solutions of fractional systems of two-point BVPs by using the iterative reproducing kernel algorithm. Ukrainian Math. J. **70**(5), 687–701 (2018)
27. Abdeljawad, T.: On conformable fractional calculus. J. Comput. Appl. Math. **279**, 57–66 (2015)
28. Feng, Q.: A new approach for seeking coefficient function solutions of conformable fractional partial differential equations based on the Jacobi elliptic equation. Chin. J. Phys. **56**, 2817–2828 (2018)
29. Thabet, H., Kendre, S.: Analytical solutions for conformable space-time fractional partial differential equations via fractional differential transform. Chaos Soliton Fractals **109**, 238–245 (2018)
30. Akbulut, A., Kaplan, M.: Auxiliary equation method for time-fractional differential equations with conformable derivative. Comput. Math Appl. **75**, 876–882 (2018)
31. Chen, C., Jiang, Y.: Simplest equation method for some time-fractional partial differential equations with conformable derivative. Comput. Math Appl. **75**, 2978–2988 (2018)
32. Kumar, D., Seadawy, A., Joardar, A.: Modified Kudryashov method via new exact solutions for some conformable fractional differential equations arising in mathematical biology. Chin. J. Phys. **56**, 75–85 (2018)

Identities for the Hermite-Based Fubini Polynomials

Burak Kurt(✉)

Mathematics of Department, Akdeniz University, 07070 Antalya, Turkey
burakkurt@akdeniz.edu.tr

Abstract. In this paper, we define the Hermite-based Fubini type polynomials. We investigate the properties of Fubini type numbers which defined by Muresan [15]. The desire of this paper is to construct a new relations and recurrence relations for Hermite-based Fubini type numbers and polynomials. We give some identities for this polynomial.

Keywords: Fubini polynomials · Hermite polynomials · Generating function · Hermite-based polynomials and numbers

2010 Mathematics Subject Classification: 11B65 · 11B75 · 33B10

1 Introduction and Notation

In this section, we give some useful and well-known polynomials which have applications in almost all branches of mathematics and mathematical physics.

The Bernoulli polynomials and the generalized Apostol-Bernoulli polynomials are defined by way of the following generating functions [8–25] respectively:

$$\sum_{n=0}^{\infty} B_n(x)\frac{t^n}{n!} = \frac{t}{e^t - 1}e^{xt}, \ |t| < 2\pi \tag{1.1}$$

and

$$\sum_{n=0}^{\infty} B_n(x;\lambda)\frac{t^n}{n!} = \frac{t}{\lambda e^t - 1}e^{xt}, \ \{|t| < \ |\log(-\lambda)| \text{ if } \lambda \neq 1, \ |t| < 2\pi \text{ if } \lambda = 1 \text{ and } \lambda \in \mathbb{C}\}. \tag{1.2}$$

From (1.1) and (1.2), we write the Apostol-Bernoulli numbers and the Bernoulli numbers as, respectively:

$$B_n(\lambda) = B_n(0;\lambda), \ B_n = B_n(0).$$

The Euler polynomials and the generalized Apostol-Euler polynomials are defined by way of the following generating functions [8–25] respectively:

$$\sum_{n=0}^{\infty} E_n(x)\frac{t^n}{n!} = \frac{2}{e^t + 1}e^{xt}, \ |t| < \pi \tag{1.3}$$

M. Zeki Sarıkaya et al. (Eds.): ICMRS 2019, LNNS 123, pp. 123–128, 2020.
https://doi.org/10.1007/978-3-030-43002-3_11

and

$$\sum_{n=0}^{\infty} E_n(x;\lambda)\frac{t^n}{n!} = \frac{2}{\lambda e^t+1}e^{xt},\ \{|t| < \ |\log(-\lambda)| \text{ if } \lambda \neq 1,\ |t| < \pi \text{ if } \lambda = 1 \text{ and } \lambda \in \mathbb{C}\}. \tag{1.4}$$

The Stirling numbers of the second kind are defined by way of the following generating function

$$\frac{(e^t-1)^k}{k!} = \sum_{n=0}^{\infty} S_2(n,k)\frac{t^n}{n!}, \tag{1.5}$$

$$S_2(n,n) = 1,\ n \in \mathbb{N}_0;\ S_2(n,0) = 0,\ n \in \mathbb{N}_0;\ \text{for } m > n,\ S_2(n,m) = 0.$$

The 2-variable Hermite polynomials $H_n(x,y)$ [1,2,6–8] defined by

$$H_n(x,y) = n!\sum_{r=0}^{\left[\frac{n}{2}\right]} \frac{x^{n-2r}y^r}{r!\,(n-2r)!}.$$

The generating functions of Hermite polynomials are defined [1,2,6–8] as

$$\sum_{n=0}^{\infty} H_n(x,y)\frac{t^n}{n!} = e^{xt+yt^2}. \tag{1.6}$$

Khan *et al.* [6–8] and Bretti *et al.* [1] defined the Hermite-based Bernoulli polynomials as:

$$\sum_{n=0}^{\infty} {}_HB_n(x,y)\frac{t^n}{n!} = \frac{t}{e^t-1}e^{xt+yt^2}. \tag{1.7}$$

Khan *et al.* [6–8] and Bretti *et al.* [1] defined the Hermite-based Euler polynomials as:

$$\sum_{n=0}^{\infty} {}_HE_n(x,y)\frac{t^n}{n!} = \frac{2}{e^t+1}e^{xt+yt^2},\ |t| < \pi. \tag{1.8}$$

We consider the Hermite-based Apostol-Bernoulli and Apostol-Euler polynomials as:

$$\sum_{n=0}^{\infty} {}_H\mathcal{B}_{n,\mu}(x,y)\frac{t^n}{n!} = \frac{t}{\mu e^t-1}e^{xt+yt^2} \tag{1.9}$$

where $\mu \in \mathbb{R}$, $|t| < 2\pi$ if $|\mu| \neq 1$; $|t| < 2\pi$ if $\mu = 1$ and

$$\sum_{n=0}^{\infty} {}_H\mathcal{E}_{n,\mu}(s,k)\frac{t^n}{n!} = \frac{2}{\mu e^t+1}e^{st+kt^2} \tag{1.10}$$

where $\mu \in \mathbb{R}$, $|t| < \pi$ if $\mu = 1$; $|t| < |\log(-\mu)|$ when $|\mu| \neq 1$.

Kim *et al.* [4] and Kim *et al.* [5] proved some relations and recurrences relations for the Fubini polynomials and two variable Fubini polynomials.

Kilar *et al.* [3] proved some relations. They properties for the Fubini numbers $W_g(n)$. They gave some beautiful relations and identities.

The generalized Fubini numbers $f_{n,k}$ are defined by way of the following generating function [15, p. 398]

$$F_k(t) = \sum_{n=0}^{\infty} f_{n,k} \frac{t^n}{n!} = \frac{e^t - 1}{k + 1 - e^t}. \tag{1.11}$$

The next section; we define Hermite-based Fubini polynomials. We give some identities by using generating function of Fubini polynomials.

2 Hermite-Based Fubini-Type Polynomials

In this section, we prove explicit relations for the Hermite-based Fubini-type polynomials. Also, we give some equalities for this polynomial.

We consider the Hermite-based Fubini type polynomials as

$$\sum_{m=1}^{\infty} {}_H f_{m,k}(x,y) \frac{t^m}{m!} = \frac{e^t - 1}{k + 1 - ke^t} e^{xt+yt^2}. \tag{2.1}$$

From (2.1), we have the following equalities:

i. $m \geq 2$,

$$\frac{\partial}{\partial x} {}_H f_{m,k}(x,y) = m \ {}_H f_{m-1,k}(x,y)$$

ii. $m \geq 3$,

$$\frac{\partial}{\partial y} {}_H f_{m,k}(x,y) = (m-1)\, m \ {}_H f_{m-1,k}(x,y)$$

iii. $m \geq 2$,

$$\begin{aligned} &{}_H f_{m+1,k}(x,y) \\ &= \frac{-1}{1+k} \left\{ {}_H B_{m-1}\left(x+1, y : \frac{k}{1+k}\right) - {}_H f_{m+1,k}(x,y) \right. \\ &\left. - \frac{k}{1+k} \sum_{s=0}^{m+1} \binom{m}{s} B_{m+1-s}\left(1, \frac{k}{1+k}\right) \left({}_H f_{s,k}(x,y)\right) \right\} + x f_{m,k} + 2ym f_{m-1,k}. \end{aligned}$$

Theorem 1. *For $n \geq 2$, we get*

$$ {}_H f_{n-1,k}(x,y) = \frac{1}{n(k+1)} \left\{ \left({}_H \mathcal{B}_n(x,y)\right) - \left({}_H \mathcal{B}_n(x+1,y)\right) \right\}.$$

Proof. From (2.1),

$$\sum_{n=1}^{\infty} {}_H f_{n,k}(x,y) \frac{t^n}{n!} = \frac{e^{(x+1)t+yt^2}}{1+k-ke^t} - \frac{e^{xt+yt^2}}{1+k-ke^t}$$

$$= \frac{1}{t}\frac{1}{1+k}\left\{\frac{-t}{\frac{k}{k+1}e^t-1}e^{(x+1)t+yt^2} - \frac{t}{\frac{k}{k+1}e^t-1}e^{xt+yt^2}\right\}$$

$$\sum_{n=1}^{\infty} {}_Hf_{n,k}(x,y)\frac{t^{n+1}}{n!} = \frac{1}{1+k}\left\{\sum_{n=0}^{\infty}\left({}_H\mathcal{B}_n(x,y) - {}_H\mathcal{B}_n(x+1,y)\right)\frac{t^n}{n!}\right\}.$$

Comparing the coefficients both sides of $\frac{t^n}{n!}$, we omit it.

We write again (2.1) as

$$\sum_{n=1}^{\infty} {}_Hf_{n,k}(x,y)\frac{t^n}{n!} = \frac{e^t-1}{k+1-ke^t}e^{xt+yt^2}.$$

From here, we write as

$$\frac{e^t-1}{k+1-ke^t}e^{yt^2} = e^{-xt}\sum_{n=1}^{\infty} {}_Hf_{n,k}(x,y)\frac{t^n}{n!}. \tag{2.2}$$

We change t by $t+z$ in (2.2) and rewrite the generating function as

$$\frac{e^{t+u}-1}{k+1-ke^{t+z}}e^{y(t+z)^2} = e^{-x(t+z)}\sum_{m=1}^{\infty}\sum_{n=1}^{\infty} {}_Hf_{n+m,k}(x,y)\frac{t^n}{n!}\frac{z^m}{m!}.$$

Replacing x by v in the above equation, we get

$$\sum_{m=1}^{\infty}\sum_{n=1}^{\infty} {}_Hf_{n+m,k}(x,y)\frac{t^n}{n!}\frac{u^m}{m!} = e^{(t+z)(v-x)}\sum_{m=1}^{\infty}\sum_{n=1}^{\infty} {}_Hf_{n+m,k}(x,y)\frac{t^n}{n!}\frac{z^m}{m!}$$

which on using formula [25, Srivastava p. 52.].

$$\sum_{Z=0}^{\infty} f(Z)\frac{(x+y)^Z}{Z!} = \sum_{m,n=0}^{\infty} f(n+m)\frac{t^n}{n!}\frac{y^m}{m!}$$

in the right hand sides, becomes

$$\sum_{m=1}^{\infty}\sum_{n=1}^{\infty} {}_Hf_{n+m,k}(x,y)\frac{t^n}{n!}\frac{z^m}{m!} = \sum_{p=0}^{\infty}\sum_{q=0}^{\infty}(v-x)^{p+q}\frac{t^p}{p!}\frac{z^q}{q!}\sum_{m=1}^{\infty}\sum_{n=1}^{\infty} {}_Hf_{n+m,k}(x,y)\frac{t^n}{n!}\frac{z^m}{m!}$$

$$= \left[\sum_{q=0}^{\infty}(v-x)^q\frac{z^q}{q!} + \sum_{p=0}^{\infty}(v-x)^p\frac{t^p}{p!} + \sum_{p=1}^{\infty}\sum_{q=1}^{\infty}(v-x)^{p+q}\frac{t^p}{p!}\frac{z^q}{q!}\right]$$
$$\times\sum_{m=1}^{\infty}\sum_{n=1}^{\infty} {}_Hf_{n+m,k}(x,y)\frac{t^n}{n!}\frac{u^m}{m!}.$$

By using Cauchy product and comparing the coefficients both sides of $\frac{t^n}{n!}\frac{z^m}{m!}$, we have the ensuing theorem.

Theorem 2. *Whereas $m \geq 2$ and $n \geq 2$,*

$$\left({}_Hf_{n+m,k}(x,y)\right) = \sum_{p=1}^{n}\binom{n}{p}\sum_{q=1}^{m}\binom{m}{q}(v-x)^{p+q}\left({}_Hf_{n+m-p-q,k}(x,y)\right).$$

References

1. Bretti, G., Natalini, P., Ricci, P.E.: Generalizations of the Bernoulli and Appell polynomials. Abstr. Appl. Anal. **2004**(7), 613–623 (2004)
2. Dattoli, G., Cesarano, C., Lorenzutta, S.: Bernoulli numbers and polynomials from a more general paint of view. Rendiconti di Math. Ser. VII **22**, 193–202 (2002)
3. Kılar, N., Simsek, Y.: A new family of Fubini type numbers and polynomials associated with Apostol-Bernoulli numbers and polynomials. J. Korean Math. Soc. **54**(3), 1605–1621 (2017)
4. Kim, T., Kim, D.S., Jang, G.-W.: A note degenerate Fubini polynomials. Proc. Jangjeon Math. Soc. **20**(4), 521–531 (2017)
5. Kim, D.S., Kim, T., Kwon, H.I., Park, J.-W.: Two variable higher order Fubini polynomials. J. Korean Math. Soc. **55**(4), 975–986 (2018)
6. Khan, S., Yasmin, G., Khan, R., Hassan, N.A.: Hermite-based Appell polynomials: properties and applications. J. Math. Anal. Appl. **351**, 756–764 (2009)
7. Khan, S., Al-Saad, M., Yasmin, G.: Some properties of Hermite-based Sheffer polynomials. Appl. Math. Comput. **207**, 2160–2183 (2010)
8. Khan, W.A., Araci, S., Acikgoz, M., Haroon, H.: A new class of partially degenerate Hermite-Genocchi polynomials. J. Nonlinear Sci. Appl. **10**, 5072–5081 (2017)
9. Kurt, B.: Notes on unified q-Apostol type polynomials. Filomat **30**, 921–927 (2016)
10. Kurt, B., Simsek, Y.: On the generalized Apostol type Frobenius-Euler polynomials. Adv. Differ. Equ. **2013**(1), 1–9 (2013)
11. Kurt, B., Simsek, Y.: Frobenius-Euler type polynomials related to Hermite-Bernoulli polynomials. Numer. Anal. Appl. Math. **1389**, 385–388 (2011)
12. Gaboury, S., Kurt, B.: Some relations involving Hermite-based Apostol-Genocchi polynomials. Appl. Math. Sci. **6**(82), 4091–4102 (2012)
13. Lu-Q, D., Srivastava, H.M.: Some series identities involving the generalized Apostol type and related polynomials. Comput. Math. Appl. **62**, 3591–3602 (2011)
14. Luo, M.-Q., Srivastava, H.M.: Some generalizations of the Apostol-Bernoulli and Apostol-Euler polynomials. J. Math. Anal. Appl. **308**(1), 290–302 (2005)
15. Muresan, M.: A Concrete Approach to classical Analysis. Canadian Mathematical Society. Springer, New York (2009)
16. Ozarslan, M.A.: Hermite-based unified Apostol-Bernoulli, Euler and Genocchi polynomials. Adv. Differ. Equ. **2013**, 116 (2013)
17. Pathan, M.A., Khan, W.: Some implicit summation formulas and symmetric identities for the generalized Hermite-Euler polynomials. East-West J. Math. **16**(1), 92–109 (2014)
18. Srivastava, H.M.: Some generalization and basic (or $-q$) extension of the Bernoulli, Euler and Genocchi polynomials. Appl. Math. Inf. Sci. **5**(3), 390–444 (2011)
19. Srivastava, H.M., Kurt, B., Simsek, Y.: [Corrigendum] Some families of Genocchi type polynomials and their interplation function. Integr. Transform. Spec. Funct. **23**, 919–938 (2012)
20. Srivastava, H.M., Kurt, B., Kurt, V.: Identities and relations involving the modified degenerate hermite-based Apostol-Bernoulli and Apostol-Euler polynomials. Revista de la Real Academia de Ciencias Exactas, Físicas y Naturales. Serie A. Matemáticas **113**(2), 1299–1313 (2019)
21. Srivastava, H.M., Masjed-Jamei, M., Reza Beyki, M.: A parametric type of the Apostol-Bernoulli, Apostol-Euler and Apostol-Genocchi polynomials. Appl. Math. Inf. Sci. **12**, 907–916 (2018)

22. Srivastava, H.M., Manocha, H.L.: A Treatise on Generating Functions. Wiley, Hoboken (1984)
23. Srivastava, H.M., Ozarslan, M.A., Yilmaz, B.: Some families of differential equations associated with the Hermite-based Appell polynomials and other classes of Hermite-based polynomials. Filomat **28**(4), 695–708 (2014)
24. Srivastava, H.M., Ozarslan, M.A., Kaanoğlu, C.: Some generalized Lagrange-based Apostol-Bernoulli, Apostol-Euler and Apostol-Genocchi polynomials. Russ. J. Math. Phys. **20**(1), 110–120 (2013)
25. Srivastava, H.M., Choi, J.: Series Associated with the Zeta and Related Functions. Kluwer Academic Publishers, Dordrect (2001)

On Subclasses of Uniformly Convex Functions Generated by Touchard Polynomials Associated with Conic Regions

Khalifa AlShaqsi(✉)

Nizwa College of Technology, MOM, Nizwa, Oman
khalifa.alshaqsi@nct.edu.om

Abstract. In this paper, we introduce the new operator $\mathcal{I}(n,m)f(z)$ defined by convolution and Touchard polynomials. We obtain several sufficient conditions for this convolution operator belonging to various subclasses of uniformly convex and starlike functions.

Keywords: Analytic functions · Convex functions · Convolution · Touchard polynomials

1 Introduction and Preliminary

Let $\mathbf{U} = \{z : z \in \mathbb{C} \text{ and } |z| < 1\}$ be the open unit disk in the plane and $\mathcal{A}$ be the class of functions f of the form:

$$f(z) = z + \sum_{k=2}^{\infty} a_k z^k, \tag{1}$$

which are analytic in $\mathbf{U}$ and satisfy the normalization conditions $f(0) = 0, f'(0) - 1 = 0$. Also, let $\mathcal{S}$ the class of all functions in $\mathcal{A}$ which are univalent in $\mathbf{U}$.

For $f \in \mathcal{A}$ such that

$$\operatorname{Re}\left\{\frac{zf'(z)}{f(z)}\right\} > \alpha \qquad (0 \leq \alpha < 1,\, z \in \mathbf{U}),$$

is called starlike functions of order α, denoted by $\mathcal{S}^*(\alpha)$. Similarly, For $f \in \mathcal{A}$ such that

$$\operatorname{Re}\left\{1 + \frac{zf''(z)}{f'(z)}\right\} > \alpha \qquad (z \in \mathbf{U}),$$

is called convex functions of order α, denoted by $\mathcal{C}(\alpha)$.

The classes $\mathcal{S}^*(\alpha)$ and $\mathcal{C}(\alpha)$ studied by Roberston [10] and Silverman [11], respectively.

M. Zeki Sarıkaya et al. (Eds.): ICMRS 2019, LNNS 123, pp. 129–135, 2020.
https://doi.org/10.1007/978-3-030-43002-3_12

In 1997, Bharti et al. [2] introduced the class $\lambda - \mathcal{UCV}(\alpha)$. A function $f \in \mathcal{A}$ in the class $\lambda - \mathcal{UCV}(\alpha)$ if

$$\operatorname{Re}\left\{1+\frac{zf''(z)}{f'(z)}\right\} \geq \lambda\left|\frac{zf''(z)}{f'(z)}\right| + \alpha \qquad (z \in \mathbf{U}), \tag{2}$$

where $\lambda \geq 0$ and $0 \leq \alpha < 1$. Clearly that $\lambda - \mathcal{UCV}(0) \equiv \lambda - \mathcal{UCV}$ and $0 - \mathcal{UCV}(0) \equiv \mathcal{UCV}$ respectively, studied by Kanas and Wisniowaska [6] and Goodman [3].

The class $k - \mathcal{S}_p(\alpha)$ obtained by using the Alexander transform:

$$f \in \lambda - \mathcal{UCV}(\alpha) \Leftrightarrow zf'(z) \in \mathcal{S}_p(\alpha).$$

The reader can refer to [4, 5, 7, 8, 12] for more results on these directions and references therein.

In 2004, Swaminathan [13] studied the class $\mathcal{P}_\gamma^\nu(\beta)$. A function $f \in \mathcal{A}$ in the class $\mathcal{P}_\gamma^\nu(\beta)$ for $\beta < 1$, $0 \leq \gamma < 1$, $\nu \in \mathbf{C} \setminus \{0\}$ if

$$\left|\frac{(1-\gamma)\frac{f(z)}{z} + \gamma f'(z) - 1}{2\nu(1-\beta) + (1-\gamma)\frac{f(z)}{z} + \gamma f'(z) - 1}\right| < 1, \quad (z \in \mathbf{U}).$$

Next, for the function $f \in \mathcal{A}$ Ponnusamy and Rønning [9] introduced and studied the classes $\mathcal{S}_\delta^*$ and $\mathcal{C}_\delta$ for $\delta > 0$ as follows

$$\mathcal{S}_\delta^* = \left\{\left|\frac{zf'(z)}{f(z)} - 1\right| < \delta\right\} \qquad (z \in \mathbf{U}), \tag{3}$$

$$\mathcal{C}_\delta = \left\{\left|\frac{zf''(z)}{f'(z)}\right| < \delta\right\} \qquad (z \in \mathbf{U}), \tag{4}$$

so that, obviously,

$$f(z) \in \mathcal{C}_\delta \Leftrightarrow zf'(z) \in \mathcal{S}_\delta^*.$$

Recently, the author [1] introduce a series with Touchard polynomials coefficients after the second force as following:

$$F(n, m; z) = z + \sum_{k=2}^{\infty} \frac{(k-1)^n m^{k-1}}{(k-1)!} e^{-m} z^k, \ (n \geq 0, \ m > 0). \tag{5}$$

Recall that the Hadamard product (or convolution) of f given by (1) and g given by

$$g(z) = z + \sum_{k=2}^{\infty} b_k z^k.$$

is

$$f(z) * g(z) = (f * g)(z) = z + \sum_{k=2}^{\infty} a_k b_k z^k.$$

Now, for $n \geq 0$ and $m > 0$, we introduce the operator $\mathcal{I} : \mathcal{A} \to \mathcal{A}$ as following:

$$\mathcal{I}f(z) = \mathcal{I}(n,m)f(z) = z + \sum_{k=2}^{\infty} \frac{(k-1)^n m^{k-1}}{(k-1)!} e^{-m} a_k z^k.$$

Finally, We need the following lemmas to establish our main results.

Lemma 1 ([2]). *If $f \in \mathcal{A}$, satisfies*

$$\sum_{k=2}^{\infty} k\{k(\lambda+1) - (\lambda+\alpha)\}|a_k| \leq 1-\alpha, \tag{6}$$

then $f \in \lambda - \mathcal{UCV}(\alpha)$.

Remark 1 ([2]). If $f(z)$ given by

$$f(z) = z - \sum_{k=2}^{\infty} a_k z^k, \ (a_k \geq 0), \tag{7}$$

in $\mathcal{A}$ is in $\lambda - \mathcal{UCV}(\alpha)$ if and only if the inequality (6) holds.

Lemma 2 ([2]). *If $f \in \mathcal{A}$, satisfies*

$$\sum_{k=2}^{\infty} \{k(\lambda+1) - (\lambda+\alpha)\}|a_k| \leq 1-\alpha, \tag{8}$$

then $f \in \lambda - \mathcal{S}_p(\alpha)$.

Remark 2 ([2]). If function f given by (7) in $\mathcal{A}$ is in $\lambda - \mathcal{S}_p(\alpha)$ if and only if the inequality (8) holds.

Lemma 3 ([6]). *Let f given by (1) in $\mathcal{S}$. If for some $\lambda(\lambda \geq 0)$, the inequality:*

$$\sum_{k=2}^{\infty} k(k-1)|a_k| \leq \frac{1}{\lambda+2}, \tag{9}$$

satisfies, then $f \in \lambda - \mathcal{UCV}(\alpha)$. Where the number $1/(\lambda+2)$ cannot be increased.

Lemma 4 ([9]). *If f given by (1) in $\mathcal{A}$, satisfies*

$$\sum_{k=2}^{\infty} (k+\delta-1)|a_k| \leq \delta, \ (\delta > 0), \tag{10}$$

then $f \in \mathcal{S}_\delta^*$.

Remark 3 ([9]). If function f given by (7) in $\mathcal{A}$ is in $\mathcal{S}_\delta^*$ if and only if the inequality (10) holds.

Lemma 5 ([9]). *If f given by (1) in $\mathcal{A}$, satisfies*

$$\sum_{k=2}^{\infty} k(\delta + k - 1)|a_k| \leq \delta, \ (\delta > 0), \tag{11}$$

then $f \in \mathcal{C}_\delta$.

Lemma 6 ([13]). *If $f \in \mathcal{P}_\gamma^\nu$ where function f given by (1), then:*

$$|a_k| \leq \frac{2|\nu|(1-\beta)}{1+\gamma(k-1)}. \tag{12}$$

2 Main Results

Theorem 1. *Let $f(z) \in \mathcal{P}_\gamma^\nu$ where $f(z)$ given by (1), satisfies*

$$\sum_{i=0}^{n} \binom{n}{k} \Big[(1-\alpha)(1-e^{-m}) + (\lambda+1)m\Big] \leq \frac{\gamma(1-\alpha)}{2|\nu|(1-\beta)}, \tag{13}$$

then $\mathcal{I}f(z) \in \lambda - \mathcal{UCV}(\alpha)$.

Proof. Since

$$\mathcal{I}f(z) = z + \sum_{k=2}^{\infty} \frac{(k-1)^n m^{k-1}}{(k-1)!} e^{-m} a_k z^k \qquad (n \geq 0, \, m > 0).$$

To prove that $\mathcal{I}f(z) \in \lambda - \mathcal{UCV}(\alpha)$, then from Lemma 1, it show that

$$\sum_{k=2}^{\infty} k\{k(\lambda+1) - (\lambda+\alpha)\}|A_k| \leq 1 - \alpha,$$

where

$$|A_k| = \frac{(k-1)^n m^{k-1}}{(k-1)!} e^{-m} a_k, \ (k \geq 2).$$

Now, by applying Lemma 6 we have

$$\sum_{k=2}^{\infty} k\{k(\lambda+1) - (\lambda+\alpha)\} \frac{(k-1)^n m^{k-1}}{(k-1)!} e^{-m} |a_k|$$
$$\leq 2|\nu|(1-\beta) \sum_{k=2}^{\infty} k\{k(\lambda+1) - (\lambda+\alpha)\} \frac{(k-1)^n m^{k-1}}{(k-1)!} e^{-m} \frac{1}{1+\gamma(k-1)}$$

$$\le 2|\nu|(1-\beta)e^{-m}\sum_{k=2}^{\infty}\{k(\lambda+1)-(\lambda+\alpha)\}\frac{(k-1)^n m^{k-1}}{(k-1)!}e^{-m}\frac{1}{\gamma}$$
$$=\frac{2|\nu|(1-\beta)}{\gamma}e^{-m}\left[\sum_{k=2}^{\infty}\{k(\lambda+1)-(\lambda+\alpha)\}\frac{(k-1)^n m^{k-1}}{(k-1)!}e^{-m}\right]$$
$$=\frac{2|\nu|(1-\beta)}{\gamma}e^{-m}\sum_{i=0}^{n}\binom{n}{k}\left[(\lambda+1)\sum_{k=2}^{\infty}\frac{m^{n-1}}{(n-2)!}+(1-\alpha)\sum_{k=2}^{\infty}\frac{m^{n-1}}{(n-2)!}\right]$$
$$=\frac{2|\nu|(1-\beta)}{\gamma}e^{-m}\sum_{i=0}^{n}\binom{n}{k}\left[(\lambda+1)me^{m}+(1-\alpha)(e^{m}-1)\right]$$
$$=\frac{2|\nu|(1-\beta)}{\gamma}\sum_{i=0}^{n}\binom{n}{k}\left[(\lambda+1)m+(1-\alpha)(1-e^{-m})\right]$$
$$\le 1-\alpha, \text{ by (13)},$$

which completes the proof of theorem. □

Theorem 2. *Let $f(z)$ given by (1) and $f(z)\in\mathcal{P}_{\gamma}^{\nu}$. If the inequality*

$$\sum_{i=0}^{n}\binom{n}{k}\left[(1-e^{-m})+\frac{\delta-1}{m}(1-e^{-m}-me^{-m})\right]\le\frac{\gamma(1-\alpha)}{2|\nu|(1-\beta)},$$

is satisfied then, $\mathcal{I}f(z)\in\lambda-\mathcal{S}_p(\alpha)$.

Proof. By using Lemma 2 the proof of Theorem 2 is lines similar to the proof of Theorem 1, so we omitted the proof of Theorem 2. □

Theorem 3. *Let $f(z)$ given by (1) and $f(z)\in\mathcal{P}_{\gamma}^{\nu}$. If the inequality*

$$\frac{2|\nu|(1-\beta)}{\gamma}\sum_{i=0}^{n}\binom{n}{k}\left[(1-e^{-m})+\frac{\delta-1}{m}(1-e^{-m}-me^{-m})\right]\le\delta, \qquad (14)$$

is satisfied then $\mathcal{I}f(z)\in\lambda-\mathcal{S}_{\delta}^{*}$.

Proof. By using Lemma 4, it is sufficient to show that

$$\sum_{k=2}^{\infty}(k+\delta-1)|A_k|\le\delta,$$

where

$$|A_k|=\frac{(k-1)^n m^{k-1}}{(k-1)!}e^{-m}a_k,\ \ (k\ge 2).$$

Since $f(z)\in\mathcal{P}_{\gamma}^{\nu}$ using Lemma 6 and $1+\gamma(k-1)\ge\gamma k$, then

$$\sum_{k=2}^{\infty}(k+\delta-1)\}\frac{(k-1)^n m^{k-1}}{(k-1)!}e^{-m}\frac{2|\nu|(1-\beta)}{1+\gamma(k-1)}.$$

Now

$$\begin{aligned}
&\sum_{k=2}^{\infty}(k+\delta-1)\}\frac{(k-1)^n m^{k-1}}{(k-1)!}e^{-m}|a_k| \\
&\leq \sum_{k=2}^{\infty}(k+\delta-1)\}\frac{(k-1)^n m^{k-1}}{(k-1)!}e^{-m}\frac{2|\nu|(1-\beta)}{1+\gamma(k-1)} \\
&\leq 2|\nu|(1-\beta)e^{-m}\sum_{k=2}^{\infty}(k+\delta-1)\frac{(k-1)^n m^{k-1}}{k!}e^{-m} \\
&= \frac{2|\nu|(1-\beta)}{\gamma}e^{-m}\sum_{i=0}^{n}\binom{n}{k}\left[\sum_{k=2}^{\infty}\frac{m^{n-1}}{(n-1)!}+\frac{(\delta-1)}{m}\sum_{k=2}^{\infty}\frac{m^n}{n!}\right] \\
&= \frac{2|\nu|(1-\beta)}{\gamma}\sum_{i=0}^{n}\binom{n}{k}\left[((1-e^{-m})+\frac{(\delta-1)}{m}(1-e^{-m}-me^{-m})\right] \\
&\leq \delta, \text{ by (14).}
\end{aligned}$$

The proof of theorem is completed. □

Theorem 4. *Let $f(z) \in \mathcal{P}_\gamma^\nu$ where $f(z)$ given by (1). If the inequality*

$$\frac{2|\nu|(1-\beta)}{\gamma}\sum_{i=0}^{n}\binom{n}{k}\left[m+\delta(1-e^{-m})\right] \leq \delta,$$

is satisfied then, $\mathcal{I}f(z) \in \lambda - \mathcal{C}_\delta$.

Proof. By using Lemma 5 the proof of Theorem 3 is lines similar to the proof of Theorem 1, so we omitted the proof of Theorem 3. □

Acknowledgement. The work here is supported by TRC-Oman research grant: BFP/RGP/CBS/18/054.

References

1. AlShaqsi, K.: On inclusion results of certain subclasses of analytic functions associated with generating function. AIP Conf. Proc. **1830**(1), 1–6 (2017). https://doi.org/10.1063/1.4980979
2. Bharati, R., Parvatham, R., Swaminathan, A.: On subclasses of uniformly convex functions and corresponding class of starlike funcitons. Tamkang J. Math. **28**, 17–32 (1997). https://journals.math.tku.edu.tw/index.php/TKJM
3. Goodman, A.W.: Uniformly convex functions. Ann. Polon. Math. **56**, 87–92 (1991). https://doi.org/10.4064/ap-56-1-87-92
4. Goodman, A.W.: On uniformly starlike functions. J. Math. Anal. Appl. **155**, 364–370 (1991). https://doi.org/10.1016/0022-247x(91)90006-1
5. Rønning, F.: Uniformly convex functions and a corresponding class of starlike functions. Proc. Am. Math. Soc. **118**, 189–196 (1993). https://doi.org/10.2307/2160026

6. Kanas, S., Wisinowaska, A.: Conic regions and k-uniform convexity. J. Comput. Appl. Math. **105**, 327–336 (1999). https://doi.org/10.1016/s0377-0427(99)00018-7
7. Kanas, S., Wisinowaska, A.: Conic regions and k-starlike functions. Rev. Roum. Math. Pure Appl. **45**, 647–657 (2000). http://imar.ro/journals/Revue-Mathematique
8. Ma, W., Minda, D.: Uniformly convex functions. Ann. Polon. Math. **57**, 165–175 (1992). https://doi.org/10.4064/ap-57-2-165-175
9. Ponnusamy, S., Rønning, F.: Starlikeness properties for convolutions involving hypergeometric series. Ann. Univ. Mariae Curie-Sklodawska SK. Todawaska L.II.1 **16**, 141–155 (1998). https://www.researchgate.net/publication/266042830
10. Robertson, M.S.: On the theory of univalent functions. Ann. Math. **37**, 374–408 (1936). https://doi.org/10.2307/1968451
11. Silverman, H.: Univalent functions with negative coefficients. Proc. Am. Math. Soc. **51**, 109–116 (1975). https://doi.org/10.1006/jmaa.1997.5882
12. AlShaqsi, K., Darus, M.: On classes of uniformly starlike and convex functions with negative coefficients. Acta Math. Aca. Pae. Nyí. **24**, 355–365 (2008). https://www.emis.de/journals/AMAPN/
13. Swaminathan, A.: Certain sufficiency conditions on Gaussian hypergeometric functions. J. Inequal. Pure Appl. Math. **5**(4), 1–10 (2004). http://www.kurims.kyoto-u.ac.jp/EMIS/journals/JIPAM/article428.html?sid=428

Curvature Tensors of Screen Semi-invariant Half-Lightlike Submanifolds of a Semi-Riemannian Product Manifold with Quarter-Symmetric Non-metric Connection

Oguzhan Bahadır(✉)

Department of Mathematics, Faculty of Arts and Sciences,
K.S.U. Kahramanmaras, Kahramanmaras, Turkey
oguzbaha@gmail.com

Abstract. In this article, The curvature tensors of the half-lightlike submanifolds of the semi-Riemann product manifold was investigated. We introduced the half-lightlike submanifolds class, which we would call screen semi-invariant. Some special distributions of screen semi-invariant half-lightlike submanifolds have been identified by us. Moreover, we obtain Curvature tensor and Ricci tensor for quarter symmetric non-metric connection.

Keywords: Half-lightlike submanifold · Product manifolds · Screen semi-invariant submanifolds · Quarter symmetric non-metric connection

Mathematics Subject Classification: 53C15 · 53C25 · 53C40

1 Introduction

Recently, one of the important topics in differential geometry is the theory of degenerate submanifolds of the semi-Riemannian manifold, which has many applications in physics and astronomy. In [10] and [9], this concept was especially introduced by Duggal and Bejancu. Duggal and Sahin presented the book (see [19]). Also, in [11–25] many authors studied some special lightlike submanifolds with respect to different spaces or different type connections. Different types of connections, which are generalized of Riemannian connection, are studied in different structures by the following authors [1–8].

In this article, we calculate the Ricci and the curvature tensors of half-lightlike submanifolds of a semi-Riemannian product manifold with respect to quarter-symmetric non-metric connection. In the second part, we give some references for the basic concepts, In the third part, screen semi-invariant half-lightlike submanifolds was introduced by us. Furthermore, we defined some distributions

M. Zeki Sarıkaya et al. (Eds.): ICMRS 2019, LNNS 123, pp. 136–146, 2020.
https://doi.org/10.1007/978-3-030-43002-3_13

for screen semi-invariant half-lightlike submanifolds so that they are orthogonal to each other. In the chapter four, firstly we obtain Gauss-Codazzi equations of half-lightlike submanifold with respect to the quarter symmetric non-metric connection determined by the product structure. Later we compute the Ricci tensor of this type manifold with quarter symmetric non-metric connection.

We use the book "Differential Geometry of Lightlike Submanifolds" [19] and the article [24] for basic concepts and writing language of the article.

2 Screen Semi-invariant Lightlike Submanifolds

Let (Ψ, h) be a half-lightlike submanifold of a semi-Riemannian product manifold $(\widetilde{\Psi}, \widetilde{h})$ For any $X \in \Gamma(T\Psi)$ we can write

$$FX = fX + wX, \tag{2.1}$$

where f and w are the projections on of $\Gamma(T\widetilde{\Psi})$ onto $T\Psi$ and $trT\Psi$, respectively, From (2.1), we can write

$$FX = fX + w_1(X)N + w_2(X)u, \tag{2.2}$$

where $w_1(X) = \widetilde{g}(FX, \xi)$, $w_2(X) = \epsilon\widetilde{g}(FX, u)$.

Definition 2.1. Let (Ψ, h) be a half-lightlike submanifold of a semi-Riemannian product manifold $(\widetilde{\Psi}, \widetilde{h})$. If $FRad\ T\Psi \subset S(T\Psi)$, $Fltr(T\Psi) \subset S(T\Psi)$ and $F(S(T\Psi^{\perp})) \subset S(T\Psi)$ then we say that Ψ is a screen semi-invariant half-lightlike submanifold.

If $FS(T\Psi) = S(T\Psi)$, then we say that Ψ is a screen invaryant half-lightlike submanifold.

Now, let Ψ be a screen semi-invariant half-lightlike submanifold of a semi-Riemannian product manifold $(\widetilde{\Psi}, \widetilde{h})$. If we set $L_1 = FRad\ T\Psi$, $L_2 = Fltr(T\Psi)$ and $L_3 = F(S(T\Psi^{\perp}))$, thus we get

$$S(T\Psi) = L_0 \perp \{L_1 \oplus L_2\} \perp L_3, \tag{2.3}$$

where L_0 is a $(n-3)$–dimensional invariant distribution. Thus we obtain the following decomposition:

$$TM = L_0 \perp \{L_1 \oplus L_2\} \perp L_3 \perp Rad\ T\Psi, \tag{2.4}$$
$$T\widetilde{\Psi} = L_0 \perp \{L_1 \oplus L_2\} \perp L_3 \perp S(T\Psi^{\perp}) \perp \{Rad\ T\Psi \oplus ltr(T\Psi)\}. \tag{2.5}$$

Let (Ψ, h) be a screen semi-invariant half-lightlike submanifold of a semi-Riemannian product manifold $(\widetilde{\Psi}, \widetilde{h})$. If we set

$$L = L_0 \perp L_1 \perp Rad\ T\Psi \quad L^{\perp} = L_2 \perp L_3,$$

then we can write

$$T\Psi = L \oplus L^{\perp}.$$

Remark 2.2. *L is a invariant and $L^{\perp}$ is anti-invariant distribution with respect to F on Ψ, respectively.*

Example 2.3. *We consider same example in [25] and let Ψ be submanifold of $\widetilde{\Psi}$ with the help of following equations;*

$$\begin{aligned} x_1 &= t_1 + t_2 - t_3, \\ x_2 &= t_1 + t_2 + t_3 + \sqrt{2}\arctan t_4, \\ x_3 &= \sqrt{2}(t_1 + t_2 + t_3) + \arctan t_4, \\ x_4 &= t_5, \\ x_5 &= t_1 - t_2 + t_3, \\ x_6 &= \arctan t_4, \\ x_7 &= t_1 - t_2 - t_3, \end{aligned}$$

where t_i are real parameters. Then we have

$$T\Psi = Span\{U_1, U_2, U_3, U_4, U_5\},$$

where

$$\begin{aligned} U_1 &= \frac{\partial}{\partial x_1} + \frac{\partial}{\partial x_2} + \sqrt{2}\frac{\partial}{\partial x_3} + \frac{\partial}{\partial x_5} + \frac{\partial}{\partial x_7}, \\ U_2 &= \frac{\partial}{\partial x_1} + \frac{\partial}{\partial x_2} + \sqrt{2}\frac{\partial}{\partial x_3} - \frac{\partial}{\partial x_5} - \frac{\partial}{\partial x_7}, \\ U_3 &= -\frac{\partial}{\partial x_1} + \frac{\partial}{\partial x_2} + \sqrt{2}\frac{\partial}{\partial x_3} + \frac{\partial}{\partial x_5} - \frac{\partial}{\partial x_7}, \\ U_4 &= \frac{\sqrt{2}}{(1+t_4^2)}\frac{\partial}{\partial x_2} + \frac{1}{(1+t_4^2)}\frac{\partial}{\partial x_3} + \frac{1}{(1+t_4^2)}\frac{\partial}{\partial x_6}, \\ U_5 &= \frac{\partial}{\partial x_4}. \end{aligned}$$

Then the vector U_1 is a degenerate vector, Ψ is a $1-$ lightlike submanifold of $\widetilde{\Psi}$. We set $\xi = U_1$, then we easily see that $Rad\ T\Psi = Span\{\xi\}$ and $S(T\Psi) = Span\{U_2, U_3, U_4, U_5\}$. Then we obtain that

$$ltr(T\Psi) = Span\{N = -\frac{\partial}{\partial x_1} + \frac{\partial}{\partial x_2} + \sqrt{2}\frac{\partial}{\partial x_3} - \frac{\partial}{\partial x_5} + \frac{\partial}{\partial x_7}\},$$

and

$$S(T\Psi^{\perp}) = Span\{u = \sqrt{2}\frac{\partial}{\partial x_2} + \frac{\partial}{\partial x_3} + \frac{\partial}{\partial x_6}\}.$$

Thus Ψ is a half-lightlike submanifold of $\widetilde{\Psi}$. Furthermore, we get

$$F\xi = U_2, \quad FN = U_3, \quad Fu = (1+t_4^2)U_4, \quad FU_5 = U_5.$$

If we set $L_0 = Span\{U_5\}$, $L_1 = Span\{U_2\}$, $L_2 = Span\{U_3\}$, $L_3 = Span\{U_4\}$, then Ψ is a screen semi-invariant half-lightlike submanifold of $\widetilde{\Psi}$.

Example 2.4. *[25] Let Ψ be a submanifold of $\widetilde{\Psi} = R_1^5$ with the help of the following equations*

$$y_1 = y_3,\ y_5 = \sqrt{1 - \{y_2^2 + y_4^2\}}.$$

Then we get

$$TM = Span\{\xi = \partial y_1 + \partial y_3,\ Y_1 = y_5 \partial y_2 - y_2 \partial y_5,\ Y_2 = y_5 \partial y_4 - y_4 \partial y_5\}.$$

It follows that Ψ is $1-$ lightlike. We obtain

$$N = \frac{1}{2}(-\partial y_1 + \partial y_3),$$

and

$$v = y_2 \partial y_2 + y_4 \partial y_4 + \sqrt{1 - \{y_2^2 + y_4^2\}}\ \partial y_5.$$

In where $ltr(T\Psi) = Span\{N\}$, $Rad\ T\Psi = Span\{\xi\}$, $S(T\Psi^{\perp}) = Span\{v\}$ and $S(T\Psi) = Span\{Y_1, Y_2\}$ [19].

If we set $F(y_1, y_2, y_3, y_4, y_5) = (y_1, -y_2, -y_3, -y_4, -y_5)$, then $F^2 = I$ and F is a product structure on R_1^5. Thus Ψ is a screen invariant of $\widetilde{\Psi}$.

Example 2.5. *[25] $\widetilde{\Psi} = R_1^1 \times R_1^3$ is a semi-Riemannian product manifold with metric tensor $\widetilde{h} = \pi^* h_1 + \sigma^* h_2$ and $F(y_1, y_2, y_3, y_4) = (y_1, y_2, -y_4, -y_3)$, where h_1 and h_2 are standard metric tensors of R_1^1, R_1^3 and (y_1, y_2, y_3, y_4) is the standard coordinate system of $R_1^1 \times R_1^3 \equiv R_2^4$, π and σ are projections on $\widetilde{\Psi}$. We consider in $\widetilde{\Psi}$ the submanifold Ψ given by the equations;*

$$\begin{aligned} y_1 &= 2t_1 - 4t_2, \\ y_2 &= t_1 + 7t_2, \\ y_3 &= 2t_1 - t_2, \\ y_4 &= t_1 + t_2. \end{aligned}$$

where t_i are real parameters.

Thus we have

$$T\Psi = Span\{U_1, U_2\},$$

where

$$\begin{aligned} U_1 &= 2\frac{\partial}{\partial y_1} + \frac{\partial}{\partial y_2} + 2\frac{\partial}{\partial y_3} + \frac{\partial}{\partial y_4}, \\ U_2 &= -4\frac{\partial}{\partial y_1} - 7\frac{\partial}{\partial y_2} - \frac{\partial}{\partial y_3} + \frac{\partial}{\partial y_4}. \end{aligned}$$

Then we can see that the vector U_1 is a degenerate vector on Ψ and Ψ is a $1-$ lightlike submanifold of $\widetilde{\Psi}$. Thus we can easily say that $RadT\ \Psi = Span\{\xi\}$ and $S(T\Psi) = Span\{U_2\}$. Futhermore, we compute that

$$ltr(T\Psi) = Span\{N = 2\frac{\partial}{\partial y_1} + \frac{\partial}{\partial y_2} - \frac{\partial}{\partial y_3} - 2\frac{\partial}{\partial y_4}\}$$

and

$$S(TM^{\perp}) = Span\{V = \frac{\partial}{\partial x_1} + 2\frac{\partial}{\partial x_3} - 2\frac{\partial}{\partial x_4}\}.$$

Then we get $F\xi = N$. Thus Ψ is a screen invariant half-lightlike submanifold of $\widetilde{\Psi}$.

3 Curvature Tensor with Respect to Quarter-Symmetric Non-metric Connection

Let (Ψ, h, F) be a semi-Riemannian product manifold and $\widetilde{\nabla}$ be the Levi-Civita connection on Ψ. If we set

$$\widetilde{D}_U V = \widetilde{\nabla}_U V + \pi(V)FU \tag{3.1}$$

for any $U, V \in \Gamma(T\widetilde{\Psi})$, then $\widetilde{D}$ is a linear connection on $\widetilde{\Psi}$, where π is a 1-form on $\widetilde{\Psi}$. Let $\widetilde{T}$ be the torsion tensor with respect to quarter-symmetric non-metric connection on $\widetilde{\Psi}$. Then we have

$$\widetilde{T}(U, V) = \pi(V)FU - \pi(U)FV, \tag{3.2}$$

and

$$(\widetilde{D}_U \widetilde{h})(V, W) = -\pi(V)\widetilde{g}(FU, W) - \pi(W)\widetilde{h}(FU, V), \tag{3.3}$$

for any $U, V, W \in \Gamma(T\widetilde{\Psi})$. Therefore $\widetilde{D}$ is a quarter-symmetric non-metric connection on $\widetilde{\Psi}$.

The Gauss and Weingarten formulas for the connection $\widetilde{D}$ are given in the following format, respectively,

$$\widetilde{D}_U V = D_U V + \widetilde{D}_1(U, V)N + \widetilde{D}_2(U, V)v, \tag{3.4}$$
$$\widetilde{D}_U N = -\widetilde{A}_N U + \widetilde{p}_1(U)N + \widetilde{p}_2(U)v, \tag{3.5}$$
$$\widetilde{D}_U v = -\widetilde{A}_v U + \widetilde{\varepsilon}_1(U)N + \widetilde{\varepsilon}_2(U)v. \tag{3.6}$$

for any $U, V \in \Gamma(T\Psi)$, From (3.1), (3.4), (3.5) and (3.6) we obtain

$$D_U V = \nabla_U V + \pi(V)fU, \tag{3.7}$$
$$\widetilde{D}_1(U, V) = D_1(U, V) + \pi(V)w_1(U), \tag{3.8}$$
$$\widetilde{D}_2(U, V) = D_2(U, V) + \pi(V)w_2(U), \tag{3.9}$$
$$\widetilde{A}_N U = A_N U - \pi(N)fU, \tag{3.10}$$
$$\widetilde{p}_1(U) = p_1(U) + \pi(N)w_1(U), \tag{3.11}$$
$$\widetilde{p}_2(U) = p_2(U) + \pi(N)w_2(U), \tag{3.12}$$
$$\widetilde{A}_v U = A_v U - \pi(v)fU, \tag{3.13}$$
$$\widetilde{\varepsilon}_1(U) = \varepsilon_1(U) + \pi(v)w_1(U), \tag{3.14}$$
$$\widetilde{\varepsilon}_2(U) = \varepsilon_2(U) + \pi(v)w_2(U). \tag{3.15}$$

for any $U, V \in \Gamma(T\Psi)$.

Moreover for $\xi \in \Gamma(RadT\Psi)$ we can write

$$D_U PV = D^*_U PV + E^*(U, PV)\xi, \tag{3.16}$$

$$D_U \xi = -\widetilde{A}^*_\xi U - \widetilde{p}_1(U)\xi, \tag{3.17}$$

where $D^*_U PV$ $\widetilde{A}^*_\xi U \in \Gamma(S(T\Psi))$, $E^*(U, PV) = \widetilde{h}(D_U PV, N)$ and $\widetilde{p}_1(U) = -\widetilde{h}(D_U \xi, N)$.

Now we will compute curvature tensor of half-lightlike submanifolds for the quarter-symmetric non-metric connection:

By using (3.4), (3.5) and (3.6) we get

$$\begin{aligned}\widetilde{R}^D(U,V)W = {} & R^D(U,V)W - \widetilde{D}_1(V,W)\widetilde{A}_N U - \widetilde{D}_2(V,W)\widetilde{A}_v U + \widetilde{D}_1(U,W)\widetilde{A}_N V \\ & + \widetilde{D}_2(U,W)\widetilde{A}_v V + [(D_U\widetilde{D}_1)(V,W) - (D_V\widetilde{D}_1)(U,W) + \widetilde{D}_1(V,W)\widetilde{p}_1(U) \\ & + \widetilde{D}_2(V,W)\widetilde{\varepsilon}_1(U) - \widetilde{D}_1(U,W)\widetilde{p}_1(V) - \widetilde{D}_2(U,W)\widetilde{\varepsilon}_1(V) \\ & - \pi(V)\widetilde{D}_1(fU,W) + \pi(U)\widetilde{D}_1(fV,W)]N \\ & + [(D_U\widetilde{D}_2)(V,W) - (D_V\widetilde{D}_2)(U,W) + \widetilde{D}_1(V,W)\widetilde{p}_2(U) \\ & + \widetilde{D}_2(V,W)\widetilde{\varepsilon}_2(U) - \widetilde{D}_1(U,W)\widetilde{p}_2(V) - \widetilde{D}_2(U,W)\widetilde{\varepsilon}_2(V) \\ & - \pi(V)\widetilde{D}_2(fU,W) + \pi(U)\widetilde{D}_2(fV,W)]v,\end{aligned} \tag{3.18}$$

where R^D and $\widetilde{R}^D$ are curvature tensors with respect to connections D and $\widetilde{D}$, respectively.

The Gauss and codazzi equation of half-lightlike submanifold with respect to quarter-symmetric non-metric connection is as follows:

$$\begin{aligned}g(\widetilde{R}^D(U,V)PW, PZ) = {} & h(R^D(U,V)PW, PZ) - \widetilde{D}_1(V,PW)h(\widetilde{A}_N U, PZ) \\ & - \widetilde{D}_2(V,PW)h(\widetilde{A}_v U, PZ) + \widetilde{D}_1(U,PW)h(\widetilde{A}_N V, PZ) \\ & + \widetilde{D}_2(U,PW)h(\widetilde{A}_v U, PZ),\end{aligned} \tag{3.19}$$

$$\begin{aligned}g(\widetilde{R}^D(U,V)PW, \xi) = {} & (D_U\widetilde{D}_1)(V,PW) - (D_V\widetilde{D}_1)(U,PW) + \widetilde{D}_1(V,PW)\widetilde{p}_1(U) \\ & + \widetilde{D}_2(V,PW)\widetilde{\varepsilon}_1(U) - \widetilde{D}_1(U,PW)\widetilde{p}_1(V) - \widetilde{D}_2(U,PW)\widetilde{\varepsilon}_1(V) \\ & - \pi(V)\widetilde{D}_1(fU,PW) + \pi(U)\widetilde{D}_1(fV,PW),\end{aligned} \tag{3.20}$$

$$\begin{aligned}g(\widetilde{R}^D(U,V)PW, v) = {} & \epsilon\{(D_U\widetilde{D}_2)(V,PW) - (D_V\widetilde{D}_2)(U,PW) + \widetilde{D}_1(V,PW)\widetilde{p}_2(U) \\ & + \widetilde{D}_2(V,PW)\widetilde{\varepsilon}_2(U) - \widetilde{D}_1(U,PW)\widetilde{p}_2(V) - \widetilde{D}_2(U,PW)\widetilde{\varepsilon}_2(V) \\ & - \pi(V)\widetilde{D}_2(fU,PW) + \pi(U)\widetilde{D}_2(fW,PW)\},\end{aligned} \tag{3.21}$$

$$\begin{aligned}g(\widetilde{R}^D(U,V)PW, N) = {} & g(R^D(U,V)PW, N) - \widetilde{D}_1(V,PW)g(\widetilde{A}_N U, N) \\ & - \widetilde{D}_2(V,PW)g(\widetilde{A}_v U, N) + \widetilde{D}_1(U,PW)g(\widetilde{A}_N V, N) \\ & + \widetilde{D}_2(U,PW)g(\widetilde{A}_v V, N),\end{aligned} \tag{3.22}$$

$$g(\widetilde{R}^D(U,V)\xi,N) = g(R^D(U,V)\xi,N) - \widetilde{D}_2(V,\xi)g(\widetilde{A}_v U,N) + \widetilde{D}_2(U,\xi)g(\widetilde{A}_v V,N)$$
$$-\widetilde{D}_1(V,\xi)g(\widetilde{A}_N U,N) - \widetilde{D}_1(U,\xi)g(\widetilde{A}_N V,N), \tag{3.23}$$

$$g(\widetilde{R}^D(U,V)\xi,v) = \epsilon\{(D_U\widetilde{D}_2)(V,\xi) - (D_V\widetilde{D}_2)(U,\xi) + \widetilde{D}_1(V,\xi)\widetilde{p}_2(U)$$
$$+ \widetilde{D}_2(V,\xi)\widetilde{\varepsilon}_2(U) - \widetilde{D}_1(U,\xi)\widetilde{p}_2(V) - \widetilde{D}_2(U,\xi)\widetilde{\varepsilon}_2(V)$$
$$- \pi(V)\widetilde{D}_2(fU,\xi) + \pi(U)\widetilde{D}_2(fV,\xi)\}. \tag{3.24}$$

for any $U,V,W,Z \in \Gamma(T\Psi)$. From (3.16) and (3.17) we obtain

$$R^D(U,V)\xi = 2d\widetilde{v}_1(U,V)\xi + E_1^*(V,\widetilde{A}_\xi^*U)\xi - E_1^*(U,\widetilde{A}_\xi^*V)\xi + \widetilde{v}_1(U)\widetilde{A}_\xi^*Y$$
$$-\widetilde{v}_1(V)\widetilde{A}_\xi^*U + D_V^*\widetilde{A}_\xi^*U - D_U^*\widetilde{A}_\xi^*V + \widetilde{A}_\xi^*[U,V]. \tag{3.25}$$

Therefore we obtain

$$g(R^D(U,V)\xi,N) = 2d\widetilde{v}_1(U,V) + E_1^*(V,\widetilde{A}_\xi^*U) - E_1^*(U,\widetilde{A}_\xi^*V). \tag{3.26}$$

From (3.16) and (3.17) we obtain

$$R^D(U,V)PW = R^*(U,V)PW + \{E_1^*(U,D_V^*PW) - E_1^*(V,D_U^*PW) + UE_1^*(V,PW)$$
$$- VE_1^*(U,PW) + E_1^*(V,PW)\widetilde{v}_1(U) - E_1^*(U,PW)\widetilde{v}_1(V) - E_1^*([U,V],PW\}\xi$$
$$- E_1^*(V,PW)\widetilde{A}_\xi^*U + E_1^*(U,PW)\widetilde{A}_\xi^*V, \tag{3.27}$$

where R^* is the curvature tensor with respect to connection D^*.

From (3.27) we obtain

$$g(R^D(U,V)PW,PZ) = g(R^*(U,V)PW,PZ) - E_1^*(V,PW)g(\widetilde{A}_\xi^*U,PZ)$$
$$+ E_1^*(U,PW)g(\widetilde{A}_\xi^*V,PZ), \tag{3.28}$$

and

$$g(R^D(U,V)PW,N) = E_1^*(U,D_V^*PW) - E_1^*(V,D_U^*PW) + UE_1^*(V,PW) - YE_1^*(U,PW)$$
$$+ E_1^*(U,PW)\widetilde{v}_1(U) - E_1^*(U,PW)\widetilde{v}_1(V) - E_1^*([U,V],PW) \tag{3.29}$$

Now we will calculate Ricci tensor of half-lightlike submanifold with respect to the quarter-symmetric non-metric connection for D. We know that

$$Ric^D(U,V) = \sum_{i=1}^{m-1} \varepsilon_i g(R^D(U,E_i)V,E_i) + g(R^D(U,\xi)V,N).$$

From (3.18) den we obtain

$$Ric^D(U,V) = \sum_{i=1}^{m-1} \varepsilon_i\{g(\widetilde{R}^D(U,E_i)V,E_i) + \widetilde{D}_1(E_i,V)g(\widetilde{A}_N U,E_i)$$
$$+ \widetilde{D}_2(E_i,V)g(\widetilde{A}_v U,E_i) - \widetilde{D}_1(U,V)g(\widetilde{A}_N E_i,E_i) - \widetilde{D}_2(U,V)g(\widetilde{A}_v E_i,E_i)\}$$
$$+ g(\widetilde{R}^D(U,\xi)V,N) + \widetilde{D}_1(\xi,V)g(\widetilde{A}_N U,N) + \widetilde{D}_2(\xi,V)g(\widetilde{A}_u U,N)$$
$$- \widetilde{D}_1(U,V)g(\widetilde{A}_N\xi,N) - \widetilde{D}_2(U,V)g(\widetilde{A}_v\xi,N). \tag{3.30}$$

Let Ψ be a screen semi-invariant half-lightlike submanifold of a semi-Riemannian product manifold $\widetilde{\Psi}$. If $\widetilde{\Psi}$ is $(n+1)$–dimensional semi-Riemannian manifold of constant curvature with respect to quarter-symmetric non-metric connection for $\widetilde{D}$, then we know that

$$\widetilde{R}^D(U,V)W = c\{g(U,W)V - g(V,W)U\}.$$

Then we obtain

$$\begin{aligned}\sum_{i=1}^{m-1} \varepsilon_i \widetilde{h}(\widetilde{R}^D(U,E_i)V,E_i) &= \sum_{i=1}^{m-1} \varepsilon_i c(g(U,V)g(E_i,E_i) - g(E_i,V)g(U,E_i)) \\ &= c(m-2)g(U,V) \end{aligned} \tag{3.31}$$

and

$$\begin{aligned}\widetilde{g}(\widetilde{R}^D(U,\xi,V,N) &= c(g(U,V)g(\xi,N) - g(\xi,V)g(U,N)) \\ &= cg(U,V) \end{aligned} \tag{3.32}$$

Then from (3.31) and (3.32) we have the following lemma.

Lemma 3.1. *Let Ψ be a screen semi-invariant half-lightlike submanifold of a semi-Riemannian product manifold $\widetilde{\Psi}$. If $\widetilde{\Psi}$ is $(n+1)-$ dimensional semi-Riemannian manifold of constant curvature with respect to quarter-symmetric non-metric connection for $\widetilde{D}$, then we have the following equation on the distribution L.*

$$\begin{aligned} Ric^D(U,V) - Ric^D(V,U) = \sum_{i=1}^{m-1} \varepsilon_i\{ &\widetilde{D}_1(E_i,V)g(\widetilde{A}_N U,E_i) + \widetilde{D}_2(E_i,V)g(\widetilde{A}_v U,E_i) \\ &- \widetilde{D}_1(E_i,U)g(\widetilde{A}_N V,E_i) - \widetilde{D}_2(E_i,U)g(\widetilde{A}_v V,E_i)\} \\ &+ \widetilde{D}_1(\xi,V)g(\widetilde{A}_N U,N) + \widetilde{D}_2(\xi,V)g(\widetilde{A}_v U,N) \\ &- \widetilde{D}_1(\xi,U)g(\widetilde{A}_N V,N) - \widetilde{D}_2(\xi,U)g(\widetilde{A}_v V,N). \end{aligned} \tag{3.33}$$

Proof. *$\widetilde{D}_1$ and $\widetilde{D}_2$ is symmetric on the distribution $L_4 = L_0 \perp L_1 \perp RadT\Psi$. Thus from (3.30) claim is proved.*

Theorem 3.2. *Let Ψ be a semi-invariant half-lightlike submanifold of a semi-Riemannian product manifold $\widetilde{\Psi}$ and let $\widetilde{\Psi}$ be $(n+1)-$ dimensional semi-Riemannian manifold of constant curvature with respect to quarter-symmetric non-metric connection for $\widetilde{D}$. If Ψ is totally geodesic then the Ricci tensor is symmetric with respect to quarter-symmetric non-metric connection on the distribution L_1 if and only if*

$$E_1(U, A^*_\xi V) = E_1(V, A^*_\xi U).$$

Proof. *For any $U,V \in \Gamma(L_1)$, Since Ψ is totally geodesic we have*

$$\widetilde{D}_1(\xi,V) = D_1(\xi,V) = 0,$$

and

$$\widetilde{D}_2(\xi, V) = D_2(\xi, V) = 0.$$

Additionally we have

$$\widetilde{D}_1(E_i, U) = D_1(E_i, U) = g(A^*_\xi U, E_i).$$

Thus from (3.10) we get

$$\widetilde{D}_1(E_i, U)g(\widetilde{A}_N V, E_i) = g(\widetilde{A}_N V, A^*_\xi U) = g(A_N V, A^*_\xi U) = E_1(V, A^*_\xi U). \tag{3.34}$$

Similarly we have

$$\widetilde{D}_1(E_i, V)g(\widetilde{A}_N U, E_i) = g(\widetilde{A}_N U, A^*_\xi V) = g(A_N U, A^*_\xi V) = E_1(U, A^*_\xi V). \tag{3.35}$$

Furthermore we obtain

$$\widetilde{D}_2(E_i, U) = D_2(E_i, U) = \epsilon g(A_v U, E_i).$$

Thus from (3.13) we obtain

$$\widetilde{D}_2(E_i, U)g(\widetilde{A}_v V, E_i) = g(\widetilde{A}_v V, A_v U) = g(A_v V, A_v U). \tag{3.36}$$

Similarly we have

$$\widetilde{D}_2(E_i, V)g(\widetilde{A}_v U, E_i) = g(\widetilde{A}_v U, A_v V) = g(A_v U, A_v V). \tag{3.37}$$

(3.34), (3.35), (3.36) and (3.37) is substituted in Eq. (3.33), then we get

$$Ric^D(U, V) - Ric^D(V, U) = E_1(U, A^*_\xi V) - E_1(V, A^*_\xi U).$$

The proof is completed.

4 Conclusion

We investigate the curvature tensors half-lightlike submanifolds of a semi-Riemannian product manifold and define new class of half-lightlike submanifolds which called screen semi-invariant half-lightlike submanifolds with some different types distribution of screen semi-invariant half-lightlike submanifold. Moreover, we calculate the Curvature tensor and Ricci tensor with respect to the quarter-symmetric non-metric connection and provide the non-trivial examples, which enrich the quality of the present manuscript. These results may be fruitful and innovative for future study. In addition, other researchers can make new studies in lightlike geometry with the help of special type distributions we have found.

References

1. Hayden, H.A.: Sub-spaces of a space with torsion. Proc. London Math. Soc. **34**, 27–50 (1932)
2. Yano, K.: On semi-symmetric metric connections. Rev. Roumania Math. Pures Appl. **15**, 1579–1586 (1970)
3. Golab, S.: On semi-symmetric metric and quarter-symmetric linear connections. Tensor **29**, 249–254 (1975)
4. Agashe, N.S., Chafle, M.R.: A semi symmetric non-metric connection in a Riemannian manifold. Indian J. Pure Appl. Math. **23**, 399–409 (1992)
5. Chaturvedi, B.B., Pandey, P.N.: Semi-symetric non metric connections on a Kahler manifold. Difer. Geom.-Dinamical Syst. **10**, 86–90 (2008)
6. Tripathi, M.M.: A new connection in a Riemannian manifold. Int. J. Geo **1**, 15–24 (2008)
7. Ahmad, M., Jun, J.B., Siddiqi, M.D.: On some properties of semi-invaraint submanifolds of a nearly Trans-Sasakian manifolds admitting a quarter symmetric non-metric connection. J. Chungcheong Math. Soc. **25**(1), 73–90 (2012)
8. Ahmad, M., Siddiqi, M.D., Ojha, J.P.: Semi-invariant submanifolds of Kenmotsu manifolds immersed in an almost r-paracontact structure admitting a quarter symmetric non-metric connection. J. Math. Comput. Sci. **2**(4), 982–998 (2010)
9. Duggal, K.L., Bejancu, A.: Lightlike Submanifolds of Semi-Riemannian Manifolds and Applications. Kluwer Academic Publishers, Dordrecht (1996)
10. Duggal, K.L., Bejancu, A.: Lightlike submanifolds of codimension two. Math. J. Toyama Univ. **15**, 59–82 (1992)
11. Duggal, K.L., Jin, D.H.: Null Curves and Hypersurfaces of Semi-Riemannian Manifolds. World Scientific Publishing Co. Pte. Ltd., Singapore (2007)
12. Duggal, K.L.: Riemannian geometry of half lightlike submanifolds. Math. J. Toyama Univ. **25**, 169–179 (2002)
13. Duggal, K.L., Sahin, B.: Screen conformal half-lightlike submanifolds. Int. J. Math. Math. Sci. **68**, 3737–3753 (2004)
14. Duggal, K.L., Sahin, B.: Screen Cauchy Riemann lightlike submanifolds. Acta Math. Hungar. **106**(1–2), 137–165 (2005)
15. Duggal, K.L., Sahin, B.: Generalized Cauchy Riemann lightlike submanifolds. Acta Math. Hungar. **112**(1–2), 113–136 (2006)
16. Duggal, K.L., Sahin, B.: Lightlike submanifolds of indefinite Sasakian manifolds. Int. J. Math. Math. Sci. **2007**, 1–21 (2007). Art ID 57585 [162]
17. Duggal, K.L., Sahin, B.: Contact generalized CR-lightlike submanifolds of Sasakian submanifolds. Acta Math. Hungar. **122**(1–2), 45–58 (2009)
18. Atceken, M., Kilic, E.: Semi-invariant lightlike submanifolds of a semi- Riemannian product manifold. Kodai Math. J. **30**(3), 361–378 (2007)
19. Duggal, K.L., Sahin, B.: Differential Geometry of Lightlike Submanifolds, Birkhauser Veriag AG, Basel-Boston-Berlin (2010)
20. Jin, D.H.: Half lightlike submanifolds of a Trans-Sasakian manifold with a semi-symmetric non-metric connection. Int. J. Math. Anal. **10**(8), 381–397 (2016). HIKARI Ltd, https://doi.org/10.12988/ijma.2016.616. www.m-hikari.com
21. Jin, D.H.: Einstein half lightlike submanifolds of a Lorentzian space form with a semi-symmetric non-metric connection. J. Inequalities Appl. **2013**, 403 (2013). http://www.journalofinequalitiesandapplications.com/content/2013/1/403
22. Jin, D.H.: Geometry of half lightlike submanifolds of a semi-Riemannian space form with a semi-symmetric metric connection. J. Chungcheong Math. Soc. **24**(4), 769–780 (2011)

23. Kilic, E., Sahin, B.: Radical anti-invariant lightlike submanifolds of a semi-Riemannian product manifold. Turkish J. Math. **32**, 429–449 (2008)
24. Kilic, E., Bahadir, O.: Lightlike hypersurfaces of a semi-Riemannian product manifold and quarter-symmetric nonmetric connections. Int. J. Math. Math. Sci. **2012**, 17 (2012). Article ID 178390
25. Bahadir, O.: Screen semi invariant half-lightlike submanifolds of a semi-Riemannian product manifold. Glob. J. Adv. Res. Class. Modern Geometries **4**(2), 116–124 (2015). ISSN 2284-5569

Viscosity Modification with Inertial Forward-Backward Splitting Methods for Solving Inclusion Problems

D. Yambangwai[1], S. Suantai[2], H. Dutta[3], and W. Cholamjiak[1](✉)

[1] School of Science, University of Phayao, Phayao 56000, Thailand
damrong.sut@gmail.com, c-wchp007@hotmail.com
[2] Department of Mathematics, Faculty of Science, Chiang Mai University, Chiang Mai 50200, Thailand
suthep.s@cmu.ac.th
[3] Department of Mathematics, Gauhati University, Guwahati 781014, India
hemen_dutta08@rediffmail.com

Abstract. In this paper, we introduce the two new different modified forward-backward algorithms combining the viscosity approximation method with the inertial technical term for solving the inclusion problem. The strongly convergent theorems are established under some suitable conditions in Hilbert spaces. The application of algorithms in this study work is use to find the minimum-norm least-squares solution of an unconstrained linear system and test some numerical experiments. Moreover, the efficiency and the implementation of our methods have been shown through the examples.

Keywords: Inertial method · Inclusion problem · Maximal monotone operator · Forward-backward algorithm · Minimum-norm least-squares solution.

Mathematics Subject Classification (2010): 15A06 · 47H04 · 47H05 · 47H10 · 47J22

1 Introduction

In this work, we study the following inclusion problem: find $\hat{x}$ in a Hilbert space H such that

$$0 \in A\hat{x} + B\hat{x} \tag{1.1}$$

where $A : H \to H$ is an operator and $B : H \to 2^H$ is a multivalued operator. The solution set of (1.1) is denoted by $(A + B)^{-1}(0)$. This problem includes, as special cases, convex programming, variational inequality problem, split feasibility problem and minimization problem. To be more precise, some concrete problems in machine learning, image processing and linear inverse problem can be modeled mathematically as this formulation.

M. Zeki Sarıkaya et al. (Eds.): ICMRS 2019, LNNS 123, pp. 147–175, 2020.
https://doi.org/10.1007/978-3-030-43002-3_14

For solving the problem (1.1), the forward-backward splitting algorithm [3, 9,17,19,24,32] is usually employed and is generated by the following manner: $x_1 \in H$ and

$$x_{n+1} = (I + rB)^{-1}(x_n - rAx_n),\ n \geq 1, \tag{1.2}$$

where $r > 0$. In this case, each step of iterates involves only with A as the forward step and B as the backward step, but not the sum of operators. This method includes, as special cases, the proximal point algorithm [29] and the gradient method. In 1979, Lions and Mercier [18] proposed the following splitting iterative methods in a real Hilbert space:

$$x_{n+1} = (2J_r^A - I)(2J_r^B - I)x_n,\ n \geq 1 \tag{1.3}$$

and

$$x_{n+1} = J_r^A(2J_r^B - I)x_n + (I - J_r^B)x_n,\ n \geq 1, \tag{1.4}$$

where $J_r^T = (I + rT)^{-1}$ with $r > 0$. The first one is often called Peaceman-Rachford algorithm [25] and the second one is called Douglas-Rachford algorithm [11]. We note that both algorithms obtain only weakly convergent in general [4,18].

In particular, if $A := \nabla f$ and $B := \partial g$, where ∇f is the gradient of f and ∂g is the subdifferential of g which is defined by $\partial g(x) := \{s \in H : g(y) \geq g(x) + \langle s, y - x\rangle,\ \ \forall y \in H\}$ then problem (1.1) is reduced to the following minimization problem:

$$\min_{x \in H} f(x) + g(x) \tag{1.5}$$

and (1.2) also becomes

$$x_{n+1} = \mathrm{prox}_{rg}(x_n - r\nabla f(x_n)), n \geq 1, \tag{1.6}$$

where $r > 0$ is the step-size and $\mathrm{prox}_{rg} = (I + r\partial g)^{-1}$ is the proximity operator of g.

In 2001, Alvarez and Attouch [1] used the heavy ball method which was studied in [26,27] for maximal monotone operators by using the proximal point algorithm. This algorithm is called the inertial proximal point algorithm and it is generated by the following form:

$$\begin{cases} y_n = x_n + \theta_n(x_n - x_{n-1}) \\ x_{n+1} = (I + r_nB)^{-1}y_n,\quad n \geq 1. \end{cases} \tag{1.7}$$

They proved that if $\{r_n\}$ is non-decreasing and $\{\theta_n\} \subset [0, 1)$ with

$$\sum_{n=1}^{\infty} \theta_n\|x_n - x_{n-1}\|^2 < \infty, \tag{1.8}$$

then algorithm (1.7) converges weakly to a zero of B. Here θ_n is an extrapolation factor and the inertial is represented by the term $\theta_n(x_n - x_{n-1})$. It is remarkable that the inertial terminology greatly improves the performance of the algorithm and has a good convergence properties [12,13,23].

In 2003, Moudafi and Oliny [22] introduced the following inertial proximal point algorithm for finding a solution of the zero-finding problem of two monotone operators:

$$\begin{cases} y_n = x_n + \theta_n(x_n - x_{n-1}) \\ x_{n+1} = (I + r_n B)^{-1}(y_n - r_n A x_n), \quad n \geq 1, \end{cases} \tag{1.9}$$

where $A : H \to H$ and $B : H \to 2^H$. They proved the weak convergence theorem provided $r_n < 2/L$ with L the Lipschitz constant of A under the condition (1.8). It is observed that, for $\theta_n > 0$, the algorithm (1.9) does not take the form of a forward-backward splitting algorithm, since operator A is still evaluated at the point x_n.

Recently, Lorenz and Pock [19] introduced the following inertial forward-backward algorithm for monotone operators:

$$\begin{cases} y_n = x_n + \theta_n(x_n - x_{n-1}) \\ x_{n+1} = (I + r_n B)^{-1}(y_n - r_n A y_n), \quad n \geq 1, \end{cases} \tag{1.10}$$

where $\{r_n\}$ is a suitable sequence in $[0, 1]$. They showed that algorithm (1.10) differs from that of Moudafi and Oliny insofar that they evaluated the operator B as the inertial extrapolate y_n. The algorithms involving the inertial term mentioned above have weak convergence. However, in some applied disciplines, the strong convergence is more desirable that the weak convergence [4].

Let C be a nonempty, closed and convex subset of a normed space X. A mapping $T : C \to C$ is said to be

1. *contraction* if there exists $k \in (0, 1)$ such that $\|Tx - Ty\| \leq k\|x - y\|$ for all $x, y \in C$;
2. *nonexpansive* if $\|Tx - Ty\| \leq \|x - y\|$ for all $x, y \in C$.

The fixed point set of T is denoted by $F(T)$, that is, $F(T) = \{x \in C : x = Tx\}$.

In 1922, Banach [2] established the famous fixed point result of a contractive mapping in complete metric spaces, known as Banach's contraction principle, which is an important tool for solving the existence problem of nonlinear mappings. Since then, many generalizations of this principle have been constructed in several directions.

In 1967, Browder [5] employed the Banach's contraction principle to prove the existence of a nonexpansive mapping in Hilbert spaces.

In 1967, Halpern [14] proposed the following classical iteration for a nonexpansive mapping $T : C \to C$ in a real Hilbert space:

$$x_{n+1} = \alpha_n u + (1 - \alpha_n) T x_n, \quad n \geq 0, \tag{1.11}$$

where $\alpha_n \in (0, 1)$ and $u \in C$.

In 1992, Wittmann [35] proved that the sequence $\{x_n\}$ converges strongly to a fixed point of T if $\{\alpha_n\}$ satisfies the following conditions:

$$C1: \lim_{n\to\infty} \alpha_n = 0; \quad C2: \sum_{n=0}^{\infty} \alpha_n = \infty; \quad C3: \sum_{n=0}^{\infty} |\alpha_{n+1} - \alpha_n| < \infty.$$

In 2000, Moudafi [21] introduced the following method which is called *viscosity approximation method*: $x_0 \in C$ and

$$x_{n+1} = \alpha_n f(x_n) + (1 - \alpha_n)Tx_n, \quad n \geq 0, \tag{1.12}$$

where $\alpha_n \in (0, 1)$. He proved that the sequence $\{x_n\}$ converges strongly to a fixed point of T if $\{\alpha_n\}$ satisfies $C1$, $C2$ and $C3$.

In 2004, Xu [36] extended Moudafi's results [21] in the framework of Hilbert spaces. He presented solving some variational inequality problem by using the viscosity approximation method: $x_0 \in C$ and

$$x_{n+1} = (1 - \alpha_n)Tx_n + \alpha_n f(x_n), \quad n \geq 0, \tag{1.13}$$

where $\{\alpha_n\} \subset (0, 1)$. He proved that the sequence $\{x_n\}$ converges strongly to $\hat{x}$ under some control conditions of the sequence $\{\alpha_n\}$, where $\widehat{x}$ is the unique solution of the variational inequality

$$\langle (I - f)\widehat{x}, x - \widehat{x}\rangle \geq 0, \quad x \in F(T),$$

if $\{\alpha_n\}$ satisfies the following conditions:

$H1$: $\lim_{n\to\infty} \alpha_n = 0$; $H2$: $\sum_{n=0}^{\infty} \alpha_n = \infty$; $H3$: either $\sum_{n=0}^{\infty} |\alpha_{n+1} - \alpha_n| < \infty$ or $\lim_{n\to\infty} (\frac{\alpha_{n+1}}{\alpha_n}) = 1$.

In 2007, Takahashi and Takahashi [34] proved strong convergence theorems by viscosity approximation methods for nonexpansive mappings in Hilbert spaces, so that the strong convergence theorem is guaranteed (see also [36]).

In 2008, Yao et al. [37] considered and analyzed a new viscosity iterative scheme for approximating a common fixed point of a sequence of nonexpansive mappings in reflexive Banach spaces: $x_1 \in C$ and

$$x_{n+1} = \alpha_n f(x_n) + \beta x_n + (1 - \alpha_n - \beta)W_n x_n, \quad n \geq 1, \tag{1.14}$$

where $\{\alpha_n\} \subset (0, 1)$, β is a constant in $(0, 1)$ and W_n is the W-mapping which is generated by Shimoji and Takahashi [30]. They proved that $\{x_n\}$ converges strongly to the fixed point $\hat{x}$ of W_n, where $\hat{x}$ is the unique solution of variational inequality

$$\langle (I - f)\widehat{x}, x - \widehat{x}\rangle \geq 0, \quad x \in F(W_n),$$

if $\{\alpha_n\}$ satisfies $H1$ and $H2$.

In this work, we introduce the modified forward-backward algorithm combining the viscosity approximation method involving the inertial technique term for solving the inclusion problems such that the strong convergence is proved in the framework of Hilbert spaces. The rest of this paper is organized as follows. In Sect. 2, we recall some basic concepts and lemmas. In Sect. 3, we prove the strong convergence theorem of our proposed methods. Finally, in Sect. 4, we provide some applications including its numerical experiments by using our methods to solve the least squares problem. The efficiency and the implementation of our scheme have been explored through the examples.

2 Preliminaries and Lemmas

Let C be a nonempty, closed and convex subset of a Hilbert space H. The nearest point projection of H onto C is denoted by P_C, that is, $\|x - P_C x\| \leq \|x - y\|$ for all $x \in H$ and $y \in C$. Such P_C is called the metric projection of H onto C. We know that $\langle x - P_C x, y - P_C x\rangle \leq 0$ holds for all $x \in H$ and $y \in C$; see also [33].

Recall that a mapping $T : H \to H$ is said to be nonexpansive if, for all $x, y \in H$,

$$\|Tx - Ty\| \leq \|x - y\|. \tag{2.1}$$

A mapping $T : H \to H$ is said to be firmly nonexpansive if, for all $x, y \in H$,

$$\|Tx - Ty\|^2 \leq \|x - y\|^2 - \|(I - T)x - (I - T)y\|^2, \tag{2.2}$$

or equivalently

$$\langle Tx - Ty, x - y\rangle \geq \|Tx - Ty\|^2 \tag{2.3}$$

for all $x, y \in H$. We denote $F(T)$ by the fixed point set of T. It is known that T is firmly nonexpansive if and only if $I - T$ is firmly nonexpansive. We know that the metric projection P_C is firmly nonexpansive.

An operator $A : C \to H$ is called α-inverse strongly monotone if there exists a constant $\alpha > 0$ with

$$\langle Ax - Ay, x - y\rangle \geq \alpha\|Ax - Ay\|^2, \quad \forall x, y \in C. \tag{2.4}$$

We see that if A is α-inverse strongly monotone, then it is $\frac{1}{\alpha}$-Lipschitzian continuous.

A multivalued mapping $B : H \to 2^H$ is said to be monotone if for all $x, y \in H, f \in B(x)$, and $g \in B(y)$ implies $\langle x - y, f - g\rangle \geq 0$. A monotone mapping B is maximal if its graph $G(B) := \{(f, x) \in H \times H : f \in B(x)\}$ of B is not properly contained in the graph of any other monotone mapping. We know that a monotone mapping B is maximal if and only if for $(x, f) \in H \times H$, $\langle x - y, f - g\rangle \geq 0$ for all $(y, g) \in G(B)$ imply $f \in B(x)$. Let $J_r^B = (I + rB)^{-1}$, $r > 0$ be the resolvent of B. It is well known that J_r^B is singlevalued, $D(J_r^B) = H$ and J_r^B is firmly nonexpansive for all $r > 0$.

Theorem 2.1 [36]. *Let H be a real Hilbert space. Let $f : C \to C$ is a contraction on C and let $T : C \to C$ be a nonexpansive mapping with a fixed point. For every $t \in (0, 1)$, the unique fixed point $x_t \in C$ of the contraction $C \ni x \mapsto tf(x) + (1 - t)Tx$ converges strongly as $t \to 0$ to a fixed point of T.*

In what follows, we shall use the following notation:

$$T_r^{A,B} = J_r^B(I - rA) = (I + rB)^{-1}(I - rA), \quad r > 0. \tag{2.5}$$

Lemma 2.2 [17]. *Let H be a real Hilbert space. Let $A : H \to H$ be an α-inverse strongly monotone operator and $B : H \to 2^H$ a maximal monotone operator. Then the following results hold*

(i) *For $r > 0$, $F(T_r^{A,B}) = (A + B)^{-1}(0)$;*

(ii) *For $0 < s \leq r$ and $x \in H$, $\|x - T_s^{A,B}x\| \leq 2\|x - T_r^{A,B}x\|$.*

Lemma 2.3 [17]. *Let H be a real Hilbert space and A be an α-inverse strongly monotone operator. For each $r > 0$, we have*

$$\begin{aligned}\|T_r^{A,B}x - T_r^{A,B}y\|^2 &\leq \|x - y\|^2 - r(2\alpha - r)\|Ax - Ay\|^2 \\ &\quad - \|(I - J_r^B)(I - rA)x - (I - J_r^B)(I - rA)y\|^2, \end{aligned} \tag{2.6}$$

for all $x, y \in B_r = \{z \in H : \|z\| \leq r\}$.

Lemma 2.4 [20]. *Let $\{a_n\}$ and $\{c_n\}$ be sequences of nonnegative real numbers such that*

$$a_{n+1} \leq (1 - \delta_n)a_n + b_n + c_n, \quad n \geq 1, \tag{2.7}$$

where $\{\delta_n\} \subseteq (0, 1)$ and $\{b_n\}$ is a real sequence. Assume $\sum_{n=1}^{\infty} c_n < \infty$. Then the following results hold:

(i) *If $b_n \leq \delta_n M$ for some $M \geq 0$, then $\{a_n\}$ is a bounded sequence.*
(ii) *If $\sum_{n=1}^{\infty} \delta_n = \infty$ and $\limsup_{n\to\infty} \frac{b_n}{\delta_n} \leq 0$, then $\lim_{n\to\infty} a_n = 0$.*

Lemma 2.5 [15]. *Assume $\{s_n\}$ is a sequence of nonnegative real numbers such that*

$$s_{n+1} \leq (1 - \delta_n)s_n + \delta_n\tau_n, \quad n \geq 1 \tag{2.8}$$

and

$$s_{n+1} \leq s_n - \eta_n + \rho_n, \quad n \geq 1. \tag{2.9}$$

where $\{\delta_n\} \subseteq (0, 1)$, $\{\eta_n\}$ is a sequence of nonnegative real numbers and $\{\tau_n\}$, and $\{\rho_n\}$ are real sequences such that

(i) $\sum_{n=1}^{\infty} \delta_n = \infty$,
(ii) $\lim_{n\to\infty} \rho_n = 0$,
(iii) $\lim_{k\to\infty} \eta_{n_k} = 0$ *implies* $\limsup_{k\to\infty} \tau_{n_k} \leq 0$ *for any subsequence real numbers $\{n_k\}$ of $\{n\}$. Then $\lim_{n\to\infty} s_n = 0$.*

Proposition 2.6 [10]. *Let H be a real Hilbert space. Let $m \in \mathbb{N}$ be fixed. Let $\{x_i\}_{i=1}^m \subset X$ and $t_i \geq 0$ for all $i = 1, 2, ..., m$ with $\sum_{i=1}^m t_i \leq 1$. Then we have*

$$\|\sum_{i=1}^m t_i x_i\|^2 \leq \frac{\sum_{i=1}^m t_i\|x_i\|^2}{2 - (\sum_{i=1}^m t_i)}. \tag{2.10}$$

3 Strong Convergence Results

In this section, we introduce the modified inertial forward-backward methods combining the viscosity approximation method for inclusion problem. We then prove the strong convergence theorem in Hilbert spaces.

Theorem 3.1. *Let H be a real Hilbert space and let $f : H \to H$ be a contraction mapping. Let $A : H \to H$ be an α-inverse strongly monotone operator and $B : H \to 2^H$ a maximal monotone operator such that $S = (A+B)^{-1}(0) \neq \emptyset$. Let $\{x_n\}$ be a sequence generated by $x_0, x_1 \in H$ and*

$$\begin{cases} y_n = x_n + \theta_n(x_n - x_{n-1}), \\ x_{n+1} = \alpha_n f(y_n) + \beta_n y_n + \gamma_n J_{r_n}^B(y_n - r_n A y_n), \quad n \geq 1, \end{cases} \tag{3.1}$$

where $J_{r_n}^B = (I + r_n B)^{-1}$, $0 < r_n \leq 2\alpha$, $\{\theta_n\} \subset [0, \theta]$ with $\theta \in [0, 1)$ and $\{\alpha_n\}$, $\{\beta_n\}$ and $\{\gamma_n\}$ are sequences in $(0, 1)$ with $\alpha_n + \beta_n + \gamma_n = 1$. Assume that the following conditions hold:

(i) $\lim_{n\to\infty} \frac{\theta_n \|x_n - x_{n-1}\|}{\alpha_n} = 0$;
(ii) $\lim_{n\to\infty} \alpha_n = 0$, $\sum_{n=1}^{\infty} \alpha_n = \infty$;
(iii) $0 < \liminf_{n\to\infty} r_n \leq \limsup_{n\to\infty} r_n < 2\alpha$;
(iv) $\liminf_{n\to\infty} \gamma_n > 0$.

Then the sequence $\{x_n\}$ converges strongly to $z = P_S f(z)$.

Proof. For each $n \in \mathbb{N}$, we put $T_n = J_{r_n}^B(I - r_n A)$ and let $\{z_n\}$ be defined by $z_1 \in H$ and

$$z_{n+1} = \alpha_n f(z_n) + \beta_n z_n + \gamma_n T_n z_n. \tag{3.2}$$

Using Lemma 2.3, we see that

$$\begin{aligned} \|x_{n+1} - z_{n+1}\| &\leq \alpha_n \|f(y_n) - f(z_n)\| + \beta_n \|y_n - z_n\| + \gamma_n \|T_n y_n - T_n z_n\| \\ &\leq (1 - \alpha_n(1-k))\|y_n - z_n\| \\ &\leq (1 - \alpha_n(1-k))\|x_n + \theta_n(x_n - x_{n-1}) - z_n\|. \end{aligned} \tag{3.3}$$

By our assumptions and Lemma 2.4 (ii), we conclude that

$$\lim_{n\to\infty} \|x_n - z_n\| = 0. \tag{3.4}$$

Let $z = P_S f(z)$. Then

$$\begin{aligned} \|z_{n+1} - z\| &\leq \alpha_n \|f(z_n) - z\| + \beta_n \|z_n - z\| + \gamma_n \|T_n z_n - z\| \\ &\leq \alpha_n \|f(z_n) - f(z)\| + \alpha_n \|f(z) - z\| + (1 - \alpha_n)\|z_n - z\| \\ &\leq (1 - \alpha_n(1-k))\|z_n - z\| + \alpha_n \|f(z) - z\| \\ &= (1 - \alpha_n(1-k))\|z_n - z\| + \alpha_n(1-k)\frac{\|f(z) - z\|}{(1-k)} \\ &\leq \max\Big\{\|z_n - z\|, \frac{\|f(z) - z\|}{(1-k)}\Big\} \\ &\ \ \vdots \\ &\leq \max\Big\{\|z_1 - z\|, \frac{\|f(z) - z\|}{(1-k)}\Big\}. \end{aligned} \tag{3.5}$$

This shows that $\{z_n\}$ is bounded. Consequently, $\{x_n\}$ and $\{y_n\}$ are also bounded. We observe that

$$\begin{aligned}\|y_n - z\|^2 &= \|x_n - z + \theta_n(x_n - x_{n-1})\|^2 \\ &\le \|x_n - z\|^2 + 2\theta_n\langle x_n - x_{n-1}, y_n - z\rangle \end{aligned} \tag{3.6}$$

and

$$\begin{aligned}\|x_{n+1} - z\|^2 &= \|\alpha_n f(y_n) + \beta_n y_n + \gamma_n T_n y_n - z\|^2 \\ &\le \|\beta_n(y_n - z) + \gamma_n(T_n y_n - z)\|^2 + 2\alpha_n\langle f(y_n) - z, x_{n+1} - z\rangle \\ &= \|\beta_n(y_n - z) + \gamma_n(T_n y_n - z)\|^2 + 2\alpha_n\langle f(y_n) - f(z), x_{n+1} - z\rangle \\ &\quad + 2\alpha_n\langle f(z) - z, x_{n+1} - z\rangle \\ &\le \|\beta_n(y_n - z) + \gamma_n(T_n y_n - z)\|^2 + 2\alpha_n\|f(y_n) - f(z)\|\|x_{n+1} - z\| \\ &\quad + 2\alpha_n\langle f(z) - z, x_{n+1} - z\rangle \\ &\le \|\beta_n(y_n - z) + \gamma_n(T_n y_n - z)\|^2 + 2\alpha_n k\|y_n - z\|\|x_{n+1} - z\| \\ &\quad + 2\alpha_n\langle f(z) - z, x_{n+1} - z\rangle \\ &\le \|\beta_n(y_n - z) + \gamma_n(T_n y_n - z)\|^2 + 2\alpha_n k\|x_n - z\|\|x_{n+1} - z\| \\ &\quad + 2\alpha_n k\theta_n\|x_n - x_{n-1}\|\|x_{n+1} - z\| + 2\alpha_n\langle f(z) - z, x_{n+1} - z\rangle \\ &\le \|\beta_n(y_n - z) + \gamma_n(T_n y_n - z)\|^2 + \alpha_n k(\|x_n - z\|^2 + \|x_{n+1} - z\|^2) \\ &\quad + 2\alpha_n k\theta_n\|x_n - x_{n-1}\|\|x_{n+1} - z\| + 2\alpha_n\langle f(z) - z, x_{n+1} - z\rangle. \end{aligned} \tag{3.7}$$

Which implies that

$$\begin{aligned}\|x_{n+1} - z\|^2 \le{}& \frac{1}{1-\alpha_n k}\|\beta_n(y_n - z) + \gamma_n(T_n y_n - z)\|^2 + \frac{\alpha_n k}{1-\alpha_n k}\|x_n - z\|^2 \\ &+ \frac{2\alpha_n k}{1-\alpha_n k}\frac{\theta_n\|x_n - x_{n-1}\|}{\alpha_n}\|\|x_{n+1} - z\| + \frac{2\alpha_n}{1-\alpha_n k}\langle f(z) - z, x_{n+1} - z\rangle. \end{aligned} \tag{3.8}$$

On the other hand, by Proposition 2.6 and Lemma 2.3, we obtain

$$\begin{aligned}\|\beta_n(y_n - z) + \gamma_n(T_n y_n - z)\|^2 &\le \frac{1}{1+\alpha_n}(\beta_n\|y_n - z\|^2 + \gamma_n\|T_n y_n - z\|^2) \\ &\le \frac{\beta_n}{1+\alpha_n}\|y_n - z\|^2 + \frac{\gamma_n}{1+\alpha_n}(\|y_n - z\|^2 \\ &\quad - r_n(2\alpha - r_n)\|Ay_n - Az\|^2 \\ &\quad - \|y_n - r_n Ay_n - T_n y_n + r_n Az\|) \\ &= \frac{1-\alpha_n}{1+\alpha_n}\|y_n - z\|^2 - \frac{\gamma_n r_n(2\alpha - r_n)}{1+\alpha_n}\|Ay_n - Az\|^2 \\ &\quad - \frac{\gamma_n}{1+\alpha_n}\|y_n - r_n Ay_n - T_n y_n + r_n Az\|. \end{aligned} \tag{3.9}$$

Replacing (3.6) into (3.9), we have

$$\begin{aligned}\|\beta_n(y_n - z) + \gamma_n(T_n y_n - z)\|^2 \le{}& \frac{1-\alpha_n}{1+\alpha_n}(\|x_n - z\|^2 + 2\theta_n\langle x_n - x_{n-1}, y_n - z\rangle) \\ &- \frac{\gamma_n r_n(2\alpha - r_n)}{1+\alpha_n}\|Ay_n - Az\|^2 \\ &- \frac{\gamma_n}{1+\alpha_n}\|y_n - r_n Ay_n - T_n y_n + r_n Az\|. \end{aligned} \tag{3.10}$$

Again replacing (3.10) into (3.8), it follows that

$$\begin{aligned}\|x_{n+1}-z\|^2 \le & \frac{1-\alpha_n}{(1-\alpha_n k)(1+\alpha_n)}\left(\|x_n-z\|^2+2\theta_n\langle x_n-x_{n-1},y_n-z\rangle\right)\\ & -\frac{\gamma_n r_n(2\alpha-r_n)}{(1-\alpha_n k)(1+\alpha_n)}\|Ay_n-Az\|^2\\ & -\frac{\gamma_n}{(1-\alpha_n k)(1+\alpha_n)}\|y_n-r_nAy_n-T_ny_n+r_nAz\|\\ & +\frac{\alpha_n k}{1-\alpha_n k}\|x_n-z\|^2+\frac{2\alpha_n k}{1-\alpha_n k}\frac{\theta_n\|x_n-x_{n-1}\|}{\alpha_n}\|\|x_{n+1}-z\|\\ & +\frac{2\alpha_n}{1-\alpha_n k}\langle f(z)-z,x_{n+1}-z\rangle \qquad (3.11)\end{aligned}$$

We can check that $\frac{2\alpha_n(1-k)}{(1-\alpha_n k)(1+\alpha_n)}$ is in $(0,1)$. From (3.11), we then have

$$\begin{aligned}\|x_{n+1}-z\|^2 \le & \left(1-\frac{2\alpha_n(1-k)}{(1-\alpha_n k)(1+\alpha_n)}\right)\|x_n-z\|^2\\ & +\frac{2\alpha_n(1-k)}{(1-\alpha_n k)(1+\alpha_n)}\left(\frac{1+\alpha_n}{1-k}\langle f(z)-z,x_{n+1}-z\rangle\right.\\ & +\frac{(1-\alpha_n)\theta_n}{(1-k)\alpha_n}\langle x_n-x_{n-1},y_n-z\rangle+\frac{\alpha_n k}{1-k}\|x_n-z\|^2\\ & \left.+\frac{(1+\alpha_n)k\theta_n}{1-k}\|x_n-x_{n-1}\|\|x_{n+1}-z\|\right) \qquad (3.12)\end{aligned}$$

and also

$$\begin{aligned}\|x_{n+1}-z\|^2 \le & \|x_n-z\|^2-\frac{\gamma_n r_n(2\alpha-r_n)}{(1-\alpha_n k)(1+\alpha_n)}\|Ay_n-Az\|^2\\ & -\frac{\gamma_n}{(1-\alpha_n k)(1+\alpha_n)}\|y_n-r_nAy_n-T_ny_n+r_nAz\|\\ & +\frac{2\alpha_n}{1-\alpha_n k}\langle f(z)-z,x_{n+1}-z\rangle\\ & +\frac{2(1-\alpha_n)\theta_n}{(1-\alpha_n k)(1+\alpha_n)}\langle x_n-x_{n-1},y_n-z\rangle\\ & +\frac{2\alpha_n^2 k}{(1-\alpha_n k)(1+\alpha_n)}\|x_n-z\|^2\\ & +\frac{2\alpha_n k\theta_n}{1-\alpha_n k}\|x_n-x_{n-1}\|\|x_{n+1}-z\|. \qquad (3.13)\end{aligned}$$

For each $n \geq 1$, we get

$$\begin{aligned}
s_n &= \|x_n - z\|^2, \ \delta_n = \frac{2\alpha_n(1-k)}{(1-\alpha_n k)(1+\alpha_n)}, \\
\tau_n &= \frac{1+\alpha_n}{1-k}\langle f(z) - z, x_{n+1} - z\rangle + \frac{(1-\alpha_n)\theta_n}{(1-k)\alpha_n}\langle x_n - x_{n-1}, y_n - z\rangle \\
&\quad + \frac{\alpha_n k}{1-k}\|x_n - z\|^2 + \frac{(1+\alpha_n)k\theta_n}{1-k}\|x_n - x_{n-1}\|\|x_{n+1} - z\|, \\
\eta_n &= \frac{\gamma_n r_n(2\alpha - r_n)}{(1-\alpha_n k)(1+\alpha_n)}\|Ay_n - Az\|^2 \\
&\quad + \frac{\gamma_n}{(1-\alpha_n k)(1+\alpha_n)}\|y_n - r_n Ay_n - T_n y_n + r_n Az\|, \\
\rho_n &= \frac{2\alpha_n}{1-\alpha_n k}\langle f(z) - z, x_{n+1} - z\rangle + \frac{2(1-\alpha_n)\theta_n}{(1-\alpha_n k)(1+\alpha_n)}\langle x_n - x_{n-1}, y_n - z\rangle \\
&\quad + \frac{2\alpha_n^2 k}{(1-\alpha_n k)(1+\alpha_n)}\|x_n - z\|^2 + \frac{2\alpha_n k\theta_n}{1-\alpha_n k}\|x_n - x_{n-1}\|\|x_{n+1} - z\|.
\end{aligned}$$

Then (3.12) and (3.13) are reduced to the following:

$$s_{n+1} \leq (1-\delta_n)s_n + \delta_n\tau_n, \quad n \geq 1 \tag{3.14}$$

and

$$s_{n+1} \leq s_n - \eta_n + \rho_n, \quad n \geq 1. \tag{3.15}$$

Since $\sum_{n=1}^{\infty} \alpha_n = \infty$, it follows that $\sum_{n=1}^{\infty} \delta_n = \infty$. By the boundedness of $\{y_n\}$ and $\{x_n\}$, and $\lim_{n\to\infty} \alpha_n = 0$, we see that $\lim_{n\to\infty} \rho_n = 0$. In order to complete the proof, using Lemma 2.5, it remains to show that $\lim_{k\to\infty} \eta_{n_k} = 0$ implies $\limsup_{k\to\infty} \tau_{n_k} \leq 0$ for any subsequence $\{\eta_{n_k}\}$ of $\{\eta_n\}$. Let $\{\eta_{n_k}\}$ be a subsequence of $\{\eta_n\}$ such that $\lim_{k\to\infty} \eta_{n_k} = 0$. So, by our assumptions, we can deduce that

$$\lim_{k\to\infty} \|Ay_{n_k} - Az\| = \lim_{k\to\infty} \|y_{n_k} - r_{n_k} Ay_{n_k} - T_{n_k} y_{n_k} + r_{n_k} Az\| = 0. \tag{3.16}$$

This gives, by the triangle inequality, that

$$\lim_{k\to\infty} \|T_{n_k} y_{n_k} - y_{n_k}\| = 0. \tag{3.17}$$

Since $\liminf_{n\to\infty} r_n > 0$, there is $r > 0$ such that $r_n \geq r$ for all $n \geq 1$. In particular, $r_{n_k} \geq r$ for all $k \geq 1$. Lemma 2.2 (ii) yields that

$$\|T_r^{A,B} y_{n_k} - y_{n_k}\| \leq 2\|T_{n_k} y_{n_k} - y_{n_k}\|. \tag{3.18}$$

Then, by (3.17), we obtain

$$\limsup_{k\to\infty} \|T_r^{A,B} y_{n_k} - y_{n_k}\| \leq 0. \tag{3.19}$$

This implies that

$$\lim_{k\to\infty} \|T_r^{A,B} y_{n_k} - y_{n_k}\| = 0. \tag{3.20}$$

On the other hand, we have

$$\begin{aligned}\|T_r^{A,B} y_{n_k} - x_{n_k}\| &\leq \|T_r^{A,B} y_{n_k} - y_{n_k}\| + \|y_{n_k} - x_{n_k}\| \\ &\leq \|T_r^{A,B} y_{n_k} - y_{n_k}\| + \theta_{n_k}\|x_{n_k} - x_{n_k-1}\|. \end{aligned} \tag{3.21}$$

By condition (i) and (3.20), we get

$$\lim_{k\to\infty} \|T_r^{A,B} y_{n_k} - x_{n_k}\| = 0. \tag{3.22}$$

Let $z_t = tf(z_t) + (1-t)T_r^{A,B} z_t$, $t \in (0,1)$. Employing Theorem 2.1, we have $z_t \to P_S f(z) = z$ as $t \to 0$. So, we obtain

$$\begin{aligned}\|z_t - x_{n_k}\|^2 &= \|t(f(z_t) - x_{n_k}) + (1-t)(T_r^{A,B} z_t - x_{n_k})\|^2 \\ &\leq (1-t)^2\|T_r^{A,B} z_t - x_{n_k}\|^2 + 2t\langle f(z_t) - x_{n_k}, z_t - x_{n_k}\rangle \\ &= (1-t)^2\|T_r^{A,B} z_t - x_{n_k}\|^2 + 2t\langle f(z_t) - z_t, z_t - x_{n_k}\rangle + 2t\|z_t - x_{n_k}\|^2 \\ &\leq (1-t)^2\left(\|T_r^{A,B} z_t - T_r^{A,B} y_{n_k}\| + \|T_r^{A,B} y_{n_k} - x_{n_k}\|\right)^2 \\ &\quad + 2t\langle f(z_t) - z_t, z_t - x_{n_k}\rangle + 2t\|z_t - x_{n_k}\|^2 \\ &\leq (1-t)^2\left(\|z_t - y_{n_k}\| + \|T_r^{A,B} y_{n_k} - x_{n_k}\|\right)^2 \\ &\quad + 2t\langle f(z_t) - z_t, z_t - x_{n_k}\rangle + 2t\|z_t - x_{n_k}\|^2 \\ &\leq (1-t)^2\left(\|z_t - x_{n_k}\| + \theta_{n_k}\|x_{n_k} - x_{n_k-1}\| + \|T_r^{A,B} y_{n_k} - x_{n_k}\|\right)^2 \\ &\quad + 2t\langle f(z_t) - z_t, z_t - x_{n_k}\rangle + 2t\|z_t - x_{n_k}\|^2. \end{aligned} \tag{3.23}$$

This shows that

$$\begin{aligned}\langle z_t - f(z_t), z_t - x_{n_k}\rangle \leq \frac{(1-t)^2}{2t}&\left(\|z_t - x_{n_k}\| + \theta_{n_k}\|x_{n_k} - x_{n_k-1}\|\right. \\ &\left. + \|T_r^{A,B} y_{n_k} - x_{n_k}\|\right)^2 + \frac{(2t-1)}{2t}\|z_t - x_{n_k}\|^2. \end{aligned} \tag{3.24}$$

From condition (i), (3.22) and (3.24), we obtain

$$\begin{aligned}\limsup_{k\to\infty}\langle z_t - f(z_t), z_t - x_{n_k}\rangle &\leq \frac{(1-t)^2}{2t}M^2 + \frac{(2t-1)}{2t}M^2 \\ &= \frac{t}{2}M^2 \end{aligned} \tag{3.25}$$

for some $M > 0$ large enough. Taking $t \to 0$ in (3.25), we obtain

$$\limsup_{k\to\infty}\langle z - f(z), z - x_{n_k}\rangle \leq 0. \tag{3.26}$$

On the other hand, we have

$$\begin{aligned}\|x_{n_k+1} - x_{n_k}\| &\leq \alpha_{n_k}\|f(y_{n_k}) - x_{n_k}\| + \beta_{n_k}\|y_{n_k} - x_{n_k}\| + \gamma_{n_k}\|T_{n_k} y_{n_k} - x_{n_k}\| \\ &\leq \alpha_{n_k}\|f(y_{n_k}) - x_{n_k}\| + (1-\alpha_{n_k})\|y_{n_k} - x_{n_k}\| + \gamma_{n_k}\|T_{n_k} y_{n_k} - y_{n_k}\| \\ &\leq \alpha_{n_k}\|f(y_{n_k}) - x_{n_k}\| + (1-\alpha_{n_k})\theta_{n_k}\|x_{n_k} - x_{n_k-1}\| + \gamma_{n_k}\|T_{n_k} y_{n_k} - y_{n_k}\|. \end{aligned} \tag{3.27}$$

By (i), (ii), (3.17) and (3.27), we see that

$$\lim_{k\to\infty} \|x_{n_k+1} - x_{n_k}\| = 0. \tag{3.28}$$

Combining (3.26) and (3.28), we get that

$$\limsup_{k\to\infty} \langle z - f(z), z - x_{n_k+1}\rangle \leq 0. \tag{3.29}$$

It also follows from (i) that $\limsup_{k\to\infty} \tau_{n_k} \leq 0$. We conclude that $\lim_{n\to\infty} s_n = 0$ by Lemma 2.5. Hence $x_n \to z$ as $n \to \infty$. We thus complete the proof. □

Theorem 3.2. *Let H be a real Hilbert space and let $f : H \to H$ be a contraction mapping. Let $A : H \to H$ be an α-inverse strongly monotone operator and $B : H \to 2^H$ a maximal monotone operator such that $S = (A + B)^{-1}(0) \neq \emptyset$. Let $\{x_n\}$ be a sequence generated by $x_0, x_1 \in H$ and*

$$\begin{cases} y_n = x_n + \theta_n(x_n - x_{n-1}), \\ x_{n+1} = \alpha_n f(x_n) + \beta_n x_n + \gamma_n J^B_{r_n}(y_n - r_n A y_n), \quad n \geq 1, \end{cases} \tag{3.30}$$

where $J^B_{r_n} = (I + r_n B)^{-1}$, $0 < r_n \leq 2\alpha$, $\{\theta_n\} \subset [0, \theta]$ with $\theta \in [0, 1)$ and $\{\alpha_n\}$, $\{\beta_n\}$ and $\{\gamma_n\}$ are sequences in $(0, 1)$ with $\alpha_n + \beta_n + \gamma_n = 1$. Assume that the following conditions hold:

(i) $\lim_{n\to\infty} \frac{\theta_n \|x_n - x_{n-1}\|}{\alpha_n} = 0$;
(ii) $\lim_{n\to\infty} \alpha_n = 0$, $\sum_{n=1}^{\infty} \alpha_n = \infty$;
(iii) $0 < \liminf_{n\to\infty} r_n \leq \limsup_{n\to\infty} r_n < 2\alpha$;
(iv) $\liminf_{n\to\infty} \gamma_n > 0$.

Then the sequence $\{x_n\}$ converges strongly to $z = P_S f(z)$.

Proof. For each $n \in \mathbb{N}$, we put $T_n = J^B_{r_n}(I - r_n A)$ and let $\{z_n\}$ be defined by $z_1 \in H$ and

$$z_{n+1} = \alpha_n f(z_n) + \beta_n z_n + \gamma_n T_n z_n. \tag{3.31}$$

Using Lemma 2.3, we see that

$$\begin{aligned} \|x_{n+1} - z_{n+1}\| &\leq \alpha_n \|f(x_n) - f(z_n)\| + \beta_n \|x_n - z_n\| + \gamma_n \|T_n y_n - T_n z_n\| \\ &\leq \alpha_n k \|x_n - z_n\| + \beta_n \|x_n - z_n\| + \gamma_n \|y_n - z_n\| \\ &\leq (1 - \alpha_n(1 - k))\|x_n - z_n\| + \frac{\gamma_n \theta_n}{\alpha_n} \|x_n - x_{n-1}\|. \end{aligned} \tag{3.32}$$

By our assumptions and Lemma 2.4 (ii), we conclude that

$$\lim_{n\to\infty} \|x_n - z_n\| = 0. \tag{3.33}$$

Let $z = P_S f(z)$. From (3.5), we have $\{z_n\}$ is bounded. This implies that $\{x_n\}$ and $\{y_n\}$ are also bounded. We observe that

$$\begin{aligned}\|y_n - z\|^2 &= \|x_n - z + \theta(x_n - x_{n+1})\|^2 \\ &\leq \|x_n - z\|^2 + 2\theta_n\langle x_n - x_{n+1}, y_n - z\rangle \end{aligned} \tag{3.34}$$

and

$$\begin{aligned}\|x_{n+1} - z\|^2 &= \|\alpha_n f(x_n) + \beta_n x_n + \gamma_n T_n y_n - z\|^2 \\ &\leq \|\beta_n(x_n - z) + \gamma_n(T_n y_n - z)\|^2 + 2\alpha_n\langle f(x_n) - z, x_{n+1} - z\rangle \\ &\leq \|\beta_n(x_n - z) + \gamma_n(T_n y_n - z)\|^2 + 2\alpha_n\langle f(x_n) - f(z), x_{n+1} - z\rangle \\ &\quad + 2\alpha_n\langle f(z) - z, x_{n+1} - z\rangle \\ &\leq \|\beta_n(x_n - z) + \gamma_n(T_n y_n - z)\|^2 + 2\alpha_n\|f(x_n) - f(z)\|\|x_{n+1} - z\| \\ &\quad + 2\alpha_n\langle f(z) - z, x_{n+1} - z\rangle \\ &\leq \|\beta_n(x_n - z) + \gamma_n(T_n y_n - z)\|^2 + 2\alpha_n k\|x_n - z\|\|x_{n+1} - z\| \\ &\quad + 2\alpha_n\langle f(z) - z, x_{n+1} - z\rangle \\ &\leq \|\beta_n(x_n - z) + \gamma_n(T_n y_n - z)\|^2 + \alpha_n k(\|x_n - z\|^2 + \|x_{n+1} - z\|^2) \\ &\quad + 2\alpha_n\langle f(z) - z, x_{n+1} - z\rangle. \end{aligned} \tag{3.35}$$

This implies that

$$\begin{aligned}\|x_{n+1} - z\|^2 &\leq \frac{1}{1-\alpha_n k}\|\beta_n(x_n - z) + \gamma_n(T_n y_n - z)\|^2 + \frac{\alpha_n k}{1-\alpha_n k}\|x_n - z\|^2 \\ &\quad + \frac{2\alpha_n}{1-\alpha_n k}\langle f(z) - z, x_{n+1} - z\rangle. \end{aligned} \tag{3.36}$$

On the other hand, by Proposition 2.6 and Lemma 2.3, we obtain

$$\begin{aligned}\|\beta_n(x_n - z) + \gamma_n(T_n y_n - z)\|^2 &\leq \frac{1}{1+\alpha_n}(\beta_n\|x_n - z\|^2 + \gamma_n\|T_n y_n - z\|^2) \\ &\leq \frac{\beta_n}{1+\alpha_n}\|x_n - z\|^2 + \frac{\gamma_n}{1+\alpha_n}(\|y_n - z\|^2 \\ &\quad - r_n(2\alpha - r_n)\|Ay_n - Az\|^2 \\ &\quad - \|y_n - r_nAy_n - T_ny_n + r_nAz\|). \end{aligned} \tag{3.37}$$

Replacing (3.35) into (3.37), we have

$$\begin{aligned}\|\beta_n(x_n - z) + \gamma_n(T_n y_n - z)\|^2 &\leq \frac{1-\alpha_n}{1+\alpha_n}(\beta_n\|x_n - z\|^2 + \frac{2\gamma_n\theta_n}{1+\alpha_n}\langle x_n - x_{n+1}, y_n - z\rangle \\ &\quad - \frac{\gamma_n r_n(2\alpha - r_n)}{1+\alpha_n}\|Ay_n - Az\|^2 \\ &\quad - \frac{\gamma_n}{1+\alpha_n}\|y_n - r_nAy_n - T_ny_n + r_nAz\|). \end{aligned} \tag{3.38}$$

Replacing (3.38) into (3.36), we get

$$\begin{aligned}\|x_{n+1}-z\|^2 \le & \frac{1-\alpha_n}{(1-\alpha_n k)(1+\alpha_n)}\|x_n-z\|^2 \\ & + \frac{2\gamma_n\theta_n}{(1-\alpha_n k)(1+\alpha_n)}\langle x_n - x_{n-1}, y_n - z\rangle \\ & - \frac{\gamma_n r_n(2\alpha - r_n)}{(1-\alpha_n k)(1+\alpha_n)}\|Ay_n - Az\|^2 \\ & - \frac{\gamma_n}{(1-\alpha_n k)(1+\alpha_n)}\|y_n - r_n Ay_n - T_n y_n + r_n Az\| \\ & + \frac{\alpha_n k}{1-\alpha_n k}\|x_n - z\|^2 \\ & + \frac{2\alpha_n}{1-\alpha_n k}\langle f(z) - z, x_{n+1} - z\rangle. \end{aligned} \tag{3.39}$$

We can check that $\frac{2\alpha_n(1-k)}{(1-\alpha_n k)(1+\alpha_n)}$ is in (0,1). From (3.39), we then have

$$\begin{aligned}\|x_{n+1}-z\|^2 \le & \left(1 - \frac{2\alpha_n(1-k)}{(1-\alpha_n k)(1+\alpha_n)}\right)\|x_n - z\|^2 \\ & + \frac{2\alpha_n(1-k)}{(1-\alpha_n k)(1+\alpha_n)}\left(\frac{\alpha_n k}{(1-k)}\|x_n - z\|^2 \right. \\ & \left. + \frac{\gamma_n\theta_n}{(1-k)\alpha_n}\langle x_n - x_{n-1}, y_n - z\rangle + \frac{1+\alpha_n}{1-k}\langle f(z) - z, x_{n+1} - z\rangle\right), \end{aligned} \tag{3.40}$$

and also

$$\begin{aligned}\|x_{n+1}-z\|^2 \le & \|x_n - z\|^2 - \frac{\gamma_n r_n(2\alpha - r_n)}{(1-\alpha_n k)(1+\alpha_n)}\|Ay_n - Az\|^2 \\ & - \frac{\gamma_n}{(1-\alpha_n k)(1+\alpha_n)}\|y_n - r_n Ay_n - T_n y_n + r_n Az\| \\ & + \frac{2\alpha_n}{1-\alpha_n k}\langle f(z) - z, x_{n+1} - z\rangle + \frac{2\alpha_n^2 k}{(1-\alpha_n k)(1+\alpha_n)}\|x_n - z\|^2 \\ & + \frac{2\gamma_n\theta_n}{(1-\alpha_n k)(1+\alpha_n)}\langle x_n - x_{n-1}, y_n - z\rangle. \end{aligned} \tag{3.41}$$

For each $n \geq 1$, we get

$$S_n = \|x_{n+1} - z\|^2, \quad \delta_n = \frac{2\alpha_n(1-k)}{(1-\alpha_n k)(1+\alpha_n)},$$
$$\tau_n = \frac{\alpha_n k}{1-\alpha}\|x_n - z\|^2 + \frac{\gamma_n \theta_n}{(1-k)\alpha_n}\langle x_n - x_{n-1}, y_n - z\rangle$$
$$+ \frac{1+\alpha_n}{1-k}\langle f(z) - z, x_{n+1} - z\rangle,$$
$$\eta_n = \frac{\gamma_n r_n(2\alpha - r_n)}{(1-\alpha_n k)(1+\alpha_n)}\|Ay_n - Az\|^2$$
$$+ \frac{\gamma_n}{(1-\alpha_n k)(1+\alpha_n)}\|y_n - r_n Ay_n - T_n y_n + r_n Az\|,$$
$$\rho_n = \frac{2\alpha_n}{1-\alpha_n k}\langle f(z) - z, x_{n+1} - z\rangle + \frac{2\alpha_n^2 k}{(1-\alpha_n k)(1+\alpha_n)}\|x_n - z\|^2$$
$$+ \frac{2\gamma_n \theta_n}{(1-\alpha_n k)(1+\alpha_n)}\langle x_n - x_{n-1}, y_n - z\rangle.$$

Then (3.40) and (3.41) are reduced to the following

$$s_{n+1} \leq (1-\delta_n)s_n + \delta_n \tau_n, \quad n \geq 1 \tag{3.42}$$

and

$$s_{n+1} \leq s_n - \eta_n + \rho_n, \quad n \geq 1. \tag{3.43}$$

Since $\sum_{n=1}^{\infty} \alpha_n = \infty$, it follows that $\sum_{n=1}^{\infty} \delta_n = \infty$. By the boundedness of $\{y_n\}$ and $\{x_n\}$, and $\lim_{n\to\infty} \alpha_n = 0$, we see that $\lim_{n\to\infty} \rho_n = 0$. In order to complete the proof, using Lemma 2.5, it remains to show that $\lim_{k\to\infty} \eta_{n_k} = 0$ implies $\limsup_{k\to\infty} \tau_{n_k} \leq 0$ for any subsequence $\{\eta_{n_k}\}$ of $\{\eta_n\}$. Let $\{\eta_{n_k}\}$ be a subsequence of $\{\eta_n\}$ such that $\lim_{k\to\infty} \eta_{n_k} = 0$. So, by our assumptions, we can deduce that

$$\lim_{k\to\infty} \|Ay_{n_k} - Az\| = \lim_{k\to\infty} \|y_{n_k} - r_{n_k} Ay_{n_k} - T_{n_k} y_{n_k} + r_{n_k} Az\| = 0. \tag{3.44}$$

This gives, by the triangle inequality, that

$$\lim_{k\to\infty} \|T_{n_k} y_{n_k} - y_{n_k}\| = 0. \tag{3.45}$$

Since $\liminf_{n\to\infty} r_n > 0$, there is $r > 0$ such that $r_n \geq r$ for all $n \geq 1$. In particular, $r_{n_k} \geq r$ for all $k \geq 1$. Lemma 2.2 (ii) yields that

$$\|T_r^{A,B} y_{n_k} - y_{n_k}\| \leq 2\|T_{n_k} y_{n_k} - y_{n_k}\|. \tag{3.46}$$

Then, by (3.45), we obtain

$$\limsup_{k\to\infty} \|T_r^{A,B} y_{n_k} - y_{n_k}\| \leq 0. \tag{3.47}$$

This implies that

$$\lim_{k\to\infty} \|T_r^{A,B} y_{n_k} - y_{n_k}\| = 0. \tag{3.48}$$

On the other hand, we have

$$\begin{aligned}\|T_r^{A,B}y_{n_k} - x_{n_k}\| &\le \|T_r^{A,B}y_{n_k} - y_{n_k}\| + \|y_{n_k} - x_{n_k}\| \\ &\le \|T_r^{A,B}y_{n_k} - y_{n_k}\| + \theta_{n_k}\|x_{n_k} - x_{n_k-1}\|. \qquad (3.49)\end{aligned}$$

By condition (i) and (3.48), we get

$$\lim_{k\to\infty} \|T_r^{A,B}y_{n_k} - x_{n_k}\| = 0. \qquad (3.50)$$

Let $z_t = tf(z_t) + (1-t)T_r^{A,B}z_t$, $t \in (0,1)$. Employing Theorem 2.1, we have $z_t \to P_S f(z) = z$ as $t \to 0$. So, we obtain

$$\begin{aligned}\|z_t - x_{n_k}\|^2 &= \|t(f(z_t) - x_{n_k}) + (1-t)(T_r^{A,B}z_t - x_{n_k})\|^2 \\ &\le (1-t)^2\|T_r^{A,B}z_t - x_{n_k}\|^2 + 2t\langle f(z_t) - x_{n_k}, z_t - x_{n_k}\rangle \\ &= (1-t)^2\|T_r^{A,B}z_t - x_{n_k}\|^2 + 2t\langle f(z_t) - z_t, z_t - x_{n_k}\rangle + 2t\|z_t - x_{n_k}\|^2 \\ &\le (1-t)^2\big(\|T_r^{A,B}z_t - T_r^{A,B}y_{n_k}\| + \|T_r^{A,B}y_{n_k} - x_{n_k}\|\big)^2 \\ &\quad + 2t\langle f(z_t) - z_t, z_t - x_{n_k}\rangle + 2t\|z_t - x_{n_k}\|^2 \\ &\le (1-t)^2\big(\|z_t - y_{n_k}\| + \|T_r^{A,B}y_{n_k} - x_{n_k}\|\big)^2 \\ &\quad + 2t\langle f(z_t) - z_t, z_t - x_{n_k}\rangle + 2t\|z_t - x_{n_k}\|^2 \\ &\le (1-t)^2\big(\|z_t - x_{n_k}\| + \theta_{n_k}\|x_{n_k} - x_{n_k-1}\| + \|T_r^{A,B}y_{n_k} - x_{n_k}\|\big)^2 \\ &\quad + 2t\langle f(z_t) - z_t, z_t - x_{n_k}\rangle + 2t\|z_t - x_{n_k}\|^2. \qquad (3.51)\end{aligned}$$

This shows that

$$\begin{aligned}\langle z_t - f(z_t), z_t - x_{n_k}\rangle \le \frac{(1-t)^2}{2t}\big(&\|z_t - x_{n_k}\| + \theta_{n_k}\|x_{n_k} - x_{n_k-1}\| \\ &+ \|T_r^{A,B}y_{n_k} - x_{n_k}\|\big)^2 + \frac{(2t-1)}{2t}\|z_t - x_{n_k}\|^2. \qquad (3.52)\end{aligned}$$

From condition (i), (3.50) and (3.52), we obtain

$$\begin{aligned}\limsup_{k\to\infty}\langle z_t - f(z_t), z_t - x_{n_k}\rangle &\le \frac{(1-t)^2}{2t}M^2 + \frac{(2t-1)}{2t}M^2 \\ &= \frac{t}{2}M^2 \qquad (3.53)\end{aligned}$$

for some $M > 0$ large enough. Taking $t \to 0$ in (3.25), we obtain

$$\limsup_{k\to\infty}\langle z - f(z), z - x_{n_k}\rangle \le 0. \qquad (3.54)$$

On the other hand, we have

$$\begin{aligned}\|x_{n_k+1} - x_{n_k}\| &\le \alpha_{n_k}\|f(y_{n_k}) - x_{n_k}\| + \gamma_{n_k}\|T_{n_k}y_{n_k} - x_{n_k}\| \\ &\le \alpha_{n_k}\|f(y_{n_k}) - x_{n_k}\| + \gamma_{n_k}\|y_{n_k} - x_{n_k}\| + \gamma_{n_k}\|T_{n_k}y_{n_k} - y_{n_k}\| \\ &\le \alpha_{n_k}\|f(y_{n_k}) - x_{n_k}\| + \gamma_{n_k}\theta_{n_k}\|x_{n_k} - x_{n_k-1}\| + \gamma_{n_k}\|T_{n_k}y_{n_k} - y_{n_k}\| \\ &\qquad (3.55)\end{aligned}$$

By (i), (ii), (3.45) and (3.55), we see that

$$\lim_{k\to\infty} \|x_{n_k+1} - x_{n_k}\| = 0. \tag{3.56}$$

Combining (3.54) and (3.56), we get that

$$\limsup_{k\to\infty} \langle z - f(z), z - x_{n_k+1}\rangle \leq 0. \tag{3.57}$$

It also follows from (i) that $\limsup_{k\to\infty} \tau_{n_k} \leq 0$. We conclude that $\lim_{n\to\infty} s_n = 0$ by Lemma 2.5. Hence $x_n \to z$ as $n \to \infty$. We thus complete the proof. □

Remark 3.3. We remark here that the condition (i) is easily implemented in numerical computation since the valued of $\|x_n - x_{n-1}\|$ is known before choosing θ_n. Indeed, the parameter θ_n can be chosen such that $0 \leq \theta_n \leq \bar{\theta}_n$, where

$$\bar{\theta}_n = \begin{cases} \min\left\{\frac{\omega_n}{\|x_n - x_{n-1}\|}, \theta\right\} & \textit{if } x_n \neq x_{n-1}, \\ \theta & \textit{otherwise}, \end{cases}$$

where $\{\omega_n\}$ is a positive sequence such that $\omega_n = o(\alpha_n)$.

4 Linear System for a Meaningful Least Squares Solution

In this section, we apply our main result to solve the unconstrained linear system $Ax = b$, where $A : H \to H$ is a bounded linear operator and $b \in H$ is fixed in a Hilbert space H. There are many physical problems whose mathematical models turn out to be numerical linear systems $Ax = b$, for example, statistical problems giving rise to multivariate linear/multiple regression models are encountered in many real-world problems such as weather forecasting, psychological research, and business management. The equations in a system will be in general, contradictory to a varying extent. Consequently, there will be no solution that will satisfy the linear system. If the system happens to be consistent (noncontradictory), then there will always be a solution which will satisfy all the equations in the system. If we attempt to find a least-squares solution x_l of such a consistent system, it will always be true solution of the system and sum of the squares of the residuals $\|Ax_l - b\|^2$ will always be a numerical zero (see in [16]).

A least-squares solution of a linear system $Ax = b$ always exist and can be easily computed just by computing the true solution of the ever consistent system $A^T Ax = A^T b$. A least-squares solution x_l is that solution for which the sum of the squares of the residuals such that $\|Ax_l - b\|^2$ is the least, where $\|\cdot\|$ denotes the Euclidean norm. This solution may not be unique. However, one of the possible least-squares solutions x_{ml} such that $\|Ax_{ml} - b\|^2$ and also $\|x_{ml}\|$ are both minimum out of all possible least-squares solutions x_l which is called the minimum-norm least-squares solution (*mls*). The *mls* x_{ml} is always unique while the solution x_l may not be unique [31].

The split feasibility problem (SFP) [8] is to find a point $\hat{x}$ such that

$$\hat{x} \in C, \quad T\hat{x} \in Q,$$

where C and Q are, respectively, closed and convex subsets of Hilbert spaces H_1 and H_2, and $A : H_1 \to H_2$ is a bounded linear operator with its adjoint A^*. The solution set of the SFP will be denoted by Ω. For solving the SFP, Byrne [6] introduced the following CQ algorithm:

$$x_{n+1} = P_C(x_n - \lambda A^*(I - P_Q)Ax_n),$$

where $0 < \lambda < 2\alpha$ with $\alpha = 1/\|A\|^2$. Here $\|A\|^2$ is the spectral radius of A^*A. We know that $A^*(I - P_Q)A$ is $1/\|A\|^2$-inverse strongly monotone [7]. If we set $x_{n+1} = x_n - \lambda A^t(Ax_n - b)$, then the SFP includes as special case the linear inverse problem $Ax = b$. So we obtain the following results.

Theorem 4.1. *Let H_1 and H_2 be real Hilbert spaces and let $f : H_1 \to H_1$ be a contraction mapping. Let $A : H_1 \to H_2$ be a bounded linear operator and $b \in H_2$ such that $\Omega \neq \emptyset$. Let $\{x_n\}$ be a sequence generated by $x_0, x_1 \in H_1$ and*

$$\begin{cases} y_n = x_n + \theta_n(x_n - x_{n-1}), \\ x_{n+1} = \alpha_n f(y_n) + \beta_n y_n + \gamma_n(y_n - r_n A^t(Ay_n - b)), \quad n \geq 1, \end{cases} \tag{4.1}$$

where $0 < r_n \leq 2\alpha$, $\{\theta_n\} \subset [0, \theta]$ with $\theta \in [0, 1)$ and $\{\alpha_n\}$, $\{\beta_n\}$ and $\{\gamma_n\}$ are sequences in $(0, 1)$ with $\alpha_n + \beta_n + \gamma_n = 1$. Assume that the following conditions hold:

(i) $\lim_{n\to\infty} \frac{\theta_n\|x_n - x_{n-1}\|}{\alpha_n} = 0$;
(ii) $\lim_{n\to\infty} \alpha_n = 0$, $\sum_{n=1}^{\infty} \alpha_n = \infty$;
(iii) $0 < \liminf_{n\to\infty} r_n \leq \limsup_{n\to\infty} r_n < \frac{2}{\|A\|^2}$;
(iv) $\liminf_{n\to\infty} \gamma_n > 0$.

Then the sequence $\{x_n\}$ converges strongly to $z = P_\Omega f(z)$.

Theorem 4.2. *Let H_1 and H_2 be real Hilbert spaces and let $f : H_1 \to H_1$ be a contraction mapping. Let $A : H_1 \to H_2$ be a bounded linear operator and $b \in H_2$ such that $\Omega \neq \emptyset$. Let $\{x_n\}$ be a sequence generated by $x_0, x_1 \in H_1$ and*

$$\begin{cases} y_n = x_n + \theta_n(x_n - x_{n-1}), \\ x_{n+1} = \alpha_n f(x_n) + \beta_n x_n + \gamma_n(y_n - r_n A^t(Ay_n - b)), \quad n \geq 1, \end{cases} \tag{4.2}$$

where $0 < r_n \leq 2\alpha$, $\{\theta_n\} \subset [0, \theta]$ with $\theta \in [0, 1)$ and $\{\alpha_n\}$, $\{\beta_n\}$ and $\{\gamma_n\}$ are sequences in $(0, 1)$ with $\alpha_n + \beta_n + \gamma_n = 1$. Assume that the following conditions hold:

(i) $\lim_{n\to\infty} \frac{\theta_n\|x_n - x_{n-1}\|}{\alpha_n} = 0$;
(ii) $\lim_{n\to\infty} \alpha_n = 0$, $\sum_{n=1}^{\infty} \alpha_n = \infty$;
(iii) $0 < \liminf_{n\to\infty} r_n \leq \limsup_{n\to\infty} r_n < \frac{2}{\|A\|^2}$;
(iv) $\liminf_{n\to\infty} \gamma_n > 0$.

Then the sequence $\{x_n\}$ converges strongly to $z = P_\Omega f(z)$.

We now give an example in Euclidean space $\mathbb{R}^m$ to support Theorems 4.1 and 4.2. Let $f : \mathbb{R}^m \to \mathbb{R}^m$ be defined by $fx = \frac{x}{2}$ where $x \in \mathbb{R}^m$. We choose $r_n = \frac{1}{\|A\|^2}$, $\alpha_n = \frac{1}{n+1}$, $\beta_n = \frac{1}{n}$ and

$$\theta_n = \begin{cases} \min\left\{\frac{1}{(n+1)^2\|x_n - x_{n-1}\|}, 0.5\right\} & if\ x_n \neq x_{n-1}, \\ 0.5 & otherwise. \end{cases}$$

The different choices of the Initial data x_0 and x_1 are given as follows:

Choice 1:

$$x_0 = \left[0.8147\ 0.9058\ 0.1270\ 0.9134\ 0.6324\ 0.0975\ 0.2785\ 0.5469\ 0.9575\ 0.9649\right]^T,$$

$$x_1 = \left[0.1576\ 0.9706\ 0.9572\ 0.4854\ 0.8003\ 0.1419\ 0.4218\ 0.9157\ 0.7922\ 0.9595\right]^T,$$

Choice 2:

$$x_0 = \left[1.3147\ 1.4058\ 0.6270\ 1.4134\ 1.1324\ 0.5975\ 0.7785\ 1.0469\ 1.4575\ 1.4649\right]^T,$$

$$x_1 = \left[0.9576\ 1.7706\ 1.7572\ 1.2854\ 1.6003\ 0.9419\ 1.2218\ 1.7157\ 1.5922\ 1.7595\right]^T.$$

We provide a numerical test of a comparison between the iteration 4.1 and the iteration 4.2 in solving linear system $Ax = b$ where $A \in \mathbb{R}^{m\times n}$, $x \in \mathbb{R}^{n\times 1}$ and $b \in \mathbb{R}^{m\times 1}$. The stopping criterion consists of Cauchy sequence norm denoted as $C_n = \|x_{n+1} - x_n\| < 10^{-4}$. The solution is obtained by using the inertial forward-backward method and the standard forward-backward method is denoted by x_l.

In case of A is square and nonsingular matrix, the solution of linear system $Ax = b$ is $x_{ml} = A^{-1}b$. If A is not square matrix or is square and singular square matrix, the solution of linear system $Ax = b$ is denoted by $x_{ml} = \text{pinv}(A) * b$ (the minimum least squares solution computed by using **pinv** function in matlab).

Example 4.3. Let $A \in \mathbb{R}^{10\times 10}$, $b \in \mathbb{R}^{10\times 1}$, Rank$[A] = 10$ and Rank$[A, b] = 10$ where

$$A = \begin{bmatrix} 0.0587 & 0.4012 & 0.8330 & 0.1351 & 0.7731 & 0.2922 & 0.4129 & 0.8412 & 0.6155 & 0.0591 \\ 0.8136 & 0.7519 & 0.4364 & 0.9385 & 0.3202 & 0.9170 & 0.9641 & 0.7796 & 0.1983 & 0.2313 \\ 0.3457 & 0.0269 & 0.6357 & 0.4679 & 0.5535 & 0.5048 & 0.5991 & 0.5837 & 0.8648 & 0.9354 \\ 0.8289 & 0.3746 & 0.6651 & 0.9332 & 0.3801 & 0.5792 & 0.9413 & 0.5190 & 0.9282 & 0.8067 \\ 0.7250 & 0.1867 & 0.3558 & 0.9732 & 0.7502 & 0.0517 & 0.7610 & 0.2991 & 0.8473 & 0.0381 \\ 0.3373 & 0.3720 & 0.3185 & 0.2844 & 0.9851 & 0.2668 & 0.1109 & 0.6902 & 0.0388 & 0.9681 \\ 0.2115 & 0.3308 & 0.6898 & 0.5388 & 0.4471 & 0.7829 & 0.2851 & 0.6279 & 0.5272 & 0.9300 \\ 0.3398 & 0.0185 & 0.0263 & 0.0126 & 0.0746 & 0.5380 & 0.2126 & 0.7764 & 0.3638 & 0.0459 \\ 0.4593 & 0.6852 & 0.9399 & 0.1572 & 0.1004 & 0.9398 & 0.4148 & 0.8680 & 0.0701 & 0.3295 \\ 0.5287 & 0.6723 & 0.6265 & 0.3516 & 0.3362 & 0.5376 & 0.5017 & 0.0264 & 0.2483 & 0.7165 \end{bmatrix},$$

$$b = \left[4.6485\ 3.8201\ 5.5270\ 4.7925\ 4.7205\ 5.4100\ 5.2035\ 6.0115\ 4.7023\ 5.0606\right]^T.$$

Example 4.4. Let $A \in \mathbb{R}^{10\times 10}$, $b \in \mathbb{R}^{10\times 1}$, Rank$[A] = 9$ and Rank$[A, b] = 10$ where

$$A = \begin{bmatrix} 0.0587 & 0.4012 & 0.8330 & 0.1351 & 0.7731 & 0.2922 & 0.4129 & 0.8412 & 0.6155 & 0.0591 \\ 0.8136 & 0.7519 & 0.4364 & 0.9385 & 0.3202 & 0.9170 & 0.9641 & 0.7796 & 0.1983 & 0.2313 \\ 0.3457 & 0.0269 & 0.6357 & 0.4679 & 0.5535 & 0.5048 & 0.5991 & 0.5837 & 0.8648 & 0.9354 \\ 0.8289 & 0.3746 & 0.6651 & 0.9332 & 0.3801 & 0.5792 & 0.9413 & 0.5190 & 0.9282 & 0.8067 \\ 0.7250 & 0.1867 & 0.3558 & 0.9732 & 0.7502 & 0.0517 & 0.7610 & 0.2991 & 0.8473 & 0.0381 \\ 0.3373 & 0.3720 & 0.3185 & 0.2844 & 0.9851 & 0.2668 & 0.1109 & 0.6902 & 0.0388 & 0.9681 \\ 0.2115 & 0.3308 & 0.6898 & 0.5388 & 0.4471 & 0.7829 & 0.2851 & 0.6279 & 0.5272 & 0.9300 \\ 0.3398 & 0.0185 & 0.0263 & 0.0126 & 0.0746 & 0.5380 & 0.2126 & 0.7764 & 0.3638 & 0.0459 \\ 0.4593 & 0.6852 & 0.9399 & 0.1572 & 0.1004 & 0.9398 & 0.4148 & 0.8680 & 0.0701 & 0.3295 \\ 5.1687 & 3.7084 & 2.8889 & 5.7529 & 1.5400 & 4.2096 & 5.6390 & 3.4225 & 2.8375 & 2.3305 \end{bmatrix},$$

$$b = \begin{bmatrix} 4.6485 & 3.8201 & 5.5270 & 4.7925 & 4.7205 & 5.4100 & 5.2035 & 6.0115 & 4.7023 & 5.0606 \end{bmatrix}^T.$$

Example 4.5. Let $A \in \mathbb{R}^{10\times 9}$, $b \in \mathbb{R}^{10\times 1}$, Rank$[A] = 8$ and Rank$[A, b] = 9$ where

$$A = \begin{bmatrix} 0.0587 & 0.4012 & 0.8330 & 0.1351 & 0.7731 & 0.2922 & 0.4129 & 0.8412 & 0.6155 \\ 0.8136 & 0.7519 & 0.4364 & 0.9385 & 0.3202 & 0.9170 & 0.9641 & 0.7796 & 0.1983 \\ 0.3457 & 0.0269 & 0.6357 & 0.4679 & 0.5535 & 0.5048 & 0.5991 & 0.5837 & 0.8648 \\ 0.8289 & 0.3746 & 0.6651 & 0.9332 & 0.3801 & 0.5792 & 0.9413 & 0.5190 & 0.9282 \\ 0.7250 & 0.1867 & 0.3558 & 0.9732 & 0.7502 & 0.0517 & 0.7610 & 0.2991 & 0.8473 \\ 0.3373 & 0.3720 & 0.3185 & 0.2844 & 0.9851 & 0.2668 & 0.1109 & 0.6902 & 0.0388 \\ 0.2115 & 0.3308 & 0.6898 & 0.5388 & 0.4471 & 0.7829 & 0.2851 & 0.6279 & 0.5272 \\ 0.3398 & 0.0185 & 0.0263 & 0.0126 & 0.0746 & 0.5380 & 0.2126 & 0.7764 & 0.3638 \\ 1.7152 & 2.1056 & 2.1223 & 2.0797 & 1.8001 & 2.2723 & 2.5475 & 2.8210 & 1.3199 \\ 5.1687 & 3.7084 & 2.8889 & 5.7529 & 1.5400 & 4.2096 & 5.6390 & 3.4225 & 2.8375 \end{bmatrix},$$

$$b = \begin{bmatrix} 4.6485 & 3.8201 & 5.5270 & 4.7925 & 4.7205 & 5.4100 & 5.2035 & 6.0115 & 21.2344 & 5.0606 \end{bmatrix}^T.$$

Example 4.6. Let $A \in \mathbb{R}^{10\times 9}$, $b \in \mathbb{R}^{10\times 1}$, Rank$[A] = 8$ and Rank$[A, b] = 8$ where

$$A = \begin{bmatrix} 0.0587 & 0.4012 & 0.8330 & 0.1351 & 0.7731 & 0.2922 & 0.4129 & 0.8412 & 0.6155 \\ 0.8136 & 0.7519 & 0.4364 & 0.9385 & 0.3202 & 0.9170 & 0.9641 & 0.7796 & 0.1983 \\ 0.3457 & 0.0269 & 0.6357 & 0.4679 & 0.5535 & 0.5048 & 0.5991 & 0.5837 & 0.8648 \\ 0.8289 & 0.3746 & 0.6651 & 0.9332 & 0.3801 & 0.5792 & 0.9413 & 0.5190 & 0.9282 \\ 0.7250 & 0.1867 & 0.3558 & 0.9732 & 0.7502 & 0.0517 & 0.7610 & 0.2991 & 0.8473 \\ 0.3373 & 0.3720 & 0.3185 & 0.2844 & 0.9851 & 0.2668 & 0.1109 & 0.6902 & 0.0388 \\ 0.2115 & 0.3308 & 0.6898 & 0.5388 & 0.4471 & 0.7829 & 0.2851 & 0.6279 & 0.5272 \\ 0.3398 & 0.0185 & 0.0263 & 0.0126 & 0.0746 & 0.5380 & 0.2126 & 0.7764 & 0.3638 \\ 1.7152 & 2.1056 & 2.1223 & 2.0797 & 1.8001 & 2.2723 & 2.5475 & 2.8210 & 1.3199 \\ 5.1687 & 3.7084 & 2.8889 & 5.7529 & 1.5400 & 4.2096 & 5.6390 & 3.4225 & 2.8375 \end{bmatrix},$$

$$b = \begin{bmatrix} 4.6485 & 3.8201 & 5.5270 & 4.7925 & 4.7205 & 5.4100 & 5.2035 & 6.0115 & 14.6130 & 21.2344 \end{bmatrix}^T.$$

Example 4.7. Let $A \in \mathbb{R}^{9\times 10}$, $b \in \mathbb{R}^{9\times 1}$, Rank$[A] = 9$ and Rank$[A, b] = 9$ where

$$A = \begin{bmatrix} 0.0587 & 0.4012 & 0.8330 & 0.1351 & 0.7731 & 0.2922 & 0.4129 & 0.8412 & 0.6155 & 0.0591 \\ 0.8136 & 0.7519 & 0.4364 & 0.9385 & 0.3202 & 0.9170 & 0.9641 & 0.7796 & 0.1983 & 0.2313 \\ 0.3457 & 0.0269 & 0.6357 & 0.4679 & 0.5535 & 0.5048 & 0.5991 & 0.5837 & 0.8648 & 0.9354 \\ 0.8289 & 0.3746 & 0.6651 & 0.9332 & 0.3801 & 0.5792 & 0.9413 & 0.5190 & 0.9282 & 0.8067 \\ 0.7250 & 0.1867 & 0.3558 & 0.9732 & 0.7502 & 0.0517 & 0.7610 & 0.2991 & 0.8473 & 0.0381 \\ 0.3373 & 0.3720 & 0.3185 & 0.2844 & 0.9851 & 0.2668 & 0.1109 & 0.6902 & 0.0388 & 0.9681 \\ 0.2115 & 0.3308 & 0.6898 & 0.5388 & 0.4471 & 0.7829 & 0.2851 & 0.6279 & 0.5272 & 0.9300 \\ 0.3398 & 0.0185 & 0.0263 & 0.0126 & 0.0746 & 0.5380 & 0.2126 & 0.7764 & 0.3638 & 0.0459 \\ 0.4593 & 0.6852 & 0.9399 & 0.1572 & 0.1004 & 0.9398 & 0.4148 & 0.8680 & 0.0701 & 0.3295 \end{bmatrix},$$

$$b = \left[4.6485\ 3.8201\ 5.5270\ 4.7925\ 4.7205\ 5.4100\ 5.2035\ 6.0115\ 4.7023\right]^T.$$

Example 4.8. Let $A \in \mathbb{R}^{9\times 10}$, $b \in \mathbb{R}^{9\times 1}$, Rank$[A] = 8$ and Rank$[A, b] = 8$ where

$$A = \begin{bmatrix} 0.0587 & 0.4012 & 0.8330 & 0.1351 & 0.7731 & 0.2922 & 0.4129 & 0.8412 & 0.6155 & 0.0591 \\ 0.8136 & 0.7519 & 0.4364 & 0.9385 & 0.3202 & 0.9170 & 0.9641 & 0.7796 & 0.1983 & 0.2313 \\ 0.3457 & 0.0269 & 0.6357 & 0.4679 & 0.5535 & 0.5048 & 0.5991 & 0.5837 & 0.8648 & 0.9354 \\ 0.8289 & 0.3746 & 0.6651 & 0.9332 & 0.3801 & 0.5792 & 0.9413 & 0.5190 & 0.9282 & 0.8067 \\ 0.7250 & 0.1867 & 0.3558 & 0.9732 & 0.7502 & 0.0517 & 0.7610 & 0.2991 & 0.8473 & 0.0381 \\ 0.3373 & 0.3720 & 0.3185 & 0.2844 & 0.9851 & 0.2668 & 0.1109 & 0.6902 & 0.0388 & 0.9681 \\ 0.2115 & 0.3308 & 0.6898 & 0.5388 & 0.4471 & 0.7829 & 0.2851 & 0.6279 & 0.5272 & 0.9300 \\ 0.3398 & 0.0185 & 0.0263 & 0.0126 & 0.0746 & 0.5380 & 0.2126 & 0.7764 & 0.3638 & 0.0459 \\ 5.1687 & 3.7084 & 2.8889 & 5.7529 & 1.5400 & 4.2096 & 5.6390 & 3.4225 & 2.8375 & 2.3305 \end{bmatrix},$$

$$b = \left[4.6485\ 3.8201\ 5.5270\ 4.7925\ 4.7205\ 5.4100\ 5.2035\ 6.0115\ 21.2344\right]^T.$$

Each examples have 4 figures which consists of:

1. The error plotting C_n with $\theta_n \neq 0$ and $\theta_n = 0$ for the iteration 4.1 of choice 1 initial data.
2. The error plotting C_n with $\theta_n \neq 0$ and $\theta_n = 0$ for the iteration 4.1 of choice 1 initial data.
3. The error plotting C_n with $\theta_n \neq 0$ for iterations 4.1 and 4.2 of choice 1 initial data.
4. The error plotting C_n with $\theta_n = 0$ for iterations 4.1 and 4.2 of choice 1 initial data.

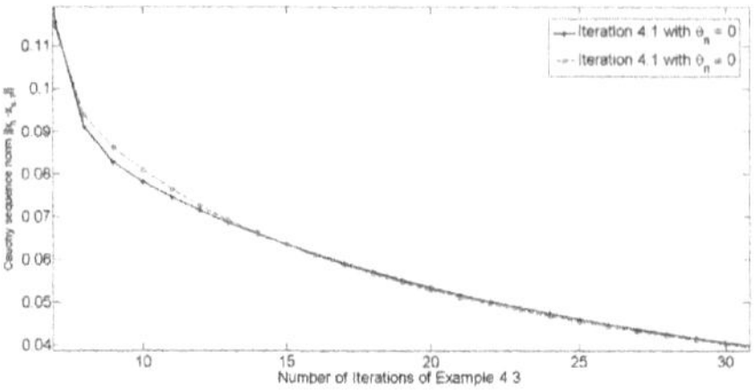

Fig. 1. The error plotting C_n with $\theta_n \neq 0$ and $\theta_n = 0$ for iteration 4.1 of choice 1 initial data

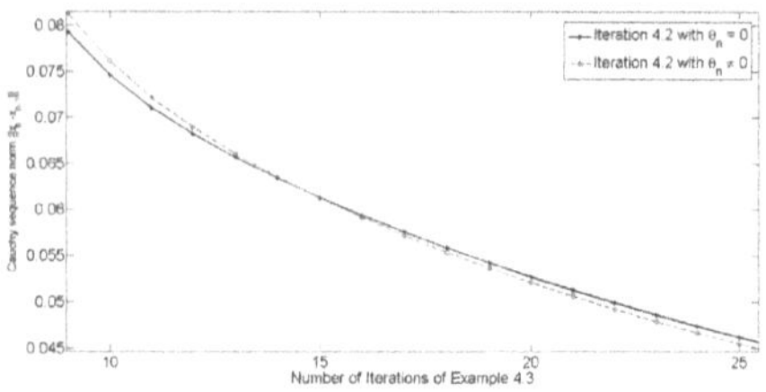

Fig. 2. The error plotting C_n with $\theta_n \neq 0$ and $\theta_n = 0$ for iteration 4.2 of choice 1 initial data

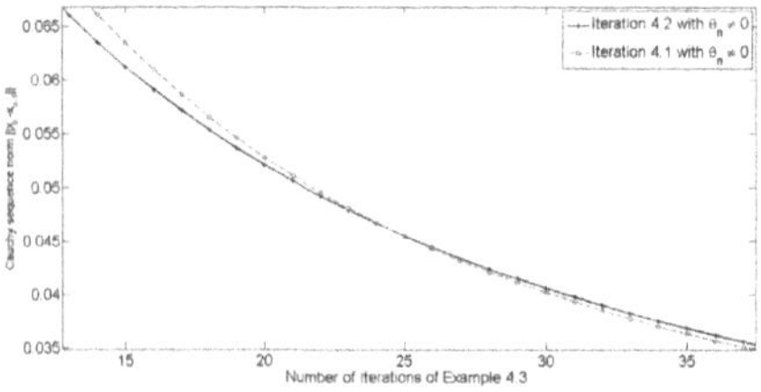

Fig. 3. The error plotting C_n with $\theta_n \neq 0$ for iterations 4.1 and 4.2 of choice 1 initial data

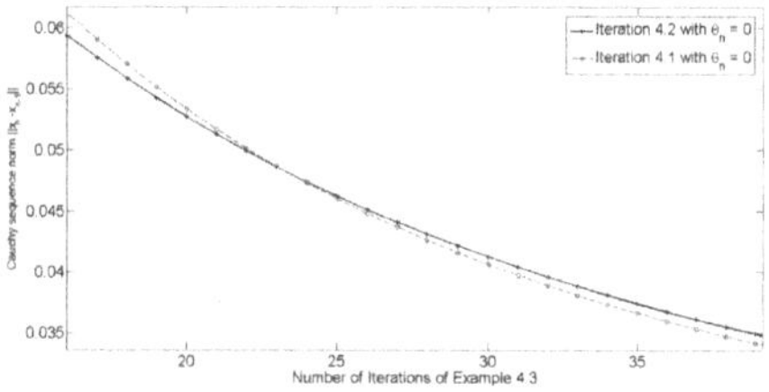

Fig. 4. The error plotting C_n with $\theta_n = 0$ for iterations 4.1 and 4.2 of choice 1 initial data

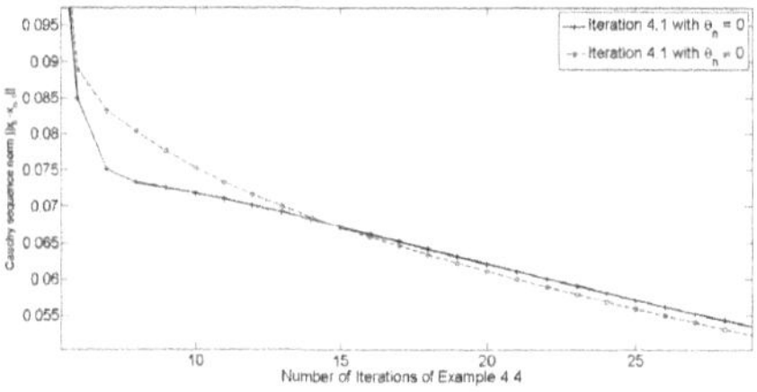

Fig. 5. The error plotting C_n with $\theta_n \neq 0$ and $\theta_n = 0$ for iteration 4.1 of choice 1 initial data

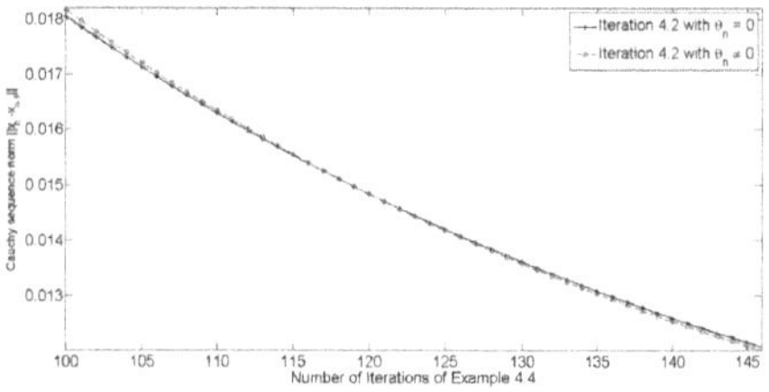

Fig. 6. The error plotting C_n with $\theta_n \neq 0$ and $\theta_n = 0$ for iteration 4.2 of choice 1 initial data

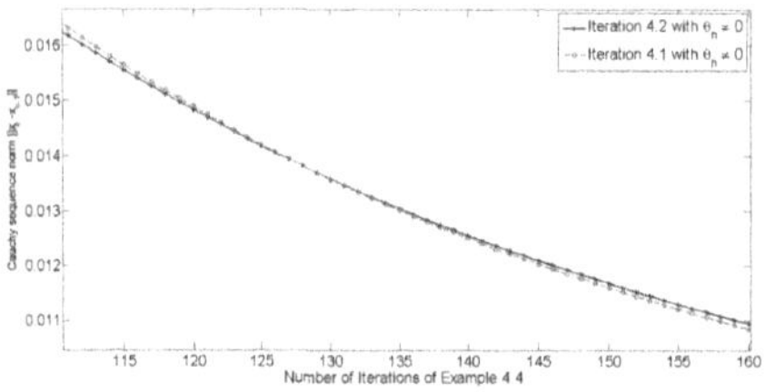

Fig. 7. The error plotting C_n with $\theta_n \neq 0$ for iterations 4.1 and 4.2 of choice 1 initial data

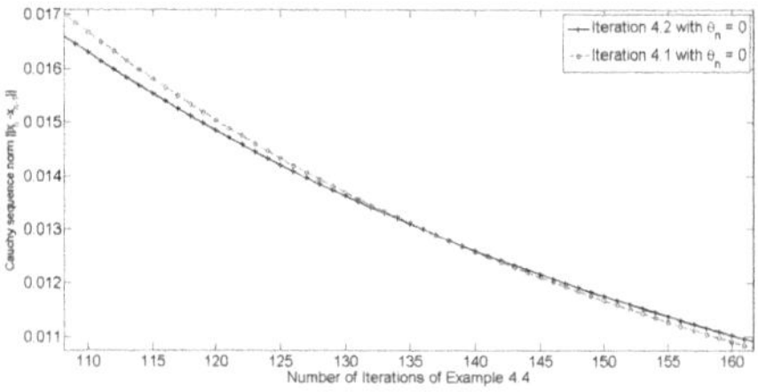

Fig. 8. The error plotting C_n with $\theta_n = 0$ for iterations 4.1 and 4.2 of choice 1 initial data

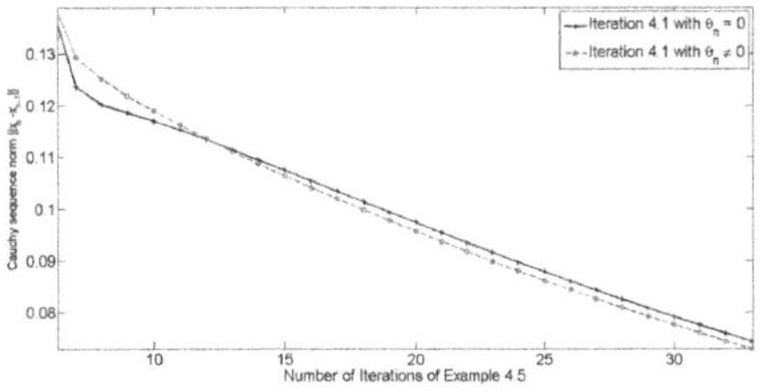

Fig. 9. The error plotting C_n with $\theta_n \neq 0$ and $\theta_n = 0$ for iteration 4.1 of choice 1 initial data

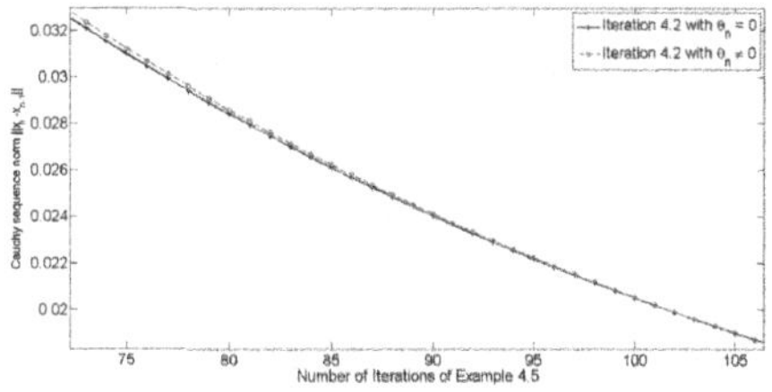

Fig. 10. The error plotting C_n with $\theta_n \neq 0$ and $\theta_n = 0$ for iteration 4.2 of choice 1 initial data

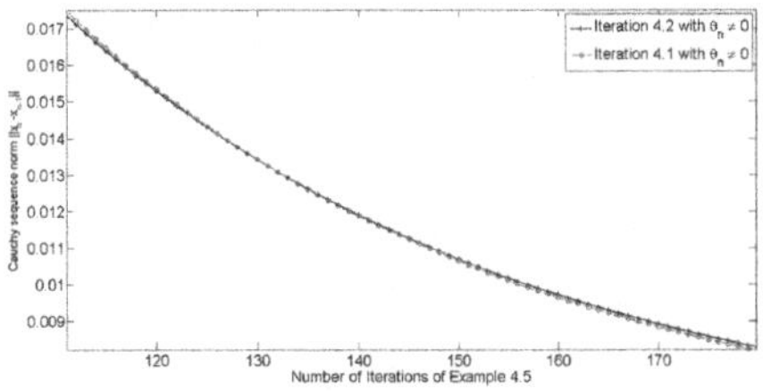

Fig. 11. The error plotting C_n with $\theta_n \neq 0$ for iterations 4.1 and 4.2 of choice 1 initial data

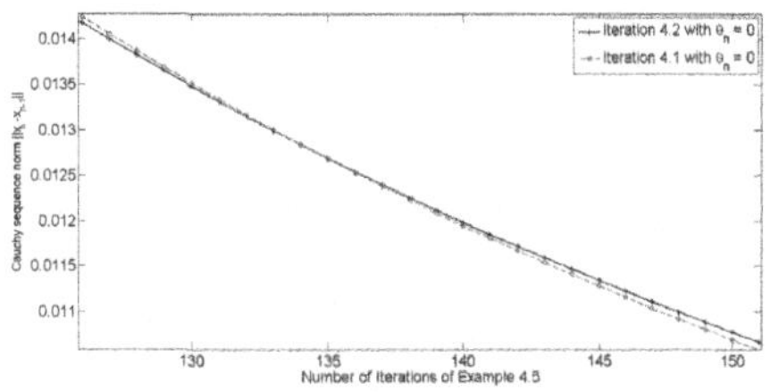

Fig. 12. The error plotting C_n with $\theta_n = 0$ for iterations 4.1 and 4.2 of choice 1 initial data

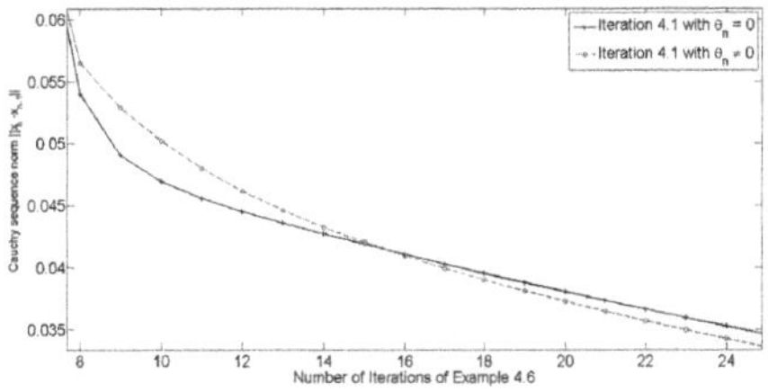

Fig. 13. The error plotting C_n with $\theta_n \neq 0$ and $\theta_n = 0$ for iteration 4.1 of choice 1 initial data

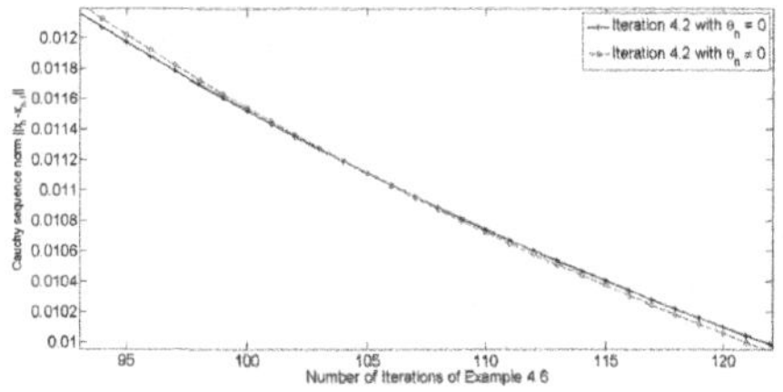

Fig. 14. The error plotting C_n with $\theta_n \neq 0$ and $\theta_n = 0$ for iteration 4.2 of choice 1 initial data

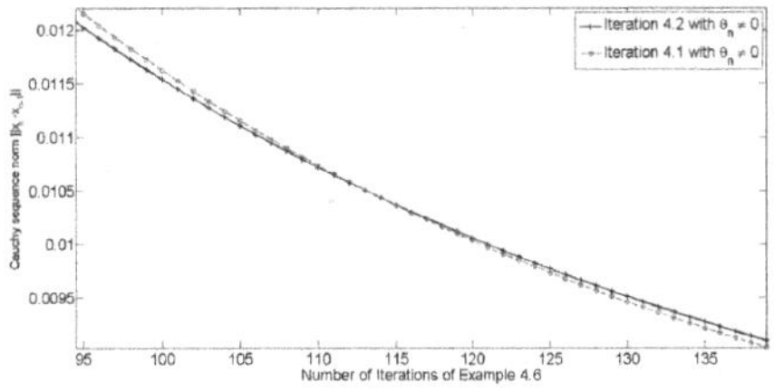

Fig. 15. The error plotting C_n with $\theta_n \neq 0$ for iterations 4.1 and 4.2 of choice 1 initial data

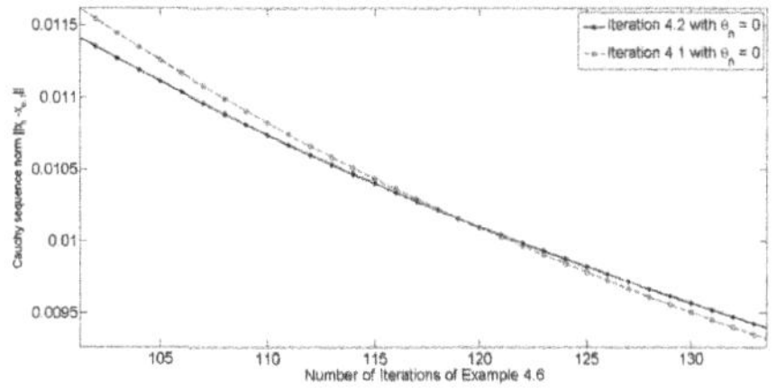

Fig. 16. The error plotting C_n with $\theta_n = 0$ for iterations 4.1 and 4.2 of choice 1 initial data

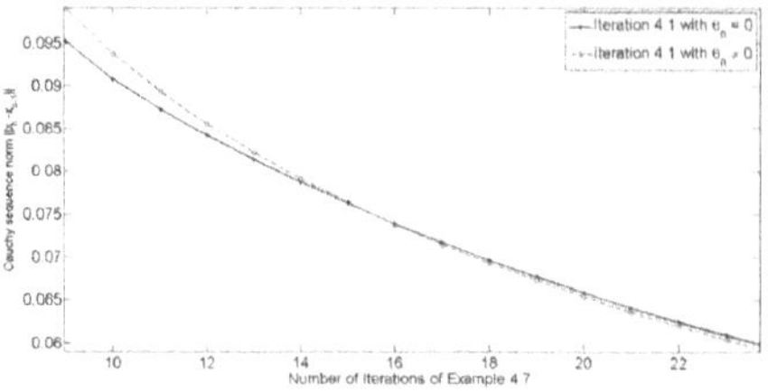

Fig. 17. The error plotting C_n with $\theta_n \neq 0$ and $\theta_n = 0$ for iteration 4.1 of choice 1 initial data

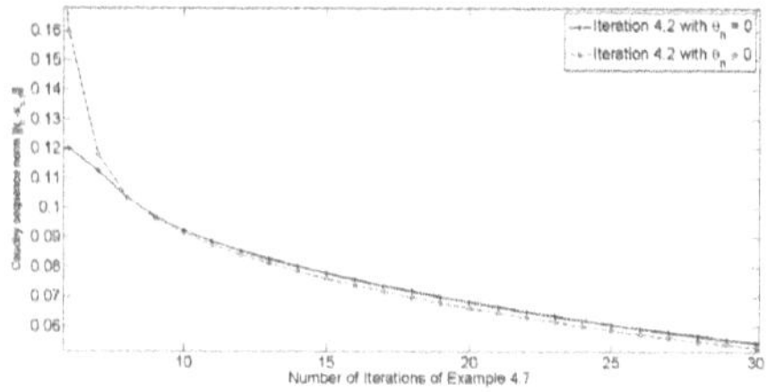

Fig. 18. The error plotting C_n with $\theta_n \neq 0$ and $\theta_n = 0$ for iteration 4.2 of choice 1 initial data

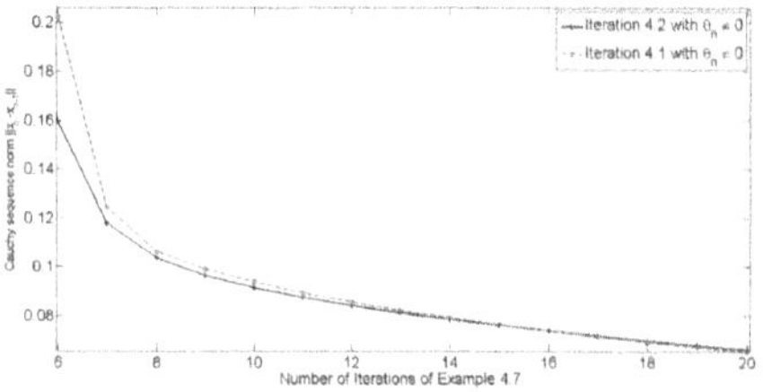

Fig. 19. The error plotting C_n with $\theta_n \neq 0$ for iterations 4.1 and 4.2 of choice 1 initial data

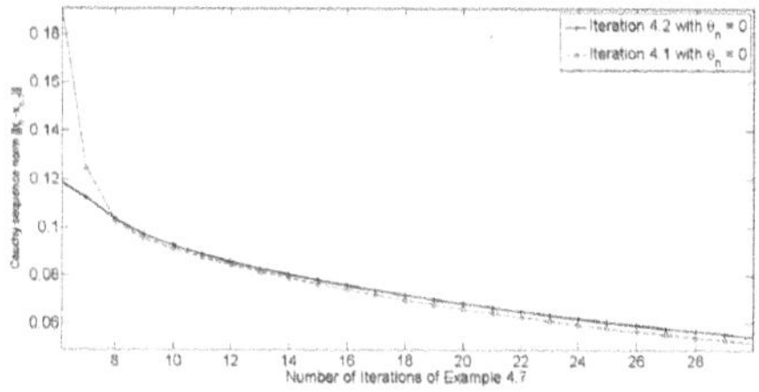

Fig. 20. The error plotting C_n with $\theta_n = 0$ for iterations 4.1 and 4.2 of choice 1 initial data

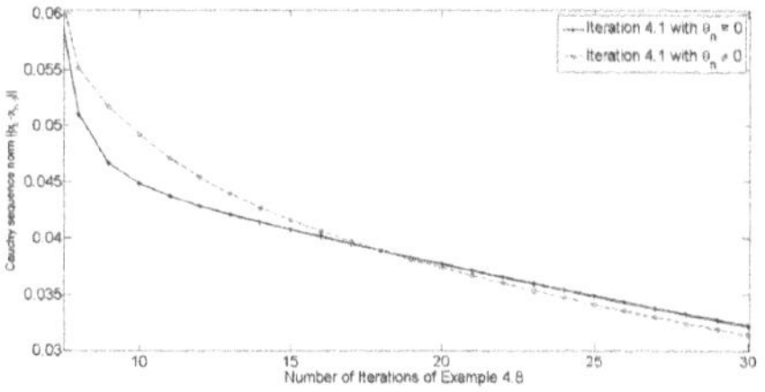

Fig. 21. The error plotting C_n with $\theta_n \neq 0$ and $\theta_n = 0$ for iteration 4.1 of choice 1 initial data

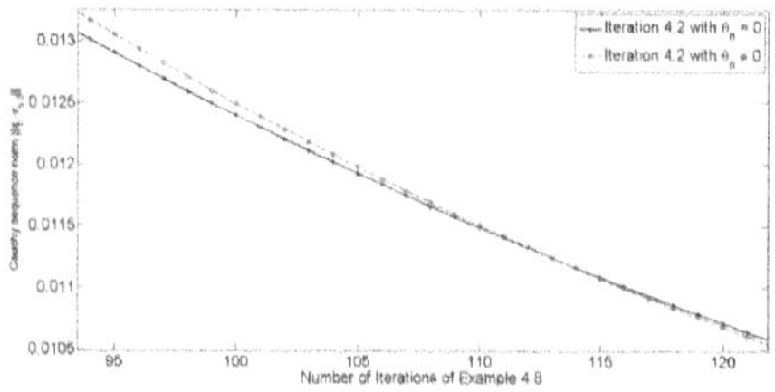

Fig. 22. The error plotting C_n with $\theta_n \neq 0$ and $\theta_n = 0$ for iteration 4.2 of choice 1 initial data

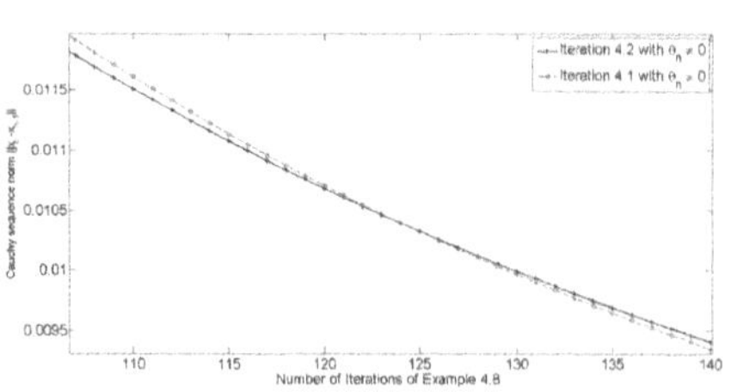

Fig. 23. The error plotting C_n with $\theta_n \neq 0$ for iterations 4.1 and 4.2 of choice 1 initial data

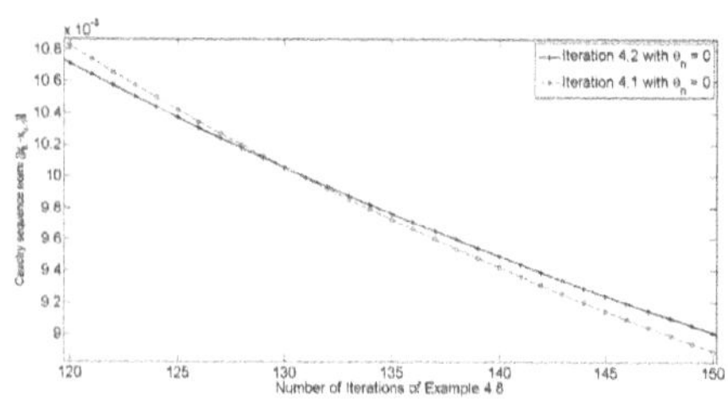

Fig. 24. The error plotting C_n with $\theta_n = 0$ for iterations 4.1 and 4.2 of choice 1 initial data

Table 1. Comparison of number of iterations for $\theta_n \neq 0$ and $\theta_n = 0$ of iterations 4.1 and 4.2 where $||x_n - x_{n-1}|| < 10^{-4}$ for Examples 4.3–4.8 with choice 1 initial data

Ex	Iterations Number of Iteration 4.1		Iterations Number of Iteration 4.2	
	$\theta_n = 0$	$\theta_n \neq 0$	$\theta_n = 0$	$\theta_n \neq 0$
1	9792	9791	9794	9793
2	33165	33164	33168	33167
3	37956	37956	37954	37954
4	8316	8313	8265	8252
5	7205	7204	7211	7209
6	10190	10190	10175	10168

Remark 4.9. 1. From Figs. 1, 2, 5, 6, 9, 10, 13, 14, 17, 18, 21 and 22, our forward-backward methods with the inertial technical term (Iteration 4.1 and 4.2) have a nice convergence speed and require small number of iterations than the standard forward-backward method ($\theta_n = 0$) for each the examples.
2. From Figs. 3, 4, 7, 8, 11, 12, 15, 16, 19, 20, 23 and 24, the iteration 4.1 has a nice convergence speed and require small number of iterations than the iteration 4.2.
3. From Table 1, we see that the number of iterations in each example are similar when it is large enough.

Remark 4.10. From Tables 2, 3, 4, 5 and 6, we see that our iterations converge to the minimum-norm least-squares solution for each of the initials.

Table 2. Solutions for Examples 4.3–4.8 and its norm by using **pinv** matlab function denoted as x_{ml}

Ex	x_{ml}	$\|\|x_{ml}\|\|$
1	[6.8901 −0.6538 −1.7113 −5.2344 6.8697 8.5983 −4.3467 −2.7384 4.8846 −1.5659]	15.8763
2	[6.8845 −15.3855 5.8147 −0.0274 7.8452 8.4295 −4.3174 1.1779 −3.1887 −3.3268]	22.1763
3	[−17.8325 −5.4899 −11.0068 7.7204 5.7341 2.2120 8.2414 10.9732 1.3064]	27.5082
4	[2.1454 −1.5970 −2.2286 0.9105 2.2872 2.0445 −3.6503 4.6407 3.4680]	8.3215
5	[5.3282 −4.2773 1.0133 0.0832 2.8741 3.1305 −4.7064 3.5732 1.7014 −0.8463]	10.2131
6	[2.1968 −1.6254 −2.1892 0.8823 2.3078 2.0769 −3.6648 4.6060 3.4517 −0.0146]	8.3210

Table 3. Comparison of $\theta_n \neq 0$ and $\theta_n = 0$ of the iteration 4.1 with 1×10^5 iterations for Examples 4.3–4.8 with choice 1 initial data

Ex	x_l, $\theta_n = 0$		x_l, $\theta_n \neq 0$	
	x_l	$\|\|x_l\|\|$	x_l	$\|\|x_l\|\|$
1	[6.8709 −0.6179 −1.7182 −5.2214 6.8330 8.5482 −4.3395 −2.7049 4.8913 −1.5470]	15.8124	[6.8709 −0.6179 −1.7182 −5.2214 6.8330 8.5482 −4.3395 −2.7049 4.8913 −1.5470]	15.8124
2	[6.6979 −14.8357 5.5655 −0.0538 7.6288 8.1547 −4.3914 1.3388 −2.8803 −3.1993]	21.4533	[6.6979 −14.8358 5.5655 −0.0537 7.6288 8.1546 −4.3914 1.3389 −2.8803 −3.1993]	21.4533
3	[−17.1161 −5.5685 −10.4967 7.1585 5.8085 2.2930 8.1357 10.6619 1.1590]	26.5631	[−17.1160 −5.5686 −10.4966 7.1586 5.8084 2.2929 8.1357 10.6620 1.1589]	26.5631
4	[2.1820 −1.6499 −2.1800 0.8517 2.3164 2.0733 −3.5894 4.5914 3.4196]	8.2634	[2.1820 −1.6500 −2.1799 0.8518 2.3163 2.0733 −3.5894 4.5915 3.4196]	8.2634
5	[5.3073 −4.2375 0.9926 0.0788 2.8584 3.1101 −4.6935 3.5835 1.7177 −0.8343]	10.1723	[5.3073 −4.2375 0.9927 0.0788 2.8584 3.1100 −4.6935 3.5836 1.7177 −0.8343]	10.1723
6	[2.1696 −1.6502 −2.1844 0.8603 2.3198 2.0774 −3.5889 4.5965 3.4195 −0.0054]	8.2668	[2.1696 −1.6503 −2.1843 0.8604 2.3197 2.0774 −3.5890 4.5965 3.4194 −0.0054]	8.2668

Table 4. Comparison of $\theta_n \neq 0$ and $\theta_n = 0$ of the iteration 4.1 with 1×10^5 iterations for Examples 4.3–4.8 with choice 2 initial data

Ex	x_l, $\theta_n = 0$		x_l, $\theta_n \neq 0$	
	x_l	$\|\|x_l\|\|$	x_l	$\|\|x_l\|\|$
1	[6.8709 −0.6179 −1.7182 −5.2214 6.8330 8.5482 −4.3395 −2.7049 4.8913 −1.5470]	15.8124	[6.8709 −0.6179 −1.7182 −5.2214 6.8330 8.5482 −4.3395 −2.7049 4.8913 −1.5470]	15.8124
2	[6.6979 −14.8357 5.5655 −0.0538 7.6288 8.1547 −4.3914 1.3388 −2.8803 −3.1993]	21.4533	[6.6979 −14.8357 5.5655 −0.0538 7.6288 8.1547 −4.3914 1.3388 −2.8803 −3.1993]	21.4533
3	[−17.1161 −5.5685 −10.4967 7.1585 5.8085 2.2930 8.1357 10.6619 1.1590]	26.5631	[−17.1161 −5.5686 −10.4967 7.1586 5.8084 2.2929 8.1357 10.6619 1.1589]	26.5631
4	[2.1820 −1.6499 −2.1800 0.8517 2.3164 2.0733 −3.5894 4.5914 3.4196]	8.2634	[2.1820 −1.6499 −2.1800 0.8518 2.3164 2.0733 −3.5894 4.5915 3.4196]	8.2634
5	[5.3073 −4.2375 0.9926 0.0788 2.8584 3.1101 −4.6935 3.5835 1.7177 −0.8343]	10.1723	[5.3073 −4.2375 0.9926 0.0788 2.8584 3.1100 −4.6935 3.5836 1.7177 −0.8343]	10.1723
6	[2.1696 −1.6502 −2.1844 0.8603 2.3198 2.0774 −3.5889 4.5965 3.4195 −0.0054]	8.2668	[2.1696 −1.6503 −2.1844 0.8604 2.3197 2.0774 −3.5890 4.5965 3.4194 −0.0054]	8.2668

Table 5. Comparison of $\theta_n \neq 0$ and $\theta_n = 0$ of the iteration 4.1 with 1×10^5 iterations for Examples 4.3–4.8 with choice 1 initial data

Ex	x_l, $\theta_n = 0$		x_l, $\theta_n \neq 0$	
	x_l	$\|\|x_l\|\|$	x_l	$\|\|x_l\|\|$
1	[6.8709 −0.6179 −1.7182 −5.2214 6.8330 8.5482 −4.3395 −2.7049 4.8913 −1.5470]	15.8124	[6.8709 −0.6179 −1.7182 −5.2214 6.8330 8.5482 −4.3395 −2.7049 4.8913 −1.5470]	15.8124
2	[6.6978 −14.8359 5.5656 −0.0535 7.6286 8.1544 −4.3914 1.3391 −2.8804 −3.1992]	21.4533	[6.6978 −14.8358 5.5656 −0.0536 7.6287 8.1545 −4.3914 1.3390 −2.8804 −3.1992]	21.4533
3	[−17.1161 −5.5683 −10.4969 7.1584 5.8086 2.2931 8.1358 10.6617 1.1591]	26.5631	[−17.1161 −5.5683 −10.4969 7.1584 5.8086 2.2931 8.1358 10.6617 1.1591]	26.5631
4	[2.1820 −1.6497 −2.1801 0.8516 2.3165 2.0734 −3.5893 4.5913 3.4197]	8.2634	[2.1820 −1.6497 −2.1802 0.8516 2.3165 2.0735 −3.5893 4.5913 3.4198]	8.2634
5	[5.3072 −4.2377 0.9928 0.0791 2.8582 3.1097 −4.6935 3.5840 1.7175 −0.8343]	10.1723	[5.3073 −4.2376 0.9927 0.0790 2.8582 3.1098 −4.6935 3.5839 1.7176 −0.8343]	10.1723
6	[2.1682 −1.6493 −2.1856 0.8609 2.3193 2.0767 −3.5885 4.5972 3.4200 −0.0050]	8.2668	[2.1684 −1.6494 −2.1855 0.8608 2.3194 2.0769 −3.5886 4.5971 3.4200 −0.0050]	8.2668

Table 6. Comparison of $\theta_n \neq 0$ and $\theta_n = 0$ of the iteration 4.1 with 1×10^5 iterations for Examples 4.3–4.8 with choice 2 initial data

Ex	x_l, $\theta_n = 0$		x_l, $\theta_n \neq 0$	
	x_l	$\|\|x_l\|\|$	x_l	$\|\|x_l\|\|$
1	[6.8709 −0.6179 −1.7182 −5.2214 6.8330 8.5482 −4.3395 −2.7049 4.8913 −1.5470]	15.8124	[6.8709 −0.6179 −1.7182 −5.2214 6.8330 8.5482 −4.3395 −2.7049 4.8913 −1.5470]	15.8124
2	[6.6979 −14.8355 5.5654 −0.0540 7.6290 8.1549 −4.3914 1.3385 −2.8801 −3.1993]	21.4533	[6.6979 −14.8355 5.5654 −0.0540 7.6290 8.1549 −4.3914 1.3385 −2.8801 −3.1993]	21.4533
3	[−17.1162 −5.5678 −10.4974 7.1580 5.8089 2.2935 8.1359 10.6613 1.1595]	26.5631	[−17.1162 −5.5678 −10.4973 7.1580 5.8089 2.2935 8.1359 10.6613 1.1595]	26.5631
4	[2.1819 −1.6492 −2.1806 0.8512 2.3168 2.0738 −3.5892 4.5908 3.4202]	8.2634	[2.1819 −1.6493 −2.1806 0.8513 2.3168 2.0738 −3.5892 4.5909 3.4201]	8.2634
5	[5.3074 −4.2372 0.9924 0.0784 2.8587 3.1104 −4.6934 3.5831 1.7179 −0.8344]	10.1723	[5.3074 −4.2372 0.9924 0.0784 2.8587 3.1104 −4.6934 3.5831 1.7179 −0.8344]	10.1723
6	[2.1683 −1.6489 −2.1859 0.8604 2.3197 2.0773 −3.5885 4.5966 3.4204 −0.0050]	8.2668	[2.1684 −1.6490 −2.1858 0.8604 2.3197 2.0773 −3.5885 4.5966 3.4203 −0.0051]	8.2668

5 Conclusion

In this work, we present the two new different modified forward-backward algorithms combining the viscosity approximation method with the inertial technical term for solving the inclusion problem. We then prove the strong convergence theorems and apply our main results to find the minimum-norm least-squares solution of the unconstrained linear system (see in Tables 2, 3, 4 and 5). Some numerical experiments show that our inertial forward-backward methods have a competitive advantage over the standard forward-backward method (see in Table 1 and Figs. 1, 2, 5, 6, 9, 10, 13, 14, 17, 18, 21 and 22). Moreover, we compare the convergence of the proposed iterations (see in Figs. 3, 4, 7, 8, 11, 12, 15, 16, 19, 20, 23 and 24).

Acknowledgement. S. Suantai would like to thank Chiang Mai University. D. Yambangwai and W. Cholamjiak would like to thank University of Phayao(Grant No. UoE62001).

References

1. Alvarez, F., Attouch, H.: An inertial proximal method for monotone operators via discretization of a nonlinear oscillator with damping. Set-Valued Anal. **9**, 3–11 (2001)
2. Banach, S.: Sur les oprations dans les ensembles abstraits et leur application aux quations intgrales. Fund. Math. **3**, 133–181 (1922)
3. Bauschke, H.H., Combettes, P.L.: Convex Analysis and Monotone Operator Theory in Hilbert Spaces. CMS Books in Mathematics. Springer, New York (2011)
4. Bauschke, H.H., Combettes, P.L.: A weak-to-strong convergence principle for Fejér-monotone methods in Hilbert spaces. Math. Oper. Res. **26**, 248–264 (2001)
5. Browder, F.E.: Convergence of approximants to fixed points of non-expansive maps in Banach spaces. Arch. Rational Mech. Anal. **24**, 82–90 (1967)

6. Byrne, C.: Iterative oblique projection onto convex sets and the split feasibility problem. Inverse Prob. **18**, 441–453 (2002)
7. Byrne, C.: A unified treatment of some iterative algorithm in signal processing and image reconstruction. Inverse Prob. **20**, 103–120 (2004)
8. Censor, Y., Elfving, T.: A multiprojection algorithm using Bregman projection in a product space. Numer. Algorithms **71**, 915–932 (2016)
9. Combettes, P.L., Wajs, V.R.: Signal recovery by proximal forward-backward splitting. Multiscale Model. Simul. **4**, 1168–1200 (2005)
10. Cholamjiak, P.: A generalized forward-backward splitting method for solving quasi inclusion problems in Banach spaces. Numer. Algorithms **8**, 221–239 (1994)
11. Douglas, J., Rachford, H.H.: On the numerical solution of the heat conduction problem in 2 and 3 space variables. Trans. Amer. Math. Soc. **82**, 421–439 (1956)
12. Dang, Y., Sun, J., Xu, H.: Inertial accelerated algorithms for solving a split feasibility problem, J. Ind. Manag. Optim. https://doi.org/10.3934/jimo.2016078
13. Dong, Q.L., Yuan, H.B., Cho, Y.J., Rassias, Th.M.: Modified inertial Mann algorithm and inertial CQ-algorithm for nonexpansive mappings. Optim. Lett. https://doi.org/10.1007/s11590-016-1102-9
14. Halpern, B.: Fixed points of nonexpanding maps. Bull. Am. Math. Soc. **73**, 957–961 (1967)
15. He, S., Yang, C.: Solving the variational inequality problem defined on intersection of finite level sets. Abstr. Appl. Anal. **2013** (2013). Art ID. 942315
16. Lakshmikantham, V., Sen, S.K.: Computational Error and Complexity in Science and Engineering. Elsevier, Amsterdam (2005)
17. López, G., Martín-Marquez, V., Wang, F., Xu, H.K.: Forward-backward splitting methods for accretive operators in Banach spaces. Abstr. Appl. Anal. **2012** (2012). Art ID 109236
18. Lions, P.L., Mercier, B.: Splitting algorithms for the sum of two nonlinear operators. SIAM J. Numer. Anal. **16**, 964–979 (1979)
19. Lorenz, D., Pock, T.: An inertial forward-backward algorithm for monotone inclusions. J. Math. Imaging Vis. **51**, 311–325 (2015)
20. Maingé, P.E.: Inertial iterative process for fixed points of certain quasi-nonexpansive mappings. Set-Valued Anal. **15**, 67–79 (2007)
21. Moudafi, A.: Viscosity approximation methods for fixed-points problems. J. Math. Anal. Appl. **241**, 46–55 (2000)
22. Moudafi, A., Oliny, M.: Convergence of a splitting inertial proximal method for monotone operators. J. Comput. Appl. Math. **155**, 447–454 (2003)
23. Nesterov, Y.: A method for solving the convex programming problem with convergence rate $O(1/k^2)$. Dokl. Akad. Nauk SSSR **269**, 543–547 (1983)
24. Passty, G.B.: Ergodic convergence to a zero of the sum of monotone operators in Hilbert space. J. Math. Anal. Appl. **72**, 383–390 (1979)
25. Peaceman, D.H., Rachford, H.H.: The numerical solution of parabolic and elliptic differentials. J. Soc. Indust. Appl. Math. **3**, 28–41 (1955)
26. Polyak, B.T.: Introduction to Optimization. Optimization Software, New York (1987)
27. Polyak, B.T.: Some methods of speeding up the convergence of iterative methods. Zh. Vychisl. Mat. Mat. Fiz. **4**, 1–17 (1964)
28. Reich, S.: Strong convergence theorems for resolvents of accretive operators in Banach spaces. J. Math. Anal. Appl. **75**, 287–292 (1980)
29. Rockafellar, R.T.: Monotone operators and the proximal point algorithm. SIAM J. Control. Optim. **14**, 877–898 (1976)

30. Shimoji, K., Takahashi, W.: Strong convergence to common fixed points of infinite nonexpasnsive mappings and applications. Taiwanese J. Math. **5**, 387–404 (2001)
31. Sen, S.K., Shaykhian, G.A.: Preprocessing inconsistent linear system for a meaningful least squares solution. J. Neural Parallel Sci. Comput. **19**, 211–228 (2011)
32. Tseng, P.: A modified forward-backward splitting method for maximal monotone mappings. SIAM J. Control. Optim. **38**, 431–446 (2000)
33. Takahashi, W.: Nonlinear Functional Analysis. Yokohama Publishers, Yokohama (2000)
34. Takahashi, S., Takahashi, W.: Viscosity approximation methods for equilibrium problems and fixed point problems in Hilbert spaces. J. Math. Anal. Appl. **331**, 506–515 (2007)
35. Wittmann, R.: Approximation of fixed points of nonexpansive mappings. Arch. Math. (Basel) **58**, 486–491 (1992)
36. Xu, H.K.: Viscosity approximation methods for nonexpansive mappings. J. Math. Anal. Appl. **298**, 279–291 (2004)
37. Yao, Y., Yao, J.C., Zhou, H.Y.: Approximation methods for common fixed points of infinite countable family of nonexpansive mappings. Comput. Math. Appl. **53**, 1380–1389 (2007)

Convergence Theorems for Two Quasi-nonexpansive Multivalued Mappings by Modifying S-Iterations

W. Cholamjiak[1(✉)], K. Moonduang[1], N. Jantharasena[1], and H. Dutta[2]

[1] School of Science, University of Phayao, Phayao 56000, Thailand
c-wchp007@hotmail.com, khuanrudee.joy29@gmail.com,
natthamon041039@gmail.com

[2] Department of Mathematics, Gauhati University, Guwahati 781014, India

Abstract. In this paper, we introduce new iterative schemes by the modified S-iterations with the inertial technical term for two quasi-nonexpansive multivalued mappings in a Hilbert space. We establish the weak convergence theorem under suitable conditions. Furthermore, we use the shrinking projection method with the S-iteration for obtaining the strong convergence theorem.

Keywords: Weak convergence · Strong convergence · Multivalued mapping · S-iteration · Shrinking projection method · Hausdorff metric

Mathematics Subject Classification (2010): 47H04 · 47H10 · 54H25.

1 Introduction

Let H be a real Hilbert space with inner product $\langle \cdot, \cdot \rangle$ and induced norm $\| \cdot \|$, respectively. Let $CB(H)$ denote the family of nonempty closed bounded subsets of H and $K(H)$ denote the family of nonempty compact subsets of H. The Hausdorff metric on $CB(H)$ is defined by the following:

$$H(A, B) = \max\left\{ \sup_{x \in A} d(x, B),\ \sup_{y \in B} d(y, A) \right\}$$

for all $A, B \in CB(H)$ where $d(x, B) = \inf_{b \in B} \|x - b\|$.

A singlevalued mapping $T : H \to H$ is called *nonexpansive* if

$$\|Tx - Ty\| \leq \|x - y\|$$

for all $x, y \in H$. A multivalued mapping $T : H \to CB(H)$ is called *nonexpansive* if

$$H(Tx, Ty) \leq \|x - y\|$$

M. Zeki Sarıkaya et al. (Eds.): ICMRS 2019, LNNS 123, pp. 176–189, 2020.
https://doi.org/10.1007/978-3-030-43002-3_15

for all $x, y \in H$. An element $z \in H$ is called a fixed point of $T : H \to H$ if $z = Tz$. An element $z \in H$ is called a fixed point of $T : H \to CB(H)$ if $z \in Tz$. The fixed point set of T is denoted by $F(T)$. If $F(T) \neq \emptyset$ and

$$H(Tx, Tp) \leq \|x - p\|$$

for all $x \in H$ and $p \in F(T)$, then T is *quasi-nonexpansive.* We denote $x_n \rightharpoonup x$ to indicate that the sequence x_n converges weakly to x and $x_n \to x$ implies that x_n converges strongly to x. Let $T : H \to CB(H)$ be a multivalued mapping, $I - T$ (I is an identity mapping) is said to be demiclosed at $y \in H$ if $\{x_n\}_{n=1}^{\infty} \subset H$, such that $x_n \rightharpoonup x$ and $\{x_n - z_n\} \to y$, where $z_n \in Tx_n$ imply $\{x - y\} \in Tx$.

Since 1956, many authors have been intensively studied and considered fixed point theorems and the existence of fixed point of multivalued mapping (see, for examples [3,6,7,10,12,13,15,22,23,32,36]).

In 1953, Mann [24] proposed the following iterative procedure for finding a fixed point of a nonexpansive mapping T in a Hilbert space H:

$$x_{n+1} = \alpha_n x_n + (1 - \alpha_n)Tx_n, \forall n \in \mathbb{N}, \tag{1.1}$$

where the initial point x_1 is taken in C arbitrarily and $\{\alpha_n\}$ is a sequence in [0,1]. We know that Mann's iteration has the only weak convergence theorem (see, for example, [2,26]).

In 1974, Ishikawa [19] proposed the following iterative scheme which is a modification of the Mann's iterative algorithm (1.1):

$$\begin{cases} x_0 & \in C \quad \textit{chosen arbitrarily}, \\ x_{n+1} & = \alpha_n x_n + (1 - \alpha_n)Tz_n, n \geq 0, \\ z_n & = \beta_n x_n + (1 - \beta_n)Tx_n \end{cases} \tag{1.2}$$

where α_n and β_n are suitable control sequences in [0, 1].

In 2007, Agarwal et al. [1] introduced and studied the S-iteration process for a class of nearly asymptotically nonexpansive mappings in Banach spaces and this iteration has a better convergence rate than Ishikawa iteration for a class of contractions in metric spaces. The sequence $\{x_n\}$ is generated by

$$\begin{cases} x_0 & \in H \quad \textit{chosen arbitrarily}, \\ x_{n+1} & = \alpha_n z_n + (1 - \alpha_n)Tz_n, \\ z_n & = \beta_n x_n + (1 - \beta_n)Tx_n. \end{cases} \tag{1.3}$$

where $\{\alpha_n\}$ and $\{\beta_n\}$ are appropriate sequences in $[0, 1]$. However, Ishikawa iteration [19] and S-iteration [34] also have only weak convergence theorem even in a Hilbert space.

For obtaining strong convergence theorem, Takahashi [35] introduced the following modification of the Mann's iteration method (1.1) which just involved one closed and convex set for a family of nonexpansive mappings T_n:

$$\begin{cases} u_0 & \in H \quad chosen \;\; arbitrarily, \\ C_1 & = C, u_1 = P_{C_1}x_0, \\ y_n & = \alpha_n u_n + (1-\alpha_n)T_n u_n, \\ C_{n+1} & = \{z \in C_n : \|y_n - z\| \le \|u_n - z\|\}, \\ u_{n+1} & = P_{C_{n+1}}x_0. \end{cases} \tag{1.4}$$

They showed that if $\alpha_n \le a$ for all $n \ge 1$ and for some $a \in (0,1)0$, then the sequence u_n converges strongly to common fixed points of T_n.

In 2008, Kohsaka and Takahashi [22] presented the activation of nonspreading mapping in Banach space. They proved fixed point theorems for a single nonspreading mapping and also a common fixed point theorems for a commutative family of nonspreading mapping in Banach spaces. Let H be a Hilbert space. A mapping $T : H \to H$ is called *nonspreading* if

$$2\|Tx - Ty\|^2 \le \|x - Ty\|^2 + \|y - Tx\|^2$$

for all $x, y \in H$. Recently, Iemoto and Takahashi [18] showed that $T : H \to H$ is nonspreading if and only if

$$\|Tx - Ty\|^2 \le \|x - y\|^2 + 2\langle x - Ty, y - Ty\rangle, \; \forall x, y \in H.$$

Further, Takahashi [35] introduced a class of nonlinear mapping which is called *hybrid* as follows:

$$\|Tx - Ty\|^2 \le \|x - y\|^2 + \langle x - Tx, y - Ty\rangle,$$

for all $x, y \in H$. They showed that a mapping $T : H \to H$ is hybrid if and only if

$$3\|Tx - Ty\|^2 \le \|x - y\|^2 + \|y - Tx\|^2 + \|x - Ty\|^2,$$

for all $x, y \in H$.

Recently, in 2013, Liu [23] introduced the following class of multivalued mappings: A mapping $T : H \to CB(H)$ is called *nonspreading* if

$$2\|u_x - u_y\|^2 \le \|u_x - y\|^2 + \|u_y - x\|^2$$

for $u_x \in Tx$ and $u_y \in Ty$ for all $x, y \in H$. In addition, he proved a weak convergence theorem for finding a common fixed point of a finite family of nonspreading and nonexpansive multivalued mappings.

Very recently, Cholamjiak and Cholamjiak [8] introduced a new concept of multivalued mappings in Hilbert spaces by using Hausdorff metric. A multivalued mapping $T : H \to CB(H)$ is called *hybrid* if

$$3H(Tx, Ty)^2 \le \|x - y\|^2 + \|y - Tx\|^2 + d(y, Tx)^2 + d(x, Ty)^2$$

for all $x, y \in H$. It was shown that if T is hybrid and $F(T) \ne \emptyset$, then T is quasi-nonexpansive. Moreover, they gave some examples of a hybrid multivalued mapping which is not nonexpansive (see Cholamjiak and Cholamjiak [8]) and proved some properties and the existence of fixed points of these mappings.

Furthermore, they also proved weak and strong convergence theorems for a finite family of hybrid multivalued mappings in Hilbert spaces.

In 2001, Alvarez and Attouch [2] modified the heavy ball method which was studied in Polyak [28,29] for maximal monotone operators by using the inertial tern $\theta_n(x_n - x_{n-1})$ where θ_n is an extrapolation factor. It is remarkable that the inertial terminology greatly improves the performance of the algorithm and has a good convergence properties [2,11,14,26].

Motivated and inspired by the above works, we introduce the modified S-iterative scheme with the inertial technical term for approximating a common fixed point of two quasi-nonexpansive multivalued mappings in a Hilbert space. We also prove a weak convergence theorem under some mind conditions. Moreover, we use shrinking projection methods with the modified the S-iteration for obtaining strong convergence theorems.

2 Preliminaries and Lemmas

We now give some basic results for the proof. Let C be a nonempty, closed, and convex subset of a Hilbert space H. The nearest point projection of H onto C is denoted by P_C, that is, $\|x - P_C x\| \leq \|x - y\|$ for all $x \in H$ and $y \in C$. Such P_C is called the *metric projection* of H onto C. We know that the metric projection P_C is firmly nonexpansive, that is

$$\|P_C x - P_C y\|^2 \leq \langle P_C x - P_C y, x - y\rangle$$

for all $x, y \in H$. Furthermore, $\langle x - P_C x, y - P_C x\rangle \leq 0$ holds for all $x \in H$ and $y \in C$; see [35].

Lemma 2.1. *[2] Let $\{\psi_n\}$, $\{\delta_n\}$, and $\{\alpha_n\}$ be the sequences in $[0, +\infty)$, such that $\psi_{n+1} \leq \psi_n + \alpha_n(\psi_n - \psi_{n-1}) + \delta_n$ for all $n \geq 1$, $\sum_{n=1}^{\infty} \delta_n < +\infty$ and there exists a real number $\alpha \in (0, 1)$ with $\alpha_n \in (0, \alpha)$ for all $n \geq 1$. Then, the followings hold:*

(i) $\sum_{n\geq 1}[\psi_n - \psi_{n-1}]_+ < +\infty$, *where* $[t]_+ = max\{t, 0\}$.

(ii) *there exists* $\psi^* \in [0, +\infty)$, *such that* $\lim_{n\to+\infty} \psi_n = \psi^*$.

Lemma 2.2. *[5] Let C be a nonempty closed and convex subset of a uniformly convex space X. Let T be a nonexpansive mapping with $F(T) \neq \emptyset$. If $\{x_n\}$ is a sequence in C, such that $x_n \rightharpoonup x$ and $(I - T)x_n \to y$, then $(I - T)x = y$. In particular, if $y = 0$, then $x \in F(T)$.*

Lemma 2.3. *[34] Let X be a Banach space satisfying Opial's condition. Let $\{x_n\}$ be a sequence in X and $u, v \in X$ be such that*

$$\lim_{n\to\infty} \|x_n - u\| \ \ \textit{and} \ \ \lim_{n\to\infty} \|x_n - v\| \ \ \textit{exist.}$$

If $\{x_{n_k}\}$ and $\{x_{m_k}\}$ are subsequences of $\{x_n\}$ which converge weakly to u and v, respectively, then $u = v$.

Proposition 2.4. *[9] Let $q > 1$ and let X be a real smooth Banach space with the generalized duality mapping j_q. Let $m \in N$ be fixed. Let $\{x_i\}_{i=1}^m \subset X$ and $t_i \geq 0$ for all $i = 1, 2, ..., m$ with $\sum_{i=1}^m t_i \leq 1$. Then, we have*

$$\|\sum_{i=1}^{m} t_i x_i\|^q \leq \frac{\sum_{i=1}^m t_i \|x_i\|^q}{q - (q-1)(\sum_{i=1}^m t_i)}$$

Condition(A). In a Hilbert space H. A multi-valued mapping $T : H \to CB(H)$ is said to satisfy *Condition (A)* if $\|x - p\| = d(x, Tp)$ for all $x \in H$ and $p \in F(T)$.

Lemma 2.5. *[8] Let H be a real Hilbert space and $T : H \to K(H)$ be a hybrid multivalued mapping. If $F(T) \neq \emptyset$, then T is quasi-nonexpansivemultivalued mapping.*

Lemma 2.6. *[8] Let H be a real Hilbert space and $T : H \to K(H)$ be a hybrid multivalued mapping with $F(T) \neq \emptyset$. Then $F(T)$ is closed.*

Lemma 2.7. *[8] Let H be a real Hilbert space and $T : H \to K(H)$ be a hybrid multivalued mapping with $F(T) \neq \emptyset$. If T satisfies Condition (A), then $F(T)$ is convex.*

Lemma 2.8. *[8] (I-T is demiclosed at 0) Let H be a real Hilbert space and $T : H \to K(H)$ be a hybrid multivalued mapping. Let $\{x_n\}$ be a sequence in H, such that $x_n \rightharpoonup p$ and $lim_{n\to\infty}\|x_n - y_n\| = 0$ for some $y_n \in Tx_n$. Then, $p \in Tp$.*

3 Main Results

In this section, we modify the S-iteration for two quasi-nonexpansive mappings and prove the weak convergence theorem in Hilbert spaces. By using the shrinking projection method, we obtain the strong convergence theorems for two quasi-nonexpansive mappings.

Theorem 3.1. *Let H be a real Hilbert space and $T_1, T_2 : H \to CB(H)$ be quasi-nonexpasive mappings satisfying Condition (A). Assume that $S = F(T_1) \cap F(T_2) \neq \emptyset$ and $I - T_i$ is demiclosed at 0 for all $i = 1, 2$. Let $\{x_n\}, \{y_n\}$ and $\{z_n\}$ be sequences generated by $x_0, x_1 \in H$ and*

$$\begin{cases} y_n = x_n + \theta_n(x_n - x_{n-1}), \\ z_n \in \alpha_n y_n + (1 - \alpha_n)T_1 y_n, \\ x_{n+1} \in \beta_n z_n + (1 - \beta_n)T_2 z_n, n \geq 1 \end{cases} \tag{3.1}$$

where $\{\theta_n\} \subset [0, \theta]$ for some $\theta \in [0, 1)$, $\{\alpha_n\}$ and $\{\beta_n\}$ are sequence in $(0, 1)$. Assume that the following conditions hold:

(i) $\sum_{n=1}^{\infty}\theta_n\|x_n - x_{n-1}\| < \infty$;
(ii) $0 < \liminf_{n\to\infty}\alpha_n \leq \limsup_{n\to\infty}\alpha_n < 1$;
(iii) $0 < \liminf_{n\to\infty}\beta_n \leq \limsup_{n\to\infty}\beta_n < 1$;
Then, the sequence $\{x_n\}$ converges weakly to $q \in S$.

Proof. Let $p \in S$. Since T_1 satisfies Condition (A), for $w_n \in T_1 y_n$ such that

$$z_n = \alpha_n y_n + (1-\alpha_n) w_n,$$

We have

$$\begin{aligned}
\|z_n - p\| &\le \|\alpha_n y_n + (1-\alpha_n) w_n - p\| \\
&\le \alpha_n \|y_n - p\| + (1-\alpha_n)\|w_n - p\| \\
&= \alpha_n \|y_n - p\| + (1-\alpha_n) d(w_n, T_1 p) \\
&\le \alpha_n \|y_n - p\| + (1-\alpha_n) H(T_1 y_n, T_1 p) \\
&\le \|y_n - p\| \\
&\le \|x_n - p\| + \theta_n \|x_n - x_{n-1}\|. \qquad (3.2)
\end{aligned}$$

By Lemmas 2.3 and (3.2), for $u_n \in T_2 z_n$ such that

$$x_{n+1} = \beta_n z_n + (1-\beta_n) u_n,$$

we have

$$\begin{aligned}
\|x_{n+1} - p\| &\le \|\beta_n z_n + (1-\beta_n) u_n - p\| \\
&\le \beta_n \|z_n - p\| + (1-\beta_n)\|u_n - p\| \\
&\le \beta_n \|z_n - p\| + (1-\beta_n) d(u_n, T_2 p) \\
&\le \beta_n \|z_n - p\| + (1-\beta_n) H(T_2 z_n, T_2 p) \\
&\le \|z_n - p\| \\
&\le \|x_n - p\| + \theta_n \|x_n - x_{n-1}\|. \qquad (3.3)
\end{aligned}$$

For Lemma 2.1 and the assumption (i), we obtain $\lim_{n\to\infty} \|x_n - p\|$ exists. In particular $\{x_n\}$ is bounded and also are $\{y_n\}$ and $\{z_n\}$. Since T_1, T_2 satisfy Condition (A), we have

$$\begin{aligned}
\|x_{n+1} - p\|^2 &\le \beta_n \|z_n - p\|^2 + (1-\beta_n)\|u_n - p\|^2 - \beta_n(1-\beta_n)\|z_n - u_n\|^2 \\
&= \beta_n \|z_n - p\|^2 + (1-\beta_n) d(u_n, T_2 p)^2 - \beta_n(1-\beta_n)\|z_n - u_n\|^2 \\
&\le \beta_n \|z_n - p\|^2 + (1-\beta_n) H(T_2 z_n, T_2 p)^2 - \beta_n(1-\beta_n)\|z_n - u_n\|^2 \\
&\le \|z_n - p\|^2 - \beta_n(1-\beta_n)\|z_n - u_n\|^2 \\
&\le \alpha_n \|y_n - p\|^2 + (1-\alpha_n)\|w_n - p\|^2 - \alpha_n(1-\alpha_n)\|w_n - y_n\|^2 \\
&= \alpha_n \|y_n - p\|^2 + (1-\alpha_n) d(w_n, T_1 p)^2 - \alpha_n(1-\alpha_n)\|w_n - y_n\|^2 \\
&\le \alpha_n \|y_n - p\|^2 + (1-\alpha_n) H(T_1 y_n, T_1 p)^2 - \alpha_n(1-\alpha_n)\|w_n - y_n\|^2 \\
&\le \|y_n - p\|^2 - \alpha_n(1-\alpha_n)\|w_n - y_n\|^2 \\
&\le \|x_n - p\|^2 + 2\theta_n \langle x_n - x_{n-1}, y_n - p\rangle - \alpha_n(1-\alpha_n)\|w_n - y_n\|^2. \qquad (3.4)
\end{aligned}$$

This implies that

$$\alpha_n(1-\alpha_n)\|w_n - y_n\|^2 \le \|x_n - p\|^2 - \|x_{n+1} - p\|^2 + 2\theta_n \langle x_n - x_{n-1}, y_n - p\rangle. \qquad (3.5)$$

Since $\lim_{n\to\infty} \|x_n - p\|$ exists, it follows from (i), (ii) and (3.4) that

$$\lim_{n\to\infty} \|y_n - w_n\| = 0. \qquad (3.6)$$

This implies that

$$\lim_{n\to\infty}\|z_n - y_n\| = \lim_{n\to\infty}(1-\alpha_n)\|y_n - w_n\| = 0. \tag{3.7}$$

From the definition of $\{y_n\}$ and the assumption (i), we have

$$\lim_{n\to\infty}\|y_n - x_n\| = \lim_{n\to\infty}\theta_n\|x_n - x_{n-1}\| = 0. \tag{3.8}$$

From (3.4), we have

$$\begin{aligned}\|x_{n+1}-p\|^2 &\le \|z_n - p\|^2 - \beta_n(1-\beta_n)\|z_n - u_n\|^2\\ &\le \|x_n - p\|^2 + 2\theta_n\langle x_n - x_{n-1}, y_n - p\rangle - \beta_n(1-\beta_n)\|z_n - u_n\|^2. \end{aligned}\tag{3.9}$$

This gives that

$$\beta_n(1-\beta_n)\|z_n - u_n\|^2 \le \|x_n - p\|^2 - \|x_{n+1}-p\|^2 + 2\theta_n\langle x_n - x_{n-1}, y_n - p\rangle. \tag{3.10}$$

By the assumption (i), (ii), we obtain

$$\lim_{n\to\infty}\|z_n - u_n\| = 0. \tag{3.11}$$

Since $\lim_{n\to\infty}\|x_n - p\|$ exists, it follows from (i), (ii) and (3.11) that

$$\lim_{n\to\infty}\|x_{n+1} - z_n\| = \lim_{n\to\infty}(1-\beta_n)\|u_n - z_n\| = 0. \tag{3.12}$$

It follows from (3.7), (3.8), (3.12) that

$$\begin{aligned}\|x_{n+1} - x_n\| &\le \|x_{n+1} - z_n\| + \|z_n - y_n\| + \|y_n - x_n\|\\ &\to 0\end{aligned}\tag{3.13}$$

as $n\to\infty$. From (3.12) and (3.13), we obtain

$$\begin{aligned}\|z_n - x_n\| &\le \|z_n - x_{n+1}\| + \|x_{n+1} - x_n\|\\ &\to 0\end{aligned}\tag{3.14}$$

as $n\to\infty$. Since $\{x_n\}$ is bounded and H is reflexive, so $\omega_w(x_n) = \{x\in H : x_{n_i}\rightharpoonup x, \{x_{n_i}\}\subset\{x_n\}\}$ is nonempty. Let $q\in\omega_w(x_n)$ be an arbitrary element. Then, there exists a subsequence $\{x_{n_i}\}\subset\{x_n\}$ converging weakly to q. Let $p\in\omega_w(x_n)$ and $\{x_{n_m}\}\subset\{x_n\}$ be such that $x_{n_m}\rightharpoonup p$. From (3.14), we also have $z_{n_i}\rightharpoonup q$ and $z_{n_m}\rightharpoonup p$. From (3.8), we obtain $y_{n_i}\rightharpoonup q$ and $y_{n_m}\rightharpoonup p$. Since $I-T_i$ is demiclosed at 0 for all $i=1,2$. It follows from (3.5) and (3.11) that $p,q\in S$. Applying Lemma 2.3, we obtain $p=q$.

Theorem 3.2. *Let H be a real Hilbert space and $T_1, T_2 : H\to CB(H)$ be quasi-nonexpasive mappings satisfying Condition (A). Assume that $S = F(T_1)\cap F(T_2)\neq\emptyset$ and $I-T_i$ is demiclosed at 0 for all $i=1,2$. Let $\{x_n\}, \{y_n\}, \{z_n\}$ and $\{v_n\}$ be sequences generated by $x_0, x_1\in H, C_1 = H$ and*

$$\begin{cases} y_n &= x_n + \theta_n(x_n - x_{n-1})\\ z_n &\in \alpha_n y_n + (1-\alpha_n)T_1 y_n,\\ v_n &\in \beta_n z_n + (1-\beta_n)T_2 z_n,\\ C_{n+1} &= \{z\in C_n : \|v_n - z\|^2 \le \|x_n - z\|^2 + 2\theta_n^2\|x_n - x_{n-1}\|^2\\ &\qquad -2\theta_n\langle x_n - z, x_{n-1} - x_n\rangle\},\\ x_{n+1} &= P_{C_{n+1}}x_1, n\ge 1,\end{cases}\tag{3.15}$$

where $\{\theta_n\} \subset [0, \theta]$ *for some* $\theta \in [0, 1)$, $\{\alpha_n\}$ *and* $\{\beta_n\}$ *are sequences in* $(0, 1)$. *Assume that the following conditions hold:*

(i) $\sum_{n=1}^{\infty} \theta_n \|x_n - x_{n-1}\| < \infty$;

(ii) $0 < \liminf_{n\to\infty} \alpha_n \leq \limsup_{n\to\infty} \alpha_n < 1$;

(iii) $0 < \liminf_{n\to\infty} \beta_n \leq \limsup_{n\to\infty} \beta_n < 1$;

Then, the sequence $\{x_n\}$ *converges strongly to* $q = P_S x_1$.

Proof. We split the proof into five steps.

Step 1. Show that $P_{C_{n+1}} x_1$ is well-defined for every $x \in H$. Since T_1, T_2 satisfy Condition (A), $F(T_1) \cap F(T_2)$ is closed and convex by Lemma 2.6 and 2.7. From the definition of C_{n+1} and Lemma 2.5, C_{n+1} is closed and convex for each $n \geq 1$.

Let $z \in S$, $u_n \in T_2 z_n$ and $w_n \in T_{n1} y_n$. Since T_1 and T_2 satisfy Condition (A), we have

$$\begin{aligned}
\|v_n - z\|^2 &\leq \beta_n \|z_n - z\|^2 + (1 - \beta_n)\|u_n - z\|^2 \\
&= \beta_n \|z_n - z\|^2 + (1 - \beta_n) d(u_n, T_2 z)^2 \\
&\leq \beta_n \|z_n - z\|^2 + (1 - \beta_n) H(T_2 z_n, T_2 z)^2 \\
&\leq \|z_n - z\|^2 \\
&\leq \alpha_n \|y_n - z\|^2 + (1 - \alpha_n)(w_n - z)\|^2 \\
&= \alpha_n \|y_n - z\|^2 + (1 - \alpha_n) d(w_n, T_1 z)^2 \\
&\leq \alpha_n \|y_n - z\|^2 + (1 - \alpha_n) H(T_1 y_n, T_1 z)^2 \\
&\leq \|y_n - z\|^2 \\
&= \|x_n - z\|^2 + 2\theta_n^2 \|x_n - x_{n-1}\|^2 - 2\theta_n \langle x_n - z, x_{n-1} - x_n \rangle.
\end{aligned} \tag{3.16}$$

Therefore, we have $z \in C_{n+1}$, and thus, $S \subset C_{n+1}$. Therefore, $P_{C_{n+1}} x_1$ is well defined.

Step 2. Show that $\lim_{n\to\infty} \|x_n - x_1\|$ exists. Since S is a nonempty closed and convex subset of H, there exists a unique $v \in S$ such that

$$v = P_S x_1 \tag{3.17}$$

From $x_n = P_{C_n} x_1$, $C_{n+1} \subset C_n$ and $x_{n+1} \in C_n$, $\forall n \geq 1$, we get

$$\|x_n - x_1\| \leq \|x_{n+1} - x_1\|, \quad \forall n \geq 1. \tag{3.18}$$

On the other hand, as $S \subset C_n$, we obtain

$$\|x_n - x_1\| \leq \|v - x_1\|, \quad \forall n \geq 1. \tag{3.19}$$

It follows that the sequence $\{x_n\}$ is bounded and nondecreasing. Therefore $\lim_{n\to\infty} \|x_n - x_1\|$ exists.

Step 3. Show that $x_n \to q \in H$ as $n \to \infty$.

For $m > n$, by the definition of C_n, we see that $x_m = P_{C_m} x_1 \in C_m \subset C_n$. By Lemma 2.5, we get

$$\|x_m - x_n\|^2 \leq \|x_m - x_1\|^2 - \|x_n - x_1\|^2 \tag{3.20}$$

since $\lim_{n\to\infty}\|x_n - x_1\|$ exists, it follows from (3.20) that $\lim_{n\to\infty}\|x_m - x_n\| = 0$. Hence, $\{x_n\}$ is Cauchy sequence in H and so $x_n \to q \in H$ as $n \to \infty$.

Step 4. Show that $q \in S$. From Step 3, we have that $\lim_{n\to\infty}\|x_{n+1} - x_n\| = 0$. Since $x_{n+1} \in C_n$, we have

$$\begin{aligned}\|v_n - x_n\| &= \|v_n - x_{n+1} + x_{n+1} - x_n\| \\ &\le \|v_n - x_{n+1}\| + \|x_{n+1} - x_n\| \\ &\le \sqrt{\|x_n - x_{n+1}\|^2 + 2\theta_n^2\|x_n - x_{n-1}\|^2 - 2\theta_n\langle x_n - x_{n-1}, x_{n-1} - x_n\rangle} \\ &\quad + \|x_{n+1} - x_n\| \end{aligned} \tag{3.21}$$

By Assumption (i) and (3.21) we obtain

$$\lim_{n\to\infty}\|v_n - x_n\| = 0 \tag{3.22}$$

Since T_1, T_2 satisfy Condition(A), we have

$$\begin{aligned}\|v_n - z\|^2 &= \beta_n\|z_n - z\|^2 + (1-\beta_n)\|u_n - z\|^2 - \beta_n(1-\beta_n)\|z_n - u_n\|^2 \\ &= \beta_n\|z_n - p\|^2 + (1-\beta_n)d(u_n, T_2p)^2 - \beta_n(1-\beta_n)\|z_n - u_n\|^2 \\ &\le \beta_n\|z_n - p\|^2 + (1-\beta_n)H(T_2z_n, T_2p)^2 - \beta_n(1-\beta_n)\|z_n - u_n\|^2 \\ &\le \|z_n - p\|^2 - \beta_n(1-\beta_n)\|z_n - u_n\|^2 \\ &\le \alpha_n\|y_n - p\|^2 + (1-\alpha_n)\|w_n - p\|^2 - \alpha_n(1-\alpha_n)\|w_n - y_n\|^2 - \beta_n(1-\beta_n)\|z_n - v_n\|^2 \\ &= \alpha_n\|y_n - p\|^2 + (1-\alpha_n)d(w_n, T_1p)^2 - \alpha_n(1-\alpha_n)\|w_n - y_n\|^2 - \beta_n(1-\beta_n)\|z_n - v_n\|^2 \\ &\le \alpha_n\|y_n - p\|^2 + (1-\alpha_n)H(T_1y_n, T_1p)^2 - \alpha_n(1-\alpha_n)\|w_n - y_n\|^2 \\ &\quad - \beta_n(1-\beta_n)\|z_n - v_n\|^2 \\ &\le \|y_n - p\|^2 - \alpha_n(1-\alpha_n)\|w_n - y_n\|^2 - \beta_n(1-\beta_n)\|z_n - v_n\|^2 \\ &\le \|x_n - p\|^2 + 2\theta_n\langle x_n - x_{n-1}, y_n - p\rangle - \alpha_n(1-\alpha_n)\|w_n - y_n\|^2 \\ &\quad - \beta_n(1-\beta_n)\|z_n - v_n\|^2 \\ &\le \|x_n - p\|^2 + 2\theta_n^2\|x_n - x_{n-1}\|^2 - 2\theta_n\langle x_n - p, x_n - x_{n-1}\rangle - \alpha_n(1-\alpha_n)\|w_n - y_n\|^2 \\ &\quad - \beta_n(1-\beta_n)\|z_n - v_n\|^2. \end{aligned} \tag{3.23}$$

This implies that

$$\begin{aligned}\alpha_n(1-\alpha_n)\|y_n - w_n\|^2 + \beta_n(1-\beta_n)\|z_n - v_n\|^2 &\le \|x_n - p\|^2 + 2\theta_n^2\|x_n - x_{n-1}\|^2 \\ &\quad - 2\theta_n\langle x_n - p, x_n - x_{n-1}\rangle - \|v_n - p\|^2. \end{aligned} \tag{3.24}$$

It follows from (3.24), the assumption (i), (ii), (iii) that

$$\lim_{n\to\infty}\|y_n - w_n\| = \lim_{n\to\infty}\|z_n - u_n\| = 0. \tag{3.25}$$

This implies that

$$\lim_{n\to\infty}\|v_n - z_n\| = \lim_{n\to\infty}(1-\beta_n)\|u_n - z_n\| = 0. \tag{3.26}$$

It follows from (3.22) and (3.26) that

$$\lim_{n\to\infty}\|z_n - x_n\| = 0 \tag{3.27}$$

By the definition of $\{y_n\}$ and the assumption (i), we obtain

$$\lim_{n\to\infty} \|y_n - x_n\| = 0. \tag{3.28}$$

It follows from (3.27) and (3.28) that

$$\|z_n - y_n\| \leq \|z_n - x_n\| + \|x_n - y_n\| \to 0 \tag{3.29}$$

as $n \to \infty$. Since $x_n \to q \in H$, $y_n \to q$ by (3.28). If follows from (3.24) that $z_n \to q$. Since $I - T_i$ is demiclosed at 0 for all $i = 1, 2$ and (3.25), we have $p \in S$.

Step 5. Show that $q \in P_S x_1$. Since $x_n = P_{C_n} x_1$ and $S \subset C_n$, we obtain

$$\langle x_1 - x_n, x_n - z\rangle \geq 0, \quad \forall z \in S. \tag{3.30}$$

By taking the limit in (3.30), we obtain

$$\big\langle x_1 - q, q - z\big\rangle \geq 0, \quad \forall z \in S. \tag{3.31}$$

This shows that $q = P_S x_1$.

□

By Lemmas 2.5–2.7, we know that if $F(T) \neq \emptyset$, then a hybrid multivalued and $F(T)$ is closed and convex. We also know that $I - T$ is demiclosed at 0 by Lemma 2.8. We then obtain the following results.

Corollary 3.3. *Let H be a real Hilbert space and $T_1, T_2 : H \to K(H)$ be hybrid multivalued mappings satisfying the condition(A) such that $S = F(T_1) \cap F(T_2) \neq \emptyset$. Let $\{x_n\}, \{y_n\}$ and $\{z_n\}$ be sequences generated by $x_0, x_1 \in H$ and*

$$\begin{cases} y_n = x_n + \theta_n(x_n - x_{n-1}), \\ z_n \in \alpha_n y_n + (1 - \alpha_n)T_1 y_n, \\ x_{n+1} \in \beta_n z_n + (1 - \beta_n)T_2 z_n, n \geq 0 \end{cases} \tag{3.32}$$

where $\{\theta_n\} \subset [0, \theta]$ for some $\theta \in [0, 1)$, $\{\alpha_n\}$ and $\{\beta_n\}$ are sequence in $(0, 1)$. Assume that the following conditions hold:
(i) $\sum_{n=1}^{\infty} \theta_n \|x_n - x_{n-1}\| < \infty$;
(ii) $0 < \liminf_{n\to\infty} \alpha_n \leq \limsup_{n\to\infty} \alpha_n < 1$;
(iii) $0 < \liminf_{n\to\infty} \beta_n \leq \limsup_{n\to\infty} \beta_n < 1$.
Then, the sequence $\{x_n\}$ converges weakly to $q \in S$.

Corollary 3.4. *Let H be a real Hilbert space and $T_1, T_2 : H \to K(H)$ be hybrid multivalued mappings satisfying the condition(A) such that $S = F(T_1) \cap F(T_2) \neq \emptyset$. Let $\{x_n\}, \{y_n\}, \{z_n\}$ and $\{v_n\}$ be sequences generated by $x_0, x_1 \in H, C_1 = H$ and*

$$\begin{cases} y_n &= x_n + \theta_n(x_n - x_{n-1}) \\ z_n &\in \alpha_n y_n + (1 - \alpha_n)T_1 y_n, \\ v_n &\in \beta_n z_n + (1 - \beta_n)T_2 z_n, \\ C_{n+1} &= \{z \in C_n : \|v_n - z\|^2 \leq \|x_n - z\|^2 + 2\theta_n^2\|x_n - x_{n-1}\|^2 \\ & \quad -2\theta_n\langle x_n - z, x_{n-1} - x_n\rangle\}, \\ x_{n+1} &= P_{C_{n+1}} x_1, n \geq 1, \end{cases} \tag{3.33}$$

where $\{\theta_n\} \subset [0,\theta]$ *for some* $\theta \in [0,1)$, $\{\alpha_n\}$ *and* $\{\beta_n\}$ *are sequences in* $(0,1)$. *Assume that the following conditions hold:*

(i) $\sum_{n=1}^{\infty}\theta_n\|x_n - x_{n-1}\| < \infty$;
(ii) $0 < \liminf_{n\to\infty}\alpha_n \le \limsup_{n\to\infty}\alpha_n < 1$;
(iii) $0 < \liminf_{n\to\infty}\beta_n \le \limsup_{n\to\infty}\beta_n < 1$.

Then, the sequence $\{x_n\}$ *converges strongly to* $q = P_s x_1$.

If $Tp = \{p\}$ for all $p \in F(T)$, T satisfying the condition(A), then we obtain the following results.

Corollary 3.5. *Let* H *be a real Hilbert space and* $T_1, T_2 : H \to K(H)$ *be hybrid multivalued mappings such that* $S = F(T_1) \cap F(T_2) \neq \emptyset$. *Let* $\{x_n\}$, $\{y_n\}$ *and* $\{z_n\}$ *be sequences generated by* $x_0, x_1 \in H$ *and*

$$\begin{cases} y_n = x_n + \theta_n(x_n - x_{n-1}), \\ z_n \in \alpha_n y_n + (1-\alpha_n)T_1 y_n, \\ x_{n+1} \in \beta_n z_n + (1-\beta_n)T_2 z_n, n \ge 0 \end{cases} \tag{3.34}$$

where $\{\theta_n\} \subset [0,\theta]$ *for some* $\theta \in [0,1)$, $\{\alpha_n\}$ *and* $\{\beta_n\}$ *are sequence in* $(0,1)$. *Assume that the following conditions hold:*

(i) $\sum_{n=1}^{\infty}\theta_n\|x_n - x_{n-1}\| < \infty$;
(ii) $0 < \liminf_{n\to\infty}\alpha_n \le \limsup_{n\to\infty}\alpha_n < 1$;
(iii) $0 < \liminf_{n\to\infty}\beta_n \le \limsup_{n\to\infty}\beta_n < 1$.

If $T_1 p = \{p\}$ *and* $T_2 q = \{q\}$, *for all* $p \in F(T_1)$ *and* $q \in F(T_2)$, *then the sequence* $\{x_n\}$ *converges weakly to a common fixed point of* $\{T_1, T_2\}$.

Corollary 3.6. *Let* H *be a real Hilbert space and* $T_1, T_2 : H \to K(H)$ *be hybrid multivalued mappings such that* $S = F(T_1) \cap F(T_2) \neq \emptyset$. *Let* $\{x_n\}, \{y_n\}, \{z_n\}$ *and* $\{v_n\}$ *be sequences generated by* $x_0, x_1 \in H, C_1 = H$ *and*

$$\begin{cases} y_n & = x_n + \theta_n(x_n - x_{n-1}) \\ z_n & \in \alpha_n y_n + (1-\alpha_n)T_1 y_n, \\ v_n & \in \beta_n z_n + (1-\beta_n)T_2 z_n, \\ C_{n+1} & = \{z \in C_n : \|v_n - z\|^2 \le \|x_n - z\|^2 + 2\theta_n^2\|x_n - x_{n-1}\|^2 \\ & \quad -2\theta_n\langle x_n - z, x_{n-1} - x_n\rangle\}, \\ x_{n+1} & = P_{C_{n+1}}x_1, n \ge 1, \end{cases} \tag{3.35}$$

where $\{\theta_n\} \subset [0,\theta]$ *for some* $\theta \in [0,1)$, $\{\alpha_n\}$ *and* $\{\beta_n\}$ *are sequences in* $(0,1)$. *Assume that the following conditions hold:*

(i) $\sum_{n=1}^{\infty}\theta_n\|x_n - x_{n-1}\| < \infty$;
(ii) $0 < \liminf_{n\to\infty}\alpha_n \le \limsup_{n\to\infty}\alpha_n < 1$;
(iii) $0 < \liminf_{n\to\infty}\beta_n \le \limsup_{n\to\infty}\beta_n < 1$.

If $T_1 p = \{p\}$ *and* $T_2 q = \{q\}$, *for all* $p \in F(T_1)$ *and* $q \in F(T_2)$, *then the sequence* $\{x_n\}$ *converges strongly to* $q = P_S x_1$.

Since P_T satisfies the condition(A), then we obtain the results.

Corollary 3.7. *Let H be a real Hilbert space and $T_1, T_2 : H \to K(H)$ be hybrid multivalued mappings such that $S = F(T_1) \cap F(T_2) \neq \emptyset$. Let $\{x_n\}$, $\{y_n\}$ and $\{z_n\}$ be sequences generated by $x_0, x_1 \in H$ and*

$$\begin{cases} y_n = x_n + \theta_n(x_n - x_{n-1}), \\ z_n \in \alpha_n y_n + (1-\alpha_n)P_{T_1}y_n, \\ x_{n+1} \in \beta_n z_n + (1-\beta_n)P_{T_2}z_n, n \geq 0 \end{cases} \tag{3.36}$$

where $\{\theta_n\} \subset [0, \theta]$ for some $\theta \in [0, 1)$, $\{\alpha_n\}$ and $\{\beta_n\}$ are sequence in $(0, 1)$. Assume that the following conditions hold:

(i) $\sum_{n=1}^{\infty}\theta_n\|x_n - x_{n-1}\| < \infty$;
(ii) $0 < \liminf_{n\to\infty}\alpha_n \leq \limsup_{n\to\infty}\alpha_n < 1$;
(iii) $0 < \liminf_{n\to\infty}\beta_n \leq \limsup_{n\to\infty}\beta_n < 1$.

If P_{T_1} and P_{T_2} are hybrid multivalued mappings, then the sequence $\{x_n\}$ converges weakly to $q \in S$.

proof *By the same proof in Theorem 3.1, we have $y_n \to w_n \in P_{T_1}y_n \subseteq T_1y_n$ and $z_n \to u_n \in P_{T_2}z_n \subseteq T_2y_n$. From Lemma 2.8, we obtain this results.*

Corollary 3.8. *Let H be a real Hilbert space and $T_1, T_2 : H \to K(H)$ be hybrid multivalued mappings such that $S = F(T_1) \cap F(T_2) \neq \emptyset$. Let$\{x_n\}$, $\{y_n\}$, $\{z_n\}$ and $\{v_n\}$ be sequences generated by $x_0, x_1 \in H, C_1 = H$ and*

$$\begin{cases} y_n &= x_n + \theta_n(x_n - x_{n-1}) \\ z_n &\in \alpha_n y_n + (1-\alpha_n)P_{T_1}y_n, \\ v_n &\in \beta_n z_n + (1-\beta_n)P_{T_2}z_n, \\ C_{n+1} &= \{z \in C_n : \|v_n - z\|^2 \leq \|x_n - z\|^2 + 2\theta_n^2\|x_n - x_{n-1}\|^2 \\ & \quad -2\theta_n\langle x_n - z, x_{n-1} - x_n\rangle\}, \\ x_{n+1} &= P_{C_{n+1}}x_1, n \geq 1, \end{cases} \tag{3.37}$$

where $\{\theta_n\} \subset [0, \theta]$ for some $\theta \in [0, 1)$, $\{\alpha_n\}$ and $\{\beta_n\}$ are sequences in $(0, 1)$. Assume that the following conditions hold:

(i) $\sum_{n=1}^{\infty}\theta_n\|x_n - x_{n-1}\| < \infty$;
(ii) $0 < \liminf_{n\to\infty}\alpha_n \leq \limsup_{n\to\infty}\alpha_n < 1$;
(iii) $0 < \liminf_{n\to\infty}\beta_n \leq \limsup_{n\to\infty}\beta_n < 1$.

If P_{T_1} and P_{T_2} are hybrid multivalued mappings, then the sequence $\{x_n\}$ converges strongly to $q \in P_S x_1$.

proof *By the same proof in Theorem 3.1, we have $y_n \to w_n \in P_{T_1}y_n \subseteq T_1y_n$ and $z_n \to u_n \in P_{T_2}z_n \subseteq T_2y_n$. From Lemma 2.8, we obtain this results.*

Remark 3.9. We remark here that the condition (i) is easily implemented in numerical computation, since the value of $\|x_n - x_{n-1}\|$ is know before choosing θ_n. Indeed, the parameter θ_n can be chosen, such that $0 \leq \theta_n \leq \theta_n$, where

$$\bar{\theta_n} = \begin{cases} min\{\frac{\omega_n}{\|x_n - x_{n-1}\|}, \theta\} & if \;\; x_n \neq x_{n-1}, \\ \theta & otherwise. \end{cases} \tag{3.38}$$

Acknowledgement. The authors would like to thank University of Phayao.

References

1. Agarwal, R.P., O'Regan, D., Sahu, D.R.: Fixed Point Theory for Lipschitzian-type Mappings with Applications. Springer, Heidelberg (2009)
2. Alvarez, F., Attouch, H.: An inertial proximal method for monotone operators via discretization of a nonlinear oscillator with damping. Set-Valued Anal. **9**, 3–11 (2001)
3. Assad, N.A., Kirk, W.A.: Fixed point theorems for set-valued mappings of contractive type. Pac. J. Math. **43**, 553–562 (1972)
4. Blum, E., Oettli, W.: From optimization and variational inequalities to equilibrium problems. Math. Stud. **63**, 123–145 (1994)
5. Browder, F.E.: Nonexpansive nonlinear operators in a Banach space. Proc. Natl. Acad. Sci. U.S.A. **54**, 1041–1044 (1965)
6. Brouwer, L.E.J.: Uber Abbildung von Mannigfaltigkeiten. Math. Ann. **71**, 598 (1912)
7. Chidume, C.E., Chidume, C.O., Djitte, N., Minjibir, M.S.: Convergence theorems for fixed points of multivalued strictly pseudocontractive mappings in Hilbert spaces. Abstr. Appl. Anal. **2013** (2013)
8. Cholamjiak, P., Cholamjiak, W.: Fixed point theorems for hybrid multivalued mappings in Hilbert spaces. J. Fixed Point Theory Appl. (2016). https://doi.org/10.1007/s11784-016-0302-3
9. Cholamjiak, P.: A generalized forward–backward splitting method for solving quasi inclusion problems in Banach spaces. Numer. Algorithms **8**, 221–239 (1994)
10. Daffer, P.Z., Kaneko, H.: Fixed points of generalized contractive multi-valued mappings. J. Math. Anal. Appl. **192**, 655–666 (1995)
11. Dang, Y., Sun, J., Xu, H.: Inertial accelerated algorithms for solving a split feasibility problem. J. Ind. Manag. Optim. https://doi.org/10.3934/jimo.2016078
12. Deimling, K.: Multivalued Differential Equations. Walter de Gruyter, Berlin (1992)
13. Benavides, T.D., Gavira, B.: A fixed point property for multivalued nonexpansive mappings. J. Math. Anal. Appl. **328**, 1471–1483 (2007)
14. Dong, Q.L., Yuan, H.B., Cho, Y.J., Rassias, Th.M.: Modified inertial Mann algorithm and inertial CQ-algorithm for nonexpansive mappings. Optim. Lett. https://doi.org/10.1007/s11590-016-1102-9
15. Feng, Y., Liu, S.: Fixed point theorems for multi-valued contractive mapping and multi-valued Caristi type mappings. J. Math. Anal. Appl. **317**, 103–112 (2006)
16. Geanakoplos, J.: Nash and Walras equilibrium via Brouwer. Econ. Theory **21**, 585–603 (2003)
17. Goebel, K., Reich, S.: Uniform Convexity, Hyperbolic Geometry, and Nonexpansive Mappings, vol. 83. Marcel Dekker, New York (1984)
18. Iemoto, S., Takahashi, W.: Approximating common fixed points of nonexpansive mappings and nonspreading mappings in a Hilbert space. Nonlinear Anal. **71**, e2080–e2089 (2009)
19. Ishikawa, S.: Fixed points by a new iteration method. Proc. Am. Math. Soc. **44**, 147–150 (1974)
20. Kakutani, S.: A generalization of Brouwer's fixed point theorem. Duke Math. J. **8**, 457–459 (1941)
21. Khan, S.H., Yildirim, I., Rhoades, B.E.: A one-step iterative process for two multivalued nonexpansive mappings in Banach spaces. Comput. Math. Appl. **61**, 3172–3178 (2011)

22. Kohsaka, F., Takahashi, W.: Existence and approximation of fixed point of family nonexpansive-type mapping in Banach spaces. SIAM J. Optim. **19**, 824–835 (2008)
23. Liu, H.B.: Convergence theorems for a finite family of nonspreading and nonexpansive multivalued mappings and nonexpansive multivalued mappings and equilibrium problems with application. Theor. Math. Appl. **3**, 49–61 (2013)
24. Mann, W.R.: Mean value methods in iteration. Proc. Am. Math. Soc. **4**, 506–551 (1953)
25. Nadler, S.B.: Multi-valued contraction mappings. Pac. J. Math. **30**, 475–488 (1969)
26. Nesterov, Y.: A method for solving the convex programming problem with convergence rate O(1 /k 2). Dokl. Akad. Nauk SSSR **269**, 543–547 (1983)
27. Pietramala, P.: Convergence of approximating fixed points sets for multivalued nonexpansive mappings. Comment Math. Univ. Carolin. **32**, 697–701 (1991)
28. Polyak, B.T.: Introduction to Optimization. Optimization Software, New York (1987)
29. Polyak, B.T.: Some methods of speeding up the convergence of iterative methods. Zh. Vychisl. Mat. Mat. Fiz. **4**, 1–17 (1964)
30. Reich, S., Zaslavski, A.J.: Convergence of iterates of nonexpansive set-valued mappings. Set Valued Mapp. Appl. Nonlinear Anal. **4**, 411–420 (2002)
31. Shahzad, N., Zegeye, H.: On Mann and Ishikawa iteration schemes for multi-valued maps in Banach spaces. Nonlinear Anal. TMA **7**, 838–844 (2009)
32. Song, Y., Cho, Y.J.: Some notes on Ishikawa iteration for multi-valued mappings. Bull. Korean Math. Soc. **48**, 575–584 (2011)
33. Song, Y., Wang, H.: Convergence of iterative algorithms for multivalued mappings in Banach spaces. Nonlinear Anal. **70**, 1547–1556 (2009)
34. Suantai, S.: Weak and strong convergence criteria of Noor iterations for asymptotically nonexpansive mappings. J. Math. Anal. Appl. **311**, 506–517 (2005)
35. Takahashi, W.: Nonlinear Functional Analysis. Yokohama Publishers, Yokohama (2000)
36. Takahashi, W.: Convex Analysis and Approximation of Fixed Points. Yokohama Publishers, Yokohama (2000). (in Japanese)
37. Takahashi, W.: Introduction to Nonlinear and Convex Analysis, Yokohama (2005). (in Japanese)
38. Takahashi, W.: Fixed point theorems for new nonlinear mappings in a Hilbert space. J. Nonlinear Convex Anal. **11**, 79–88 (2005)
39. Wang, S., Gong, X., Abdou, A.A., Cho, Y.J.: Iterative algorithm for a family of split equilibrium problems and fixed point problems in Hilbert spaces with applications. Fixed Point Theory Appl. **2016**, 4 (2016)

Certain Properties of a Subclass of Multivalent Analytic Functions Using Multiplier Transformation

Laxmipriya Parida, Ashok Kumar Sahoo, and Susanta Kumar Paikray(✉)

VSS University of Technology, Sambalpur 768017, India
laxmipriya.parida94@gmail.com, ashokuumt@gmail.com,
skpaikray_math@vssut.ac.in

Abstract. Using the operator $\mathcal{Q}_p^m(c,d)$, here we create a subclass of A_p and studied inclusion and majorization properties of the subclass. Moreover, we have studied certain properties of the functions of the class A_p in connection with subordination by using the above operator. Also, associated results realised by earlier researchers are particularised under the consideration of specific parameters. Furthermore, the results which are consistent with the previously settled results are also reflected here.

Keywords: p-valent analytic functions · Inclusion relationships · Majorisation properties · Hadamard product · Subordination · Neighborhood

MSC: 30C45

1 Introductory Framework

In this entire text, we take z in the unit disk $\mathcal{U} = \{z \in \ C : |z| < 1\}$, $p, n \in \mathcal{N}$, where $\mathcal{N}$ is the set of natural numbers and $\mathcal{N}_0 = \mathcal{N} \cup \{0\}, \mathbf{I} = \{\cdots - 2, -1, 0, 1, 2 \cdots\}$. Let $\mathbb{R}$ be the set of real numbers.

The family of the functions of the form

$$f(z) = z^p + \sum_{k=n}^{\infty} a_{p+k} z^{p+k}, \tag{1}$$

which are p-valent, analytic in the unit disk $\mathcal{U}$ and we symbolize this class by $\mathrm{A}_p(n)$ and $\mathrm{A}_p(1) = \mathrm{A}_p, \mathrm{A}_1(n) = \mathrm{A}(n), \mathrm{A}_1(1) = \mathrm{A}$.

The Hadamard product (or convolution) of f and u where f is given by (1) and the function u defined in $\mathcal{U}$ by

$$u(z) = z^p + \sum_{k=n}^{\infty} b_{p+k} z^{p+k},$$

M. Zeki Sarıkaya et al. (Eds.): ICMRS 2019, LNNS 123, pp. 190–214, 2020.
https://doi.org/10.1007/978-3-030-43002-3_16

is given by

$$(f * u)(z) = z^p + \sum_{k=n}^{\infty} a_{p+k} b_{p+k} z^{p+k} = (u * f)(z). \tag{2}$$

The linear operator $\mathcal{I}^m_{p,c,d} : \mathrm{A}_p \longrightarrow \mathrm{A}_p$ defined as follows:

$$\mathcal{I}^0_{p,c,d} f(z) = f(z), \mathcal{I}^1_{p,c,d} f(z) = \mathcal{I}_{p,c,d} f(z) = \frac{cf(z) + dzf'(z)}{c + pd},$$

for $m \in \mathcal{N}_0, c \in \mathbb{R}, d > 0$ with $c + pd > 0$, was introduced and studied by Swamy [1] (see also [2]). The operator can be expressed recursively as

$$\mathcal{I}^m_{p,c,d} f(z) = \mathcal{I}_{p,c,d} \left(\mathcal{I}^{m-1}_{p,c,d} f(z) \right) (m \in \mathcal{N}_0).$$

By using definition of the operator $\mathcal{I}^m_{p,c,d}$, for $f(z) = z^p + \sum_{k=p+1}^{\infty} a_k z^k \in \mathrm{A}_p$ we can write

$$\mathcal{I}^m_{p,c,d} f(z) = z^p + \sum_{k=p+1}^{\infty} \left(\frac{c + kd}{c + pd} \right)^m a_k z^k (m \in \mathcal{N}_0). \tag{3}$$

For the negative integral values we can express the operator as

$$\mathcal{I}^{-m}_{p,c,d} f(z) = z^p + \sum_{k=p+1}^{\infty} \left(\frac{c + pd}{c + kd} \right)^m a_k z^k (m \in \mathcal{N}_0), \tag{4}$$

which is possible as $c + kd > c + pd > 0$ for $k \geq p + 1$. We can check easily

$$\mathcal{I}^{-1}_{p,c,d} f(z) = \frac{\frac{c}{d} + p}{z^{\frac{c}{d}}} \int_0^z t^{\frac{c}{d} - 1} f(t) dt = \mathcal{I}^{-1}_{p,c,d} \left(\frac{z^p}{1 - z} \right) \star f(z).$$

further, taking convolution of the operator $\mathcal{I}^{-1}_{p,c,d}$ on $\left(\frac{z^p}{1-z} \right)$, we get

$$\mathcal{I}^{-m}_{p,c,d} f(z) = \mathcal{I}^{-1}_{p,c,d} \left(\frac{z^p}{1 - z} \right) \star \mathcal{I}^{-1}_{p,c,d} \left(\frac{z^p}{1 - z} \right) \star \ldots \mathcal{I}^{-1}_{p,c,d} \left(\frac{z^p}{1 - z} \right) \star f(z).$$

Using (3) and (4), we can frame the operator given below

$$\mathcal{Q}^m_p(n; c, d) : \mathrm{A}_p(n) \longrightarrow \mathrm{A}_p(n)$$

by

$$\mathcal{Q}^m_p(n; c, d) f(z) = z^p + \sum_{k=p+n}^{\infty} \left(\frac{c + kd}{c + pd} \right)^m a_k z^k \tag{5}$$

where $m \in \mathbf{I}$, $c \in R$ and $d > 0$ with $c + pd > 0$. From (5), one can get softly

$$d\, z \left(\mathcal{Q}^m_p(n; c, d) f \right)'(z) = (c + pd) \mathcal{Q}^{m+1}_p(n; c, d) f(z) - c\, \mathcal{Q}^m_p(n; c, d) f(z) \quad (m \in \mathbf{I}). \tag{6}$$

As an easy notation, we write

$$\mathcal{Q}_p^m(1; c, d)\, f(z) = \mathcal{Q}_p^m(c, d)\, f(z). \tag{7}$$

Upon taking specific value of different parameters such as d, c, p in (7) we get various operator which are linear designed earlier by different researchers.

(i) $\mathcal{Q}_1^m(c,d)f(z) = \mathcal{I}_{c,d}^m f(z)(m \in \mathcal{N}_0)$.
(ii) $\mathcal{Q}_p^m(c,d)\, f(z) = I_{p,c,d}^m f(z)(m \in \mathcal{N}_0)$.

The above two operators were investigated by Swamy [1,2] first one for class A and second one for A_p.

(iii) $\mathcal{Q}_p^m(l + p - pt, t)f(z) = \mathcal{J}_p^m(t,l)(f \in \mathrm{A}_p, l \geq 0, t > 0; m \in \mathbf{I})$ (see Cătas [3]).
(iv) $\mathcal{Q}_p^m(c,1)f(z) = I_p^m(c)f(z) \quad (f \in \mathrm{A}_p, c > -p; m \in \mathcal{N}_0)$ (see [4–6]).
(v) $\mathcal{Q}_p^m(0,d)f(z) = \mathcal{D}_p^m f(z)(f \in \mathrm{A}_p, d > 0; m \in \mathcal{N}_0)$ (see [7–9]).
(vi) $\mathcal{Q}_1^m(r,1)f(z) = \mathcal{I}_r^m f(z)(r > 0, m \in \mathcal{N}_0)$.
This subclass was analysed by Cho and Kim [10] and Cho and Srivastava [11] seperately.
(vii) $\mathcal{Q}_1^m(1-t,t)f(z) = \mathcal{D}_t^m f(z)(t \geq 0, m \in \mathcal{N}_0)$, the class was well mentioned in [12] which in turn develop the operator $\mathcal{D}^m$ designed by Salagean [13] for $t = 0$.

We now develop the subclass of $A_p(n)$ with the operator $\mathcal{Q}_p^m(n; c, d)$, as follows

$$\mathbb{S}_{p,n}^m(c, d, \eta, A, B) = \{f \in \mathrm{A}_p(n) : f\text{satisfies}(8)\}$$

$$(1-\eta)\frac{\mathcal{Q}_p^{m+1}(n;c,d)f(z)}{\mathcal{Q}_p^m(n;c,d)f(z)} + \eta\frac{\mathcal{Q}_p^{m+2}(n;c,d)f(z)}{\mathcal{Q}_p^{m+1}(n;c,d)f(z)} \prec \frac{1+Az}{1+Bz} \tag{8}$$

where $m \in \mathbf{I}$, $B < A; A, B \in [-1,1]$, $z \in \mathcal{U}$, $d > 0, \eta \geq 0$ and $c + pd > 0$. Here we make some notational relaxation for easy understanding

(i) $\mathbb{S}_{p,1}^m(l + p - pt, t, \eta; A, B) = \mathbb{S}_p^m(t, l, \eta, A, B)$, is the subclass of $f \in \mathrm{A}_p$ obeying the subordination requirement:

$$(1-\eta)\frac{\mathcal{J}_p^{m+1}(t,l)f(z)}{\mathcal{J}_p^m(t,l)f(z)} + \eta\frac{J_p^{m+2}(t,l)f(z)}{\mathcal{J}_p^{m+1}(t,l)f(z)} \prec \frac{1+Az}{1+Bz}(t > 0, l > -p, m \in \mathbf{I} : z \in \mathcal{U}).$$

(ii) $\mathbb{S}_{p,1}^m(c,d,1,A,B) = \mathbb{S}_p^m(c,d,A,B) = \left\{f \in \mathrm{A}_p : \frac{\mathcal{Q}_p^{m+2}(c,d)f(z)}{\mathcal{Q}_p^{m+1}(c,d)f(z)} \prec \frac{1+Az}{1+Bz}\right\}$ for $z \in \mathcal{U}$.
(iii) $\mathbb{S}_{p,1}^m(c,d,\eta,A,B) = \mathbb{S}_p^m(c,d,\eta,A,B)$.
(iv) $\mathbb{S}_{p,1}^m(c,d,\eta,1-2\rho,-1) = \mathbb{S}_p^m(c,d,\eta,\rho)$, is the subclass of A_p with real part of left side of (8) for $n = 1$ is greater than ρ where $0 \leq \rho < 1$.

(v) $\mathbb{S}^m_{p,1}(c,d,0,1-2\rho,-1) = \mathbb{S}^{m-1}_p(c,d,1,1-2\rho,-1) = \mathbb{S}^m_p(c,d;\rho)$, the class of functions $f \in \mathrm{A}_p$ satisfying

$$Re\left\{\frac{\mathcal{Q}^{m+1}_p(c,d)f(z)}{\mathcal{Q}^m_p(c,d)f(z)}\right\} > \rho,$$

for ρ and z as mentioned in (iv).

(vi) $\mathbb{S}^{m-1}_{p,1}\left(c,d,1,\frac{Ad(p-\eta)+B(c+d\eta)}{c+pd},B\right)$
$= \mathbb{S}^m_{p,1}\left(c,d,0,\frac{Ad(p-\eta)+B(c+d\eta)}{c+pd},B\right) = \mathbb{S}^m_{p,c,d}(\eta;A,B)$
$= \left\{f \in \mathrm{A}_p : \frac{1}{p-\eta}\left(\frac{z\left(\mathcal{Q}^m_p(c,d)f\right)'(z)}{\mathcal{Q}^m_p(c,d)f(z)} - \eta\right) \prec \frac{1+Az}{1+Bz}\right\}$ this subclass is exactly same as defined in [1] which in turn resemblance to the subclass designed in [2] for $n=1$.

(vii) $\mathbb{S}^0_{p,1}(0,d,0,A,B) = \mathbb{S}^*_p(A,B) = \left\{f \in \mathrm{A}_p : 1 + \frac{zf''(z)}{f'(z)} \prec \frac{p(1+Az)}{1+Bz}\right\}$.

(viii) $\mathbb{S}^0_{p,1}(0,d,1,A,B) = \mathbb{S}^1_p(0,d,0;A,B) = C_p(A,B)$
$= \left\{f \in \mathrm{A}_p : 1 + \frac{zf''(z)}{f'(z)} \prec \frac{p(1+Az)}{1+Bz}\right\}$.

Motivated essentially by the works of Swamy [1], Mac Gregor [15] and Patel et al. [14] we aimed to bring out more general results in connection with the relation of containment, majorization for the class $\mathbb{S}^m_{p,n}(c,d,\eta,A,B)$. Our investigation is also open to accommodate some new results, which we obtained here using properties of subordination. As a consequence, the obtained results generalize various subclasses of the class $\mathbb{S}^m_{p,n}(c,d,\eta,A,B)$.

2 Preliminary Lemmas

Here we keep some previously proved facts, which are required to justify our main results.

Lemma 1. *If*

$$v(z) + \frac{zv'(z)}{b} \prec u(z) Re(b) \geq 0, b \neq 0), \tag{9}$$

where v has the form given below

$$v(z) = 1 + a_n z^n + a_{n+1}z^{n+1} + \dots, \tag{10}$$

and the analytic function is convex, simple (univalent) in $\mathcal{U}$, then

$$v(z) \prec \frac{b}{n} z^{-\frac{b}{n}} \int_0^z t^{\frac{b}{n}-1} u(t)dt = m(z) \prec u(z).$$

Further dominant m is the best of the subordination (9).

The function of the form

$$1 + a_1 z + a_2 z^2 + \cdots = d(z). \tag{11}$$

With the property that $\text{Re}d(z) > q, (0 \le q < 1)$ constitute a class, symbolized as $G(q)$.

Lemma 2. $\text{Re}\{d(z)\} \ge 2q - 1 + \frac{2(1-q)}{1+|z|} (0 \le q < 1)$.

For any element d of $G(q)$ and has the form (11).

This result follows from [18] and Lemma 1 is evident in [16] (see also [17, p. 71].

Lemma 3. *[24] The differential equation*

$$q(z) + \frac{zq'(z)}{\varsigma q(z) + \tau} = \frac{1 + Az}{1 + Bz}$$

has a solution in the unit disk, which is univalent and written as

$$q(z) = \begin{cases} \frac{z^{\varsigma+\tau}(1+Bz)^{\varsigma(A-B)/B}}{\varsigma \int_0^z t^{\varsigma+\tau-1}(1+Bt)^{\varsigma(A-B)/B}\,dt} - \frac{\tau}{\varsigma}, & B \neq 0 \\ \frac{z^{\varsigma+\tau}\exp(\varsigma Az)}{\varsigma \int_0^z t^{\varsigma+\tau-1}\exp(\varsigma At)dt} - \frac{\tau}{\varsigma}, & B = 0, \end{cases} \tag{12}$$

under the following conditions

(i) $-1 \le B < A \le 1$ *and* $\varsigma > 0$,
(ii) $Re(\tau) \ge -\varsigma(1-A)/(1-B)$, *where* τ *is in* $\mathbb{C}$.

Further suppose the function d given by (11) is analytic in $\mathcal{U}$ and satisfies the following subordination:

$$d(z) + \frac{zd'(z)}{\varsigma d(z) + \tau} \prec \frac{1 + Az}{1 + Bz}, \tag{13}$$

then

$$d(z) \prec q(z) \prec \frac{1 + Az}{1 + Bz}$$

and q is the best dominant of (13).

The lemma is proved in [19].

Lemma 4. *Let* $h(z,t) : \mathcal{U} \times [0,1] \to \mathbb{C}$, *holomorphic in* $\mathcal{U}$ *and for each* $0 \le t \le 1$. *The function* h *is* ν*-integrable on* $[0,1]$, *where on* $[0,1]$, ν *be a positive measure. With additional assumption* $\text{Re}\{h(\cdot,t)\} > 0$, $h(-r,t)$ *is real and* $\text{Re}\{h(z,t)\}^{-1} \ge [h(-r,t)]^{-1}$ $(|z| \le r < 1, 0 \le t \le, 1])$. *If the function* $\mathcal{H}$ *is defined in* $\mathcal{U}$ *by*

$$\mathcal{H}(z) = \int_0^1 h(z,t)\,d\nu(t),$$

then

$$\text{Re}\{\mathcal{H}(z)\}^{-1} \ge [\mathcal{H}(-r)]^{-1}.$$

Lemma 5. *For $a, b \in \mathbb{R}$ or $\mathbb{C}$, we have*

$$\int_0^1 t^{b-1}(1-t)^{e-b-1}\frac{1}{(1-tz)^a}\,dt = [\Gamma(e)]^{-1}\Gamma(b)\Gamma(e-b)\Gamma(e){}_2F_1(a,b;e;z) \quad (\mathrm{Re}(e) > \mathrm{Re}(b) > 0); \tag{14}$$

$${}_2F_1(a,b;e;z) = \frac{1}{(1-z)^a}{}_2F_1\left(a, e-b; e, z(z-1)^{-1}\right); \tag{15}$$

$${}_2F_1(a,b;e;z) = {}_2F_1(b,a;e;z); \tag{16}$$

$$(a+1){}_2F_1(1,a;a+1;z) = (a+1) + az{}_2F_1(1,a+1;a+2;z), \tag{17}$$

where and $e(e \neq 0, -1, -2, \ldots)$, and the identities (14) to (17) are well depicted in [20, Chapter 14]) for the Gaussian hypergeometric function ${}_2F_1$.

3 Inclusion Relationships

Here we present our results out of investigation based on relationships involving containment for the class $\mathbb{S}^m_{p,n}(c, d, \eta, A, B)$, in this context the declaration about the variables are as follows $c, d \in \mathbb{R}, d > 0$ and $A, B \in [-1, 1]$ with the condition $B < A$. We further declare $c + pd > 0, \eta > 0, m \in \mathbf{I}$.

Theorem 1. *If $f \in \mathbb{S}^m_p(c, d, \eta; A, B)$, then*

$$\frac{\mathcal{Q}^{m+1}_p(c,d)f(z)}{\mathcal{Q}^m_p(c,d)f(z)} \prec \frac{\eta d}{(c+pd)Q(z)} = q(z) \prec \frac{1+Az}{1+Bz} \quad (z \in \mathcal{U}), \tag{18}$$

where

$$Q(z) = \begin{cases} \displaystyle\int_0^1 t^{\frac{c+pd}{\eta d}-1}\left(\frac{1+Btz}{1+Bz}\right)^{\frac{(c+pd)(A-B)}{\eta d B}} dt, & B \neq 0 \\ \displaystyle\int_0^1 t^{\frac{c+pd}{\eta d}-1}\exp\left(\frac{(c+pd)}{\eta d}A(t-1)z\right) dt, & B = 0 \end{cases} \tag{19}$$

and q is the best dominant of (18). *Furthermore, if*

$$A \leq -\frac{\eta d}{c+pd}B \quad \text{with} \quad -1 \leq B < 0,$$

then

$$\mathbb{S}^m_p(c, d, \eta, A, B) \subset \mathbb{S}^m_p(c, d, \rho), \tag{20}$$

where

$$\rho = \left\{{}_2F_1\left(1, \frac{c+pd}{\eta d}\left(\frac{B-A}{B}\right); \frac{c+pd}{\eta d}+1; \frac{B}{B-1}\right)\right\}^{-1}.$$

The result is the best possible.

Proof. Let $f \in \mathbb{S}^m_p(c, d, \eta; A, B)$. Consider the function h defined by

$$h(z) = z\left(\frac{\mathcal{Q}^m_p(c,d)f(z)}{z^p}\right)^{\frac{d}{c+pd}} \tag{21}$$

and h is single-valued and analytic in $|z| < r_1$ by choosing r_1 is the supremum of all r with the condition that $h(z) \neq 0$ for the r satisfying $0 < |z| \leq r < 1$.

Differentiating (21) after taking logarithm then by using identity (6) and (8), we get the function μ given by

$$\mu(z) = \frac{zh'(z)}{h(z)} = \frac{\mathcal{Q}_p^{m+1}(c,d)f(z)}{\mathcal{Q}_p^m(c,d)f(z)}, \tag{22}$$

the above function $\mu(z)$ has the value 1 at $z = 0$ and in the disc $|z| < r_1$ it is analytic. Taking logarithm in (22) then differentiating with using identity (6) and (8), we derive that

$$\mu(z) + \frac{\eta d\, z\mu'(z)}{(c+pd)\mu(z)} \prec \frac{1+Az}{1+Bz} \quad (|z| < r_1). \tag{23}$$

Now, by using Lemma 3, we find that

$$\mu(z) \prec \frac{\eta d}{(c+pd)Q(z)} = q(z) \prec \frac{1+Az}{1+Bz} \quad (|z| < r_1), \tag{24}$$

where q which is seen in (12) plays as the best dominant of (18) with $\varsigma = (c+pd)/\eta d$ and $\tau = 0$. For $-1 \leq B < A \leq 1$, it can be shown easily

$$\text{Re}\left(\frac{1+Az}{1+Bz}\right) > 0,$$

by using (24), we get

$$\text{Re}\{\mu(z)\} > 0 \quad (|z| < r_1).$$

By (22) it is quite evident that h is starlike and in fact univalent in the open disk $|z| < r_1$. Hence it will be impossible for h to be vanished on the boundary of this disk. So an easy conclusion arise that $r_1 = 1$ and (22) representing for μ is analytic in the unit disk. Hence, in the light of (24), we get

$$\mu(z) \prec q(z) \prec \frac{1+Az}{1+Bz}.$$

The assertion (18) is proved.

We must show

$$inf_{z\in\mathcal{U}}\{\text{Re}(q(z))\} = q(-1). \tag{25}$$

in order to prove (20).

Upon setting

$$a = \frac{c+pd}{\eta d}\left(\frac{B-A}{B}\right), b = \frac{c+pd}{\eta d} \text{ and } e = \frac{c+pd}{\eta d} + 1,$$

we see $e > b > 0$. Using all results of Lemma 5 i.e. from (14) to (17) we desired to get from (19)

$$\begin{aligned} Q(z) &= (1+Bz)^a \int_0^1 t^{b-1}(1+Btz)^{-a}\, dt \\ &= \frac{\eta d}{c+pd}\, {}_2F_1\left(1, \frac{c+pd}{\eta d}\left(\frac{B-A}{B}\right); \frac{c+pd}{\eta d} + 1; \frac{Bz}{1+Bz}\right). \end{aligned} \tag{26}$$

Now if we are able to show

$$\operatorname{Re}\left(\frac{1}{Q(z)}\right) \geq \frac{1}{Q(-1)},$$

then we can endorse (25).

Since

$$A \leq -\frac{\eta d}{c+pd}B \quad \text{with} \quad -1 \leq B < 0$$

implies that $c > a > 0$, by using (12), we find from (26) that

$$Q(z) = \int_0^1 h(z,t)\,dt,$$

where

$$h(z,t) = \frac{1+Bz}{1+B(1-t)z}(0 \leq t \leq 1) \text{ and } d\nu(t) = \frac{\Gamma(b)}{\Gamma(a)\Gamma(c-a)}t^{a-1}t^{c-a-1}\,dt.$$

Which is a positive measure on $[0,1]$. As $h(-r,t)$ is real for $0 \leq t \leq 1, 0 \leq |z| \leq r < 1$ and for $B \in (-1,0)$ $\operatorname{Re}\{h(z,t)\} > 0$. So by virtue of Lemma 4, we see

$$\operatorname{Re}\left(\frac{1}{Q(z)}\right) \geq \frac{1}{Q(-r)} \quad (|z| \leq r < 1)$$

which, upon letting $r \to 1^-$ yields

$$\operatorname{Re}\left(\frac{1}{Q(z)}\right) \geq \frac{1}{Q(-1)}.$$

Further, by taking $A \to (-\eta d\,B/(c+pd))^+$ for the case $A = -\eta d\,B/(c+pd)$ and using (18), we get (20). The function q is the best dominant of the subordination (18), so there will be no doubt to say the above result is best possible. With this we conclude the proof of Theorem 1.

If we see the corollary-1 of Patel et al. [21], one can easily observed our Theorem 1 has two step reduction to this. For the above facts to realize, we need to set $\eta = 1$ in Theorem 1 (which well within the permissible range)then we get the following corollaries which all are best possible in the sense that there exist a member of the respective class to achieve sharpness.

Corollary 1. *If $A \leq -dB/(c+pd)(-1 \leq B < 0)$, then*

$$\mathbb{S}_p^{m-1}(c,d,A,B) \subset \mathbb{S}_p^{m}(c,d,\widetilde{\rho}),$$

where

$$\widetilde{\rho} = \left\{ {}_2F_1\left(1, \frac{c+pd}{d}\left(\frac{B-A}{B}\right); \frac{c+pd}{d}+1; \frac{B}{B-1}\right)\right\}^{-1}.$$

Now in the above Corollary 1 taking $m = c = 0$ we yield the result of Patel et al. [21] as discussed above.

Similarly Corollary-2 of Patel et al. [21], is one of the derived product of Theorem 1. Upon taking $m = -1, \eta = d = 1$ and $\widetilde{A} = \{(p+c)A - cB\}/p$, Theorem 1 produce the result as given below.

Corollary 2. *For f to be member of $\mathbb{S}_p^*(\widetilde{A}, B)$, under the conditions $c > -p$, $B \in [-1, 0)$ and $\widetilde{A} \leq (p^{-1}) \min\{p + c(1-B), -(c+1)B\} \leq 1$, we assure*

$$\mathrm{Re}\left(z^c f(z)\left[\int_0^1 t^{c-1} f(t)dt\right]^{-1}\right) > (p+c)/{}_2F_1\left(1, p\left(\frac{B-\widetilde{A}}{B}\right); c+p+1; \frac{B}{B-1}\right).$$

Corollary 3. *If $p + l > 0, A \leq \eta d(p+l)^{-1}, B \in (-1, 0)$ then for $f \in \mathbb{S}_p^m(t, l, \eta, A, B)$, we find*

$$\mathrm{Re}\left(\frac{\mathcal{J}_p^{m+1}(t,l)f(z)}{\mathcal{J}_p^m(t,l)f(z)}\right) > \left\{{}_2F_1\left(1, \frac{(p+l)(B-A)}{\eta t B}; \frac{p+l}{\eta t} + 1; \frac{B}{B-1}\right)\right\}^{-1}.$$

Now one basic question arise, that under what condition f a member of $\mathbb{S}_p^m(c, d; \kappa)$, to be a member of $\mathbb{S}_p^m(c, d, \eta, 1-2\kappa, -1)$ whatever the parameters involved here we have freedom to define in the name of condition imposed. To answer this lets follow the following discussion.

Since $f \in \mathbb{S}_p^m(c, d, \kappa)$, we have

$$\frac{\mathcal{Q}_p^{m+1}(c,d)f(z)}{\mathcal{Q}_p^m(c,d)f(z)} = \kappa + (1-\kappa)\phi(z), \tag{27}$$

where ϕ is of the form (11) and is in $G(q)$. Differentiating (27) with use of logarithm, we find

$$\begin{aligned} &\mathrm{Re}\left\{(1-\eta)\frac{\mathcal{Q}_p^{m+1}(c,d)f(z)}{\mathcal{Q}_p^m(c,d)f(z)} + \eta\frac{\mathcal{Q}_p^{m+2}(c,d)f(z)}{\mathcal{Q}_p^{m+1}(c,d)f(z)}\right\} - \kappa \\ &\geq (1-\kappa)\left\{\mathrm{Re}(\phi(z)) - \frac{\eta d}{c+pd}\frac{|z\phi'(z)|}{|\kappa + (1-\kappa)\phi(z)|}\right\}. \end{aligned} \tag{28}$$

The above process also take the view of (6).

With the help of estimation mentioned in [15] one can deduce

$$\text{(i)}|z\phi'(z)| \leq \frac{2nr^n}{1-r^{2n}}\mathrm{Re}(\phi(z)) \quad \text{and} \quad \text{(ii)}\mathrm{Re}\{\phi(z)\} \geq \frac{1-r^n}{1+r^n}. \tag{29}$$

on taking $n = 1$ in (28), the result obtain is

$$\begin{aligned} &\mathrm{Re}\left\{(1-\eta)\frac{\mathcal{Q}_p^{m+1}(c,d)f(z)}{\mathcal{Q}_p^m(c,d)f(z)} + \eta\frac{\mathcal{Q}_p^{m+2}(c,d)f(z)}{\mathcal{Q}_p^{m+1}(c,d)f(z)}\right\} - \kappa \\ &\geq (1-\kappa)\mathrm{Re}\{\phi(z)\}\left[1 - \frac{2\eta d r}{(c+pd)\{\kappa(1-r^2) + (1-\kappa)(1-r)^2\}}\right]. \end{aligned}$$

The parameter R is shown in (30). If we take $r < R$ then above real part is positive. To justify best possibility of the upper bound R we take the function h of A_p satisfying

$$\frac{\mathcal{Q}_p^{m+1}(c,d)h(z)}{\mathcal{Q}_p^{m}(c,d)h(z)} = \frac{1+(1-2\kappa)z}{1-z} \quad \left(\kappa = \frac{1-A}{1-B}\right).$$

We now summarize the above discussion in the following result.

Theorem 2. *Let $f \in \mathbb{S}_p^m(c,d;\kappa)$ and $\kappa \times (1-B) = (1-A), B \neq 1$ then*

$$f \in \mathbb{S}_p^m(c,d,\eta,1-2\kappa,-1)\, when\, |z| < \triangle = \triangle(p,c,d,\eta,\kappa),$$

where

$$\triangle = \begin{cases} \dfrac{(c+pd)(1-\kappa)+\eta d - [(c+pd)\kappa - \eta d\}^2 + 2\eta d(c+pd)]^{1/2}}{(c+pd)(1-2\kappa)}, & \kappa \neq \dfrac{1}{2} \\ \dfrac{c+pd}{2\{(c+pd)(1-\kappa)+\eta d\}}, & \kappa = \dfrac{1}{2}. \end{cases} \tag{30}$$

In Theorem 2, if we set $B = -1$ and $m = c = 0, A = 1-(2\vartheta/p)$, in Theorem 2 we see this simplified result.

Corollary 4. *For $0 \leq \vartheta < p$ and $f \in \mathbb{S}_p^*(\vartheta)$, we have*

$$Re\left[(1-\eta)\frac{zf'(z)}{f(z)} + \eta\left\{1 + \frac{zf''(z)}{f'(z)}\right\}\right] > \vartheta \quad for \quad |z| < T(p,\eta,\vartheta),$$

where

$$T(p,\eta,\vartheta) = \begin{cases} \dfrac{(p+\eta-\vartheta) - \sqrt{\vartheta^2 + 2\eta(p-\vartheta)+\eta^2}}{p-2\vartheta}, & \vartheta \neq \dfrac{p}{2} \\ \dfrac{p}{2(p+\eta-\vartheta)}, & \vartheta = \dfrac{p}{2}. \end{cases}$$

The result is the best possible.

We now define the integral operator $\mathcal{F}_{\Lambda,p}$ For a function $f \in \mathrm{A}_p(n)$, we define the integral on $\mathrm{A}_p(n)$ as follows

$$\mathcal{F}_{\Lambda,p}(f)(z) = \frac{\Lambda+p}{z^{\Lambda}}\int_0^z t^{\Lambda-1}f(t)dt \quad (\Lambda > -p). \tag{31}$$

Application of $\mathcal{F}_{\Lambda,p}$ on the function f given by (1), is as follows

$$\begin{aligned} \mathcal{F}_{\Lambda,p}(f)(z) &= z^p + \sum_{k=n}^{\infty} \frac{\Lambda+p}{\Lambda+p+k} a_{p+k} z^{p+k} \quad (z \in \mathcal{U}) \\ &= z^p_2F_1(1,\Lambda+p;\Lambda+p+1;z) \star f(z) = \mathcal{Q}_p^{-1}(n,\Lambda,1)f(z). \end{aligned} \tag{32}$$

For $f \in \mathrm{A}_p(n)$ and $\Lambda > -p$,

$$z\left(\mathcal{Q}_p^m(n;c,d)\mathcal{F}_{\Lambda,p}(f)(z)\right)' \\ = (\Lambda+p)\mathcal{Q}_p^{m+1}(n;c,d)\mathcal{F}_{\Lambda,p}(f)(z) - \Lambda\mathcal{Q}_p^m(n;c,d)\mathcal{F}_{\Lambda,p}(f)(z). \tag{33}$$

Which is derived using (5) and (32).

Theorem 3. *For $\Lambda \in \mathbb{R}$ and*

$$\Lambda \geq \frac{c(A-B)-pd(1-A)}{d(1-B)}.$$

(a) The function $\mathcal{F}_{\Lambda,p}(f) \in \mathbb{S}_p^m(c,d,A,B)$ where $\mathcal{F}_{\Lambda,p}(f)$ is given in (31) if f is member of $\mathbb{S}_p^m(c,d,A,B)$.
Furthermore,

$$\frac{\mathcal{Q}_p^{m+1}(c,d)\mathcal{F}_{\Lambda,p}(f)(z)}{\mathcal{Q}_p^m(c,d)\mathcal{F}_{\Lambda,p}(f)(z)} \prec \frac{1}{c+pd}\left(\frac{d}{Q(z)} - (\Lambda d - c)\right) = q(z) \prec \frac{1+Az}{1+Bz},$$

where

$$Q(z) = \begin{cases} \displaystyle\int_0^1 t^{\Lambda+p-1}\left(\frac{1+Btz}{1+Bz}\right)^{\frac{(c+pd)(A-B)}{dB}} dt, & B \neq 0 \\ \displaystyle\int_0^1 t^{\Lambda+p-1}\exp\left(\frac{(c+pd)}{d}A(t-1)\right), & \text{Otherwise} \end{cases} \tag{34}$$

and the dominant q is the best.
(b) If $B \in [-1,0)$ and

$$\Lambda \geq \max\left\{\frac{c+pd}{d}B^{-1}(B-A) - p - 1, \frac{c}{d} - \frac{(c+pd)(1-A)}{d(1-B)}\right\}.$$

Then

$$f \in \mathbb{S}_p^m(c,d,A,B) \Longrightarrow \mathcal{F}_{\Lambda,p}(f) \in \mathbb{S}_p^m(c,d;\tau),$$

where

$$\tau = \frac{1}{c+pd}\left[d(\Lambda+p)\left\{{}_2F_1\left(1, \frac{(c+pd)(B-A)}{dB}; \Lambda+p+1; \frac{B}{B-1}\right)\right\}^{-1} - (\Lambda d - c)\right].$$

τ is the best possible.

Proof. Setting

$$g(z) = z\left(\frac{\mathcal{Q}_p^m(c,d)\mathcal{F}_{\Lambda,p}(f)(z)}{z^p}\right)^{\frac{d}{c+pd}}. \tag{35}$$

It is evident that g analytic and single valued in $\beth = \{z : |z| < r_1\}$.

Where $r_1 = \sup\{r : g(z) \neq 0 \text{ for } z \in R = \{z : 0 < |z| \leq r < 1\}$, on differentiation after simplifying by application of logarithm, then using the identity (33) for the function $\mathcal{F}_{\Lambda,p}(f)$ (35) gives

$$\iota(z) = \frac{zg'(z)}{g(z)} = \frac{\mathcal{Q}_p^{m+1}(c,d)\mathcal{F}_{\Lambda,p}(f)(z)}{\mathcal{Q}_p^m(c,d)\mathcal{F}_{\Lambda,p}(f)(z)}, \tag{36}$$

is analytic in $\beth = \{z : |z| < r_1\}$ and $\iota(0) = 1$. With the help of the identity (6) and (33), we extract that

$$\frac{\mathcal{Q}_p^m(c,d)(f)(z)}{\mathcal{Q}_p^m(c,d)\mathcal{F}_{\Lambda,p}(f)(z)} = \frac{(c+pd)\iota(z) + (\Lambda d - c)}{d(\Lambda + p)} \quad (|z| < r_1). \tag{37}$$

Since $f \in \mathbb{S}_p^m(c,d,A,B)$, it is clear that $\mathcal{Q}_p^m(c,d)(f)(z) \neq 0$ in $0 < |z| < 1$. So, in observance of (37), we find

$$\frac{\mathcal{Q}_p^m(c,d)\mathcal{F}_{\Lambda,p}(f)(z)}{\mathcal{Q}_p^m(c,d)(f)(z)} = \frac{d(\Lambda + p)}{(c+pd)\iota(z) + (\Lambda d - c)} \quad (z \in \beth). \tag{38}$$

At this stage we take logarithm in both sides of (38) and differentiate, then in the equation we get, after following the identities (6), (33) and (36), which leads to

$$\frac{\mathcal{Q}_p^{m+1}(c,d)(f)(z)}{\mathcal{Q}_p^m(c,d)(f)(z)} = \iota(z) + \frac{z\iota'(z)}{\left(\frac{c+pd}{d}\right)\iota(z) + \left(\Lambda - \frac{c}{d}\right)} \prec \frac{1+Az}{1+Bz} \quad (z \in \beth). \tag{39}$$

Thus, by making use of Lemma 3 with $\varsigma = (c+pd)/d$ and $\tau = (\Lambda d - c)/d$ in (39), we get

$$\iota(z) = \frac{1}{c+pd}\left(\frac{d}{Q(z)} - (\Lambda d - c)\right) = q(z) \prec \frac{1+Az}{1+Bz} \quad (|z| < r_1), \tag{40}$$

here q is the best dominant and (34) representing Q.

Since for $A, B \in [-1,1], \leq B < A$,

$$\mathrm{Re}\left(\frac{1+Az}{1+Bz}\right) > 0,$$

by (39), we have $\mathrm{Re}\{\iota(z)\} > 0$ in $\beth$. Viewing (36) we see g is univalent in $\beth$. Which confirms g never be zero in $\beth$, if $r_1 < 1$. Which tends to a conclusion that $r_1 = 1$ and ι is analytic in the unit disk. Now the assertion (a) of Theorem 3 is confirmed due to (36) and (40).

To prove the assertion (b) of Theorem 3 we can follow the technique as in the proof of Theorem 1. As q is the best dominant, so we say the result obtained is best possible.

Remark 1. To get the result due to Patel et al. [14, Remark 2]. We set the following values of parameters $A = 1 - (2\eta/p)(0 \leq \eta < p), B = -1$,then $m = c = 0$ and $m = 1, c = 0$, in Theorem 3.

4 Properties Involving the Operator $\mathfrak{Q}_p^m(c,d)$

Here by using the operator $\mathfrak{Q}_p^m(c,d)f(z)$, we have studied certain properties and characteristics of functions in A_p.

Theorem 4. *For $A+\mu(1-B)\leq 1, \mu\in(0,1), \gamma\in(0,1]$ and. If*

$$(1-\mu)\left(\frac{\mathfrak{Q}_p^m(c,d)f(z)}{z^p}\right)^{\frac{1}{\gamma}}+\mu\frac{\mathfrak{Q}_p^{m+1}(c,d)f(z)}{\mathfrak{Q}_p^m(c,d)f(z)}\prec\frac{1+Az}{1+Bz}, (f\in\mathrm{A}_p), \tag{41}$$

then

$$\left(\frac{\mathfrak{Q}_p^m(c,d)f(z)}{z^p}\right)^{\frac{1}{\gamma}}\prec\frac{\mu d\gamma}{(1-\mu)(c+pd)}\left(\frac{1}{Q(z)}\right)=q(z)\prec\frac{1+\frac{A-\mu B}{1-\mu}z}{1+Bz}, \tag{42}$$

where

$$Q(z)=\begin{cases}\int_0^1 t^{\dagger(1-\mu)-1}\left(\frac{1+Btz}{1+Bz}\right)^{\dagger\left(\frac{A-B}{B}\right)}dt, & B\neq 0\\ \int_0^1 t^{\dagger(1-\mu)-1}\exp\left(\dagger A(t-1)\right)dt, & B=0\end{cases}$$

$\dagger=\dfrac{c+pd}{\mu\gamma d}$ and best dominant of (42) is q. Over and above, if

$$A\leq\min\left\{1-\mu(1-B), B\times\left(\frac{-\mu}{\dagger}+\mu\right)\right\}\quad \text{with}\quad -1\leq B<0,$$

then

$$\mathrm{Re}\left(\frac{\mathfrak{Q}_p^m(c,d)f(z)}{z^p}\right)^{\frac{1}{\gamma}}>\xi,$$

where

$$\xi=\left\{{}_2F_1\left(1,\dagger\left(B[B-1]^{-1}\right);(1-\mu)\dagger+1;B[B-1]^{-1}\right)\right\}^{-1}.$$

The value of ξ is sharp.

Proof. Let us take

$$\flat=\left(\frac{\mathfrak{Q}_p^m(c,d)f(z)}{z^p}\right)^{\frac{1}{\gamma}}\ (\gamma>0), \tag{43}$$

here we remember $\flat$ is analytic in unit disk and has the form (11).

Taking logarithmic differentiation in both sides of (43) and using (6) in the resulting equation, we deduce that

$$\psi(z)+\frac{z\psi'(z)}{\varsigma\psi(z)+\tau}\prec\frac{1+Az}{1+Bz}\quad (z\in\mathcal{U}), \tag{44}$$

where $\varsigma=(c+pd)/\mu d\gamma$, $\psi(z)=\mu+(1-\mu)\flat$, and $\tau=-(d\gamma)^{-1}(c+pd)$. Applying Lemma 3 in (44) and following the lines of proof of Theorem 1, we shall obtain the assertion of Theorem 4.

Letting $m = 1, c = 0, d = 1, A = 1 - (2\eta/p)$ and $B = -1$ in Theorem 4, we find.

Corollary 5. *If*

$$\mathrm{Re}\left\{(1-\mu)\frac{f'(z)}{z^{p-1}} + \mu\left(1 + \frac{zf''(z)}{f'(z)}\right)\right\} > \eta,$$

for $\max(1/2)\,[2p\mu, p + (p-1)\mu] \leq \eta < p$ *and* $f \in \mathrm{A}_p$,
then

$$\mathrm{Re}\left\{\frac{f'(z)}{z^{p-1}}\right\} > \frac{p}{{}_2F_1\left(1, \frac{2(p-\eta)}{\mu}; \frac{p(1-\mu)}{\mu} + 1; \frac{1}{2}\right)}.$$

The result is sharp.

Setting

$$\left(\frac{\mathcal{Q}_p^m(c,d)f(z)}{z^p}\right)^{\frac{1}{\gamma}} = \kappa + (1-\kappa)\flat \quad \left(0 < \gamma \leq 1, \kappa = (1-A)(1-B)^{-1}, B \neq 1\right),$$

where $\flat$ is of the form (10), using the estimates (29) and following the lines of proof of Theorem 2, we obtain the following result.

Theorem 5. *If* $f \in \mathrm{A}_p(n), 0 < \daleth < 1, 0 < \omega \leq 1$ *and subordination condition*

$$\frac{\mathcal{Q}_p^m(n,c,d)f(z)}{z^p} \prec \left((1+Az)(1+Bz)^{-1}\right)^{\omega},$$

is satisfied, then

$$Re\left\{(1-\daleth)\left(\frac{\mathcal{Q}_p^m(n;c,d)f(z)}{z^p}\right)^{\frac{1}{\omega}} + \daleth\frac{\mathcal{Q}_p^{m+1}(n;c,d)f(z)}{\mathcal{Q}_p^m(n;c,d)f(z)}\right\} > \daleth + (1-\daleth)\varpi,$$

for $|z| < \mathcal{W} \equiv \mathcal{W}(p, n, \daleth, c, d, \omega, \varpi)$, *where* $\mathcal{W} > 0$ *is the least root of the equation*

$$(1-\daleth)(c+pd)(1-2\varpi)r^{2n} - 2\{(1-\daleth)(c+pd)(1-\varpi) + n\daleth d\omega\}r^n + (1-\daleth)(c+pd) = 0.$$

For sharpness we can choose $f \in \mathcal{A}_p(n)$ defined by

$$\left(\frac{\mathcal{Q}_p^m(n;c,d)f(z)}{z^p}\right)^{\frac{1}{n}} = \left(\frac{1+(1-2\varpi)z^n}{1-z^n}\right)^{\omega} \left(0 < \omega \leq 1, \varpi = \frac{1-A}{1-B}\right.$$

Theorem 6. *If*

$$(1-\daleth)\frac{\mathcal{Q}_p^m(n;c,d)f(z)}{z^p} + \daleth\frac{\mathcal{Q}_p^{m+1}(n;c,d)f(z)}{z^p} \prec \frac{1+Az}{1+Bz}, \daleth > 0, f \in \mathrm{A}_p(n), \quad (45)$$

then for $z \in \mathcal{U}$,

$$\frac{\mathcal{Q}_p^m(n;c,d)f(z)}{z^p} \prec \begin{cases} \frac{A}{B} + \left(1 - \frac{A}{B}\right)(1+Bz)^{-1}\,{}_2F_1\left(1,1;\frac{c+pd}{\daleth dn}+1;\frac{Bz}{1+Bz}\right), & B \neq 0 \\ 1 + \frac{c+pd}{c+pd+\daleth dn}Az, & B = 0. \end{cases} \tag{46}$$

Further,

$$\mathrm{Re}\left[\left(\frac{\mathcal{Q}_p^m(n;c,d)f(z)}{z^p}\right)^{1/t}\right] > \sigma^{1/\Upsilon} \quad (\Upsilon \in \mathcal{N};), \tag{47}$$

where

$$\sigma = \begin{cases} \frac{A}{B} + \left(1 - \frac{A}{B}\right)(1-B)^{-1}\,{}_2F_1\left(1,1;\frac{c+pd}{\daleth dn}+1;\frac{B}{B-1}\right), & B \neq 0 \\ 1-(c+pd)(c+pd+\daleth dn)^{-1}A, & B = 0. \end{cases}$$

The result is the sharp as σ *can not increased.*

Proof. Let

$$\Gamma(z) = \frac{\mathcal{Q}_p^m(n;c,d)f(z)}{z^p}, f \in \mathrm{A}_p(n). \tag{48}$$

On simplifying right side of (48) we observed Γ take the form (10) and in the unit disk it is analytic. Now if we differentiate (48), with the application of identity (6) followed by the use of (45), the (48) takes the form

$$\Gamma(z) + \frac{z\Gamma'(z)}{(c+pd)/\daleth d} \prec \frac{1+Az}{1+Bz} \quad (z \in \mathcal{U}). \tag{49}$$

Using Lemma 1 (with $b = (c+pd)/\daleth d$) in (49), the subordination as

$$\frac{\mathcal{Q}_p^m(n;c,d)f(z)}{z^p} \prec Q(z) = \frac{c+pd}{\daleth d\, n} z^{-\frac{c+pd}{\daleth d\, n}} \int_0^z t^{\frac{c+pd}{\daleth d\, n}-1}\left(\frac{1+At}{1+Bt}\right) dt,$$

which gives assertion (46) after an incorporating application of identities (14) to (17) followed by change of variables. This complete the prove the first part of Theorem 6.

If we are able to show

$$\inf_{z \in \mathcal{U}} \{\mathrm{Re}(Q(z))\} = Q(-1), \tag{50}$$

then it will be sufficient to establish (47) Indeed, for $\Omega = \{z : |z| \leq r < 1\}$,

$$\mathrm{Re}\left(\frac{1+Az}{1+Bz}\right) \geq \frac{1-Ar}{1-Br}.$$

Setting

$$\mathcal{G}(s,z)=\frac{1+Asz}{1+Bsz}\quad(0\leq s\leq 1)\quad\text{and}\quad d\nu(s)=\frac{c+pd}{\daleth dn}s^{\frac{c+pd}{\daleth dn}-1}ds,$$

which is a positive measure on the closed interval $[0,1]$, we see

$$Q(z)=\int_0^1\mathcal{G}(s,z)d\nu(s),$$

so that

$$\text{Re}\{Q(z)\}\geq\int_0^1\frac{1-Asz}{1-Bsz}d\nu(s)=Q(-r)\quad(z\in\Omega).$$

In the above inequality, by using fundamental inequality we find

$$\text{Re}\left(\omega^{1/\Upsilon}\right)\geq(\text{Re}(\omega))^{1/\Upsilon}\quad(\text{Re}(\omega)>0;\Upsilon\in\mathcal{N}),$$

and for $\text{Re}(\omega)>0;\Upsilon\in\mathcal{N}$ taking $r\to 1^-$ it takes the form of

$$\text{Re}\left(\omega^{1/\Upsilon}\right)\geq(\text{Re}(\omega))^{1/\Upsilon}.$$

Here from (50), the recognition of (47) is clear and best possible as the best dominant is Q.

Corollary 6. *Let* $\Phi=\mathcal{J}_p^m(t,l)f(z)$,

$$(1-\daleth)\frac{\Phi}{z^p}+\daleth\frac{\Phi}{z^p}\prec\frac{1+Az}{1+Bz},f\in\text{A}_p,$$

satisfies, then

$$Re\left\{\frac{\Phi}{z^p}\right\}>\varrho,$$

where

$$\varrho=\begin{cases}\frac{A}{B}+\left(1-\frac{A}{B}\right)(1-B)^{-1}{}_2F_1\left[1,1;\frac{p+l}{\daleth tn}+1;\frac{B}{B-1}\right], & B\neq 0\\ 1-(p+\ell)(p+l+\daleth tn)^{-1}A, & B=0,\end{cases}$$

The result is the best possible.

Setting $m=-1,c=\Lambda,d=\daleth=1,A=1-2\eta\,(0\leq\eta<1)$ and $B=-1$ in Theorem 6, we obtain

Corollary 7. *If* $f\in\text{A}_p(n)$ *satisfies*

$$(1-\daleth)\frac{\mathcal{F}_{\Lambda,p}(f)(z)}{z^p}+\daleth\frac{f(z)}{z^p}\prec\frac{1+Az}{1+Bz}\quad(\daleth>0,\Lambda>-p;z\in\mathcal{U}),$$

then

$$\text{Re}\left(\frac{\mathcal{F}_{\Lambda,p}(f)(z)}{z^p}\right)>\eta+(1-\eta)\left[{}_2F_1\left(1,1;\frac{\Lambda+p}{\daleth n}+1;\frac{1}{2}\right)-1\right].$$

The result is the best possible.

Theorem 7. *If $\kappa = (1-A)/(1-B)$ and $f \in \mathrm{A}_p(n)$ satisfies the subordination condition:*

$$\frac{\mathcal{Q}_p^m(n;c,d)f(z)}{z^p} \prec \frac{1+Az}{1+Bz}, \kappa = (1-A)/(1-B), f \in \mathrm{A}_p(n) \tag{51}$$

then

$$\mathrm{Re}\left\{(1-\urcorner)\frac{\mathcal{Q}_p^m(n;c,d)f(z)}{z^p} + \urcorner\frac{\mathcal{Q}_p^{m+1}(n;c,d)f(z)}{z^p}\right\} > \kappa$$
$$for |z| < R = R(p,c,d,\urcorner,n),$$

where

$$R = \left[\frac{\sqrt{(c+pd)^2 + (\urcorner dn)^2} - \urcorner dn}{c+pd}\right]^{\frac{1}{n}}.$$

The result is the best possible.

Proof. In view of (51), we see

$$\mathrm{Re}\left\{\frac{\mathcal{Q}_p^m(n;c,d)f(z)}{z^p}\right\} > \kappa,$$

so that

$$\frac{\mathcal{Q}_p^m(n;c,d)f(z)}{z^p} = \kappa + (1-\kappa)\phi(z), \tag{52}$$

where ϕ, given by (10), has a positive real part in $\mathcal{U}$ and is analytic. Taking logarithm and differentiating in both sides of (52), followed by applying (6) we derive from the equation so obtained

$$\mathrm{Re}\left[(1-\urcorner)\frac{\mathcal{Q}_p^m(n;c,d)f(z)}{z^p} + \urcorner\frac{\mathcal{Q}_p^{m+1}(n;c,d)f(z)}{z^p}\right] - \kappa$$
$$\geq (1-\kappa)\left\{\mathrm{Re}(\phi(z)) - \frac{\urcorner dn}{c+pd}|z\phi'(z)|\right\}.$$

Following the proof of Theorem 2 and admitting the estimate (a) of (29) in the above inequality we get assertion of Theorem 7 as required.

For bound R to be best possible, we take $f \in \mathrm{A}_p(n)$ defined by

$$\frac{\mathcal{Q}_p^m(n;c,d)f(z)}{z^p} = \frac{1+(1-2\kappa)z^n}{1-z^n} \quad \left(\kappa = \frac{1-A}{1-B}; z \in \mathcal{U}\right).$$

Putting m, c, B, A as $-1, \Lambda, -1A = 1-2\eta$ respectively and $d = 1$ in Theorem 7, the result is

Corollary 8. *For $\urcorner > 0, \Lambda > -p$*

$$\mathrm{Re}\left(\frac{\mathcal{F}_{\Lambda,p}(f)(z)}{z^p}\right) > \eta (0 \leq \eta < 1),$$

holds for f *of* $\mathrm{A}_p(n)$ *then*

$$(1-\daleth)\frac{\mathcal{F}_{\Lambda,p}(f)(z)}{z^p}+\daleth\frac{f(z)}{z^p}>\eta \quad for \quad |z|<R(p,\Lambda,\daleth,n)=R,$$

where

$$R=\left[\frac{\sqrt{(\Lambda+p)^2+(\daleth n)^2}-\daleth n}{\Lambda+p}\right]^{\frac{1}{n}}.$$

The bound R *is the best possible for the function* $f\in\mathrm{A}_p(n)$ *defined by*

$$\frac{\mathcal{F}_{\Lambda,p}(f)(z)}{z^p}=\frac{1+(1-2\eta)z^n}{1-z^n},$$

for $0\leq\eta<1, \Lambda>-p$.

Theorem 8. *For* $\mathcal{F}_{\Lambda,p}(f)$ *is given by* (31), *If*

$$(1-\daleth)\frac{\mathcal{Q}_p^m(n;c,d)\mathcal{F}_{\Lambda,p}(f)(z)}{z^p}+\daleth\frac{\mathcal{Q}_p^m(n;c,d)f(z)}{z^p}\prec\frac{1+Az}{1+Bz}, \tag{53}$$

for $\daleth>0$ *and* $\Lambda>-p$,
then for $f\in\mathrm{A}_p(n)$

$$\mathrm{Re}\left\{\left(\frac{\mathcal{Q}_p^m(n;c,d)\mathcal{F}_{\Lambda,p}(f)(z)}{z^p}\right)^{1/t}\right\}>\xi^{1/t}(t\in\mathbb{N}),$$

where

$$\xi=\begin{cases}\frac{A}{B}+\left(1-\frac{A}{B}\right)(1-B)^{-1}{}_2F_1\left(1,1;\frac{\Lambda+p}{\daleth n}+1;B(B-1)^{-1}\right), & B\neq 0\\ 1-(\Lambda+p)(\Lambda+p+\daleth n)^{-1}A, & B=0.\end{cases}$$

The result is the best possible.

Proof. If we let

$$\phi(z)=\frac{\mathcal{Q}_p^m(n;c,d)\mathcal{F}_{\Lambda,p}(f)(z)}{z^p}, \tag{54}$$

the above function takes the form (10) and of course is analytic in unit disk. On differentiation of (54) followed by (33), (53) then ϕ is of the form (10) and it is analytic in $\mathcal{U}$. We get

$$\begin{aligned}\phi(z)+\frac{z\phi'(z)}{(\Lambda+p)/\daleth}&=(1-\daleth)\frac{\mathcal{Q}_p^m(n;c,d)\mathcal{F}_{\Lambda,p}(f)(z)}{z^p}+\daleth\frac{\mathcal{Q}_p^m(n;c,d)f(z)}{z^p}\\&\prec\frac{1+Az}{1+Bz}.\end{aligned}$$

We omit the proof of the rest part of the Theorem 8 as the procedure is similar to that of Theorem 6.

Letting $A = 1 - 2\eta, B = -1, m = -1, c = \Lambda$ and $\daleth = d = t = 1$ in Theorem 7, we obtain

Corollary 9. *Suppose that $f \in \mathrm{A}_p(n)$ and $\mathcal{F}_{\Lambda,p}(f)$ is given by (31). If*

$$\mathrm{Re}\left\{\frac{\mathcal{F}_{\Lambda,p}(f)(z)}{z^p}\right\} > \eta \quad (0 \leq \eta < 1, \Lambda > -p),$$

then

$$\mathrm{Re}\left(\frac{\int_0^z t^{\Lambda-1}\mathcal{F}_{\Lambda,p}(f)(t)dt}{z^{\Lambda+p}}\right) > \frac{\eta + (1-\eta)\left\{{}_2F_1\left(1,1;\frac{\Lambda+p}{\daleth n}+1;\frac{1}{2}\right)-1\right\}}{\Lambda+p}.$$

The result is sharp.

Theorem 9. *If $f_j \in \mathrm{A}_p$ satisfy the subordination condition* (45)

$$\mathcal{H}(z) = \mathcal{Q}_p^m(c,d)(f_1 \star f_2)(z),$$

for $-1 \leq B_j < A_j \leq 1\,(j = 1, 2)$ and $\daleth > 0$ then

$$\mathrm{Re}\left\{(1-\daleth)\frac{\mathcal{Q}_p^m(c,d)\mathcal{H}(z)}{z^p} + \daleth\frac{\mathcal{Q}_p^{m+1}(c,d)\mathcal{H}(z)}{z^p}\right\} > \eta \quad (0 \leq \eta < 1;), \tag{55}$$

where

$$\eta = 1 - \frac{4(A_1 - B_1)(A_2 - B_2)}{(1-B_1)(1-B_2)}\left\{1 - \frac{1}{2}{}_2F_1\left(1,1;\frac{c+pd}{\daleth d}+1;\frac{1}{2}\right)\right\},$$

and

$$\mathcal{H}(z) = \mathcal{Q}_p^m(c,d)(f_1 \star f_2)(z).$$

For $B_1 = B_2 = -1$ the result will be best.

Proof. Let us set

$$\phi_j(z) = (1-\daleth)\frac{\mathcal{Q}_p^m(c,d)f_j(z)}{z^p} + \daleth\frac{\mathcal{Q}_p^{m+1}(c,d)f_j(z)}{z^p}(j = 1, 2), \tag{56}$$

where each $\phi_j \in G(q_j)$ for $q_j = (1 - A_j)/(1 - B_j), j = 1, 2$ due to (45). The following derivation is the application of the identity (6) and (56).

$$\mathcal{Q}_p^m(c,d)f_j(z) = \frac{p+cd}{\daleth d}z^{p-\frac{p+cd}{\daleth d}}\int_0^z t^{\frac{c+pd}{\daleth d}-1}\phi_j(t)dt \quad (j = 1, 2). \tag{57}$$

Following a simple calculation and using (57), we obtain

$$\mathcal{Q}_p^m(c,d)\mathcal{H}(z) = \frac{p+cd}{\daleth d}z^{p-\frac{p+cd}{\daleth d}}\int_0^z t^{\frac{c+pd}{\daleth d}-1}\phi_0(t)dt, \tag{58}$$

where

$$\phi_0(z) = (1-\daleth)\frac{\mathcal{Q}_p^m(c,d)\mathcal{H}(z)}{z^p} + \daleth\frac{\mathcal{Q}_p^{m+1}(c,d)\mathcal{H}(z)}{z^p}$$
$$= \frac{p+cd}{\daleth d} z^{-\frac{p+cd}{\daleth d}} \int_0^z t^{\frac{c+pd}{\daleth d}-1}(\phi_1 \star \phi_2)(t)dt. \tag{59}$$

Now for each j, ϕ_j is a member of $G(q_j)$ we find from [25] that

$$(\phi_1 \star \phi_2) \in G(q_3) \quad (q_3 = 1 - 2(1-q_1)(1-q_2)),$$

and best possible bound is q_3.

Now (59), gives by using Lemma 2

$$\begin{aligned}
\operatorname{Re}\{\phi_0(z)\} &= \frac{p+cd}{\daleth d}\int_0^1 s^{\frac{p+cd}{\daleth d}-1}\operatorname{Re}\{(\phi_1 \star \phi_2)(sz)\}ds \\
&\geq \frac{p+cd}{\daleth d}\int_0^1 s^{\frac{p+cd}{\daleth d}-1}\left(2\gamma_3 - 1 + \frac{2(1-q_3)}{1+s|z|}\right)ds \\
&> \frac{c+pd}{\daleth d}\int_0^1 s^{\frac{p+cd}{\daleth d}-1}\left(2\gamma_3 - 1 + \frac{2(1-q_3)}{1+s}\right)ds \\
&= 1 - \frac{4(A_1-B_1)(A_2-B_2)}{(1-B_1)(1-B_2)}\left(1 - \frac{c+pd}{\daleth d}\int_0^1 \frac{s^{\frac{c+pd}{\daleth d}-1}}{1+s}ds\right) \\
&= 1 - 4(A_1-B_1)(A_2-B_2)\left((1-B_1)(1-B_2)\right)^{-1} \\
&\times \left[1 - \frac{1}{2}{}_2F_1\left(1,1;\frac{c+pd}{\daleth d}+1;\frac{1}{2}\right)\right] = \eta.
\end{aligned}$$

When $B_1 = B_2 = -1$, we consider the functions $f_j \in \mathrm{A}_p$ satisfying the hypothesis (45) and defined by

$$\mathcal{Q}_p^m(c,d)f_j(z) = \frac{p+cd}{\daleth d} z^{p-\frac{p+cd}{\daleth d}} \int_0^z t^{\frac{c+pd}{\daleth d}-1}\left(\frac{1+A_j t}{1-t}\right)dt \quad (j=1,2).$$

It follows from (59) and Lemma 2 that

$$\begin{aligned}
\phi_0(z) &= \frac{p+cd}{\daleth d} z^{p-\frac{p+cd}{\daleth d}} \int_0^1 t^{\frac{c+pd}{\daleth d}-1}\left(1-(1+A_1)(1+A_2) + \frac{(1+A_1)(1+A_2)}{1-sz}\right)ds \\
&= 1-(1+A_1)(1+A_2) + (1+A_1)(1+A_2)(1-z)^{-1}\,{}_2F_1\left(1,1;\frac{c+pd}{\daleth d}+1;\frac{z}{z-1}\right) \\
&\longrightarrow 1-(1+A_1)(1+A_2) + (1+A_1)(1+A_2)(1-z)^{-1}\,{}_2F_1\left(1,1;\frac{c+pd}{\daleth d}+1;\frac{1}{2}\right),
\end{aligned}$$

as $z \to -1$, and Theorem 9 is concluded.

5 Majorization Properties

For the sub family $\mathbb{S}_p^m(c,d,A,B)$, we establish properties based on majorization of functions

Theorem 10. *For $g \in \mathbb{S}_p^m(c,d,A,B), f \in \mathrm{A}_p$. If*

$$\mathcal{Q}_p^{m+1}(c,d)f(z) \ll \mathcal{Q}_p^{m+1}(c,d)g(z), \tag{60}$$

then for $(|z| < \widetilde{r}) = \widetilde{r}(p,c,d,A,B)$

$$\left|\mathcal{Q}_p^{m+1}(c,d)f(z)\right| \leq \left|\mathcal{Q}_p^{m+1}(c,d)g(z)\right|, \tag{61}$$

and $\widetilde{r}$ is the least positive root of the equation

$$(c+pd)|A|r^3 - ((c+pd)+2d|B|)r^2 - ((c+pd)|A|+2d)r + (c+pd) = 0. \tag{62}$$

Proof. As $g \in \mathbb{S}_p^m(c,d,A,B)$, we see

$$\mathcal{Q}_p^{m+1}(c,d)g(z) = \left(\frac{1+Bw(z)}{1+Aw(z)}\right)\mathcal{Q}_p^{m+2}(c,d)g(z).$$

The analytic function w in $\mathcal{U}$ has the property $|w(z)| \leq 1, w(0) = 0$. Thus,

$$\left|\mathcal{Q}_p^{m}(c,d)g(z)\right| \leq \frac{1+|B||z|}{1-|A||z|}\left|\mathcal{Q}_p^{m+1}(c,d)g(z)\right|. \tag{63}$$

Taking help of (60) we get from (61)

$$\mathcal{Q}_p^{m+1}(c,d)f(z) = \varepsilon(z)\mathcal{Q}_p^{m+1}(c,d)g(z)(z \in \mathcal{U}), \tag{64}$$

where the analytic function ε has the property $|\varepsilon(z)| \leq 1$ in $\mathcal{U}$. Both side differentiation of (64) followed by use of identity (6), we derive from the resulting equation,

$$\left|\mathcal{Q}_p^{m+2}(c,d)f(z)\right| \leq |\varepsilon(z)|\left|\mathcal{Q}_p^{m+2}(c,d)g(z)\right| + \frac{d|z|}{c+pd}|\varepsilon'(z)|\left|\mathcal{Q}_p^{m+1}(c,d)g(z)\right|. \tag{65}$$

By the application of estimation reflected in [22], our realization for $(z \in \mathcal{U})$ as follows

$$|\varepsilon'(z)| \leq 1 - |\varepsilon(z)|^2(1-|z|^2)^{-1} = 1 - \ddagger,$$

consequently by use of (63) in (65), we see

$$\left|\mathcal{Q}_p^{m+2}(c,d)f(z)\right| \leq \left\{|\varepsilon(z)| + \ddagger\frac{d(1+|B||z|)|z|}{(c+pd)(1-|A||z|)}\right\}\left|\mathcal{Q}_p^{m+2}(c,d)g(z)\right|,$$

now at $|\varepsilon(z)| = x \; (0 \leq x \leq 1)$ i.e. at boundary $|z| = r$ the inequality we obtain

$$\left|\mathcal{Q}_p^{m+1}(c,d)f(z)\right| \leq \frac{\varsigma(x)}{(1-r^2)\{(c+pd)(1-|A|r)\}}\left|\mathcal{Q}_p^{m+1}(c,d)g(z)\right|, \tag{66}$$

where

$$\varsigma(x) = -dr(1+|B|r)x^2 + (c+pd)(1-r^2)(1-|A|r)x + dr(1+|B|r).$$

The maximum value of ς is achieved at $x = 1$. If $y \in [0, \widetilde{r}]$, then the function

$$\varsigma(x) = -(|B|y+1)x^2dy + (c+pd)(1-y^2)(1-|A|y)x + (|B|y+1)dy.$$

The mapping $\varsigma(x)$ increases in $[0, 1]$, so that

$$\varsigma(1) = (c+pd)(1-y^2)(1-|A|y) \geq \varsigma(x).$$

(61) of Theorem 10 is evident using the above fact and (66).

Following are the consequences of the above theorem.

Corollary 10. *Let*

$$\mathrm{Re}\left\{\frac{\P g(z)}{\P g(z)}\right\} \prec \frac{1+Az}{1+Bz}(g \in \mathbb{S}_p^m(c,d,A,B)).$$

If for $f \in \mathrm{A}_p$, $f(z) \ll g(z)$ happens in $\mathcal{U}$, then we get following result on taking $-pt+l+p = c.t = d$ in above theorem,

$$|\P f(z)| \leq |\P g(z)|, \text{ for } \quad |z| \leq \widetilde{r}(p,l,t,A,B) = \beta,$$

where $\P = \mathcal{J}_p^{m+2}(t,l)$ and $\beta > 0$ is the least root of the equation

$$\wp|A|\alpha^3 - (\wp + t|B|)\alpha^2 - (\wp|A| + 2t)\alpha + \wp = 0, p + l = \wp.$$

Remark 2. Equation (62) of Theorem 10 looks

$$\emptyset\alpha^3 - (p+2)\alpha^2 - (\emptyset+2)\alpha + p = 0.$$

By substituting B, m, A, c as $-1, -1, 1-(2\Omega/p), 0$ respectively, this equation has three solutions and we can verify by taking $\alpha = 1$ as one solution, the other two is obtained by solving

$$\emptyset\alpha^2 - (\emptyset + p + 2)\alpha + p = 0, |p - 2\Omega| = \emptyset. \tag{67}$$

Hence least solution of (67), can be found out.

Taking view of this remark we are in a position to derive the result as given here.

Corollary 11. *Let*

$$f(z) \ll g(z), g \in \mathcal{S}_p^*(\Omega)\,(0 \leq \Omega < p),$$

and f as in Corollary 10 then absolute value of $f'(z)$ will be less than or equal to absolute value of $g'(z)$ for any z in $\mathcal{U}$ where

$$\widetilde{r}(p,\Omega) = \begin{cases} \dfrac{p+\emptyset+2-\sqrt{(p+\emptyset+2)^2-4p\emptyset}}{2\emptyset}, & \Omega \neq \dfrac{p}{2} \\ p(p+2)^{-1}, & \Omega = \dfrac{p}{2}. \end{cases}.$$

Corollary 11 is consistent with the result of MacGregor [23, p.96, Theorem 1B] on putting $p = 1$ and $\Omega = 0$.

On substituting parameters A, m, B as $1 - (2\Omega/p), -1, -1$ respectively we obtain

$$\mathcal{C}_p(\Omega) \subset \mathbb{S}_p^*(\rho), \text{ for } (p-1)/2 \leq \Omega < p, \tag{68}$$

where

$$\rho = \frac{p}{{}_2F_1\left(1, 2(p-\Omega); p+1; 2^{-1}\right)}. \tag{69}$$

The above discussion is consolidated in the corollary given below, witch by further specification of parameters i.e. $p = 1$ and $\Omega = 0$ we achieve the result reflected in [23, p.96, Theorem 1C].

Corollary 12. *For $(p-1) \leq 2\Omega < 2p$. Suppose*

$$f(z) \ll g(z), g \in \mathcal{C}_p(\Omega), f \in \mathrm{A}_p,$$

then consequence of Corollary 11 is also achieved for $|z| < \widetilde{k}(p, \rho) = \mathfrak{K}$, where

$$\mathfrak{K} = \begin{cases} \dfrac{p + \bigvee +2 - \sqrt{(p + \bigvee +2)^2 - 4p\bigvee}}{2\bigvee}, & 2\rho \neq p \\ p(p+2)^{-1}, & 2\rho = p \end{cases},$$

and $\bigvee = |p - 2\rho|$ and ρ is given by (69).

Theorem 11. *Let $0 < \mu < 1, 0 < \gamma \leq 1$ and $f \in \mathrm{A}_p(n)$ satisfies the following subordination condition*

$$\frac{\mathcal{Q}_p^m(n, c, d)f(z)}{z^p} \prec \left(\frac{1 + Az}{1 + Bz}\right)^\gamma \quad (z \in \mathcal{U}),$$

then

$$Re\left\{(1-\mu)\left(\frac{\mathcal{Q}_p^m(n; c, d)f(z)}{z^p}\right)^{\frac{1}{\gamma}} + \mu\frac{\mathcal{Q}_p^{m+1}(n; c, d)f(z)}{\mathcal{Q}_p^m(n; c, d)f(z)}\right\} > \mu + (1-\mu)\kappa$$

for $|z| < R \equiv R(p, n, \mu, c, d, \gamma, \kappa)$, where R is the smallest positive root of the equation

$$(1-\mu)(c+pd)(1-2\kappa)r^{2n} - 2\{(1-\mu)(c+pd)(1-\kappa) + n\mu d\gamma\}r^n + (1-\mu)(c+pd) = 0.$$

References

1. Swamy, S.R.: Inclusion properties of certain subclasses of analytic functions defined by a generalized multiplier transformation. Int. J. Math. Anal. **6**, 1553–1564 (2012)
2. Swamy, S.R.: Inclusion properties of certain subclasses of analytic functions. Int. Math. Forum **7**, 1751–1760 (2012)

3. Cătas, A.: On certain classes of p-valent functions defined by multiplier transformation. In: Proceedings of the International Symposium on Geometric Function Theory and Applications, Istanbul, Turkey, pp. 241-250 (2007)
4. Aghalary, R., Ali, R.M., Joshi, S.B., Ravichandran, V.: Inequalities for functions defined by certain linear operator. Int. J. Math. Sci. **4**, 267–274 (2005)
5. Kumar, S.S., Taneja, C.H., Ravichandran, V.: Classes of multivalent functions defined by Dziok-Srivastava linear operator and multiplier transformation. Kyungpook Math. J. **46**, 97–109 (2006)
6. Srivastava, H.M., Suchitra, K.B., Stephen, A.B., Sivasubramanian, S.: Inclusion and neighborhood properties of certain subclasses of multivalent functions of complex order. J. Inequalities Pure Appl. Math. **7**(5), 1–8 (2006)
7. Aouf, M.K., Mostafa, A.O.: On a subclasses of $n-p$-valent prestarlike functions. Comput. Math. Appl. **55**, 851–861 (2008)
8. Kamali, M., Orhan, H.: On a subclass of certain starlike functions with negative coefficients. Bull. Korean Math. Soc. **41**, 53–71 (2004)
9. Orhan, H., Kiziltunc, H.: A generalization on subfamily of p-valent functions with negative coefficients. Appl. Math. Comput. **155**, 521–530 (2004)
10. Cho, N.E., Kim, T.H.: Multiplier transformations and strongly close-to-convex functions. Bull. Korean Math. Soc. **40**(3), 399–410 (2003)
11. Cho, N.E., Srivastava, H.M.: Argument estimates of certain analytic functions defined by a class of multiplier transformations. Math. Comput. Model. **37**, 39–49 (2003)
12. Al-Oboudi, F.M.: On univalent functions defined by a generalized Salagean operator. Int. J. Math. Math. Sci. **27**, 1429–1436 (2004)
13. Salagean, G.S.: Subclasses of univalent functions. Lecture Notes in Math, vol. 1013, pp. 362–372. Springer, Heidelberg (1983)
14. Patel, J., Cho, N.E., Srivastava, H.M.: Certain subclasses of multivalent functions associated with a family of linear operators. Math. Comput. Model. **43**, 320–338 (2006)
15. MacGregor, T.H.: Functions whose derivative has a positive real part. Trans. Am. Math. Soc. **104**, 532–537 (1962)
16. Hallenbeck, D.J., Ruscheweyh, S.: Subordination by convex functions. Proc. Am. Math. Soc. **52**, 191–195 (1975)
17. Miller, S.S., Mocanu, P.T.: Differential Subordinations, Theory and Applications. Series on Monographs and Textbooks in Pure and Applied Mathematics, vol. 225. Marcel Dekker Inc., New York (2000)
18. Pashkouleva, D.Ž.: The starlikeness and spiral-convexity of certain subclasses of analytic functions. In: Srivastava, H.M., Owa, S. (eds.) Current Topics in Analytic Function Theory, pp. 266–273. World Scientific Publishing Company, Singapore (1992)
19. Wilken, D.R., Feng, J.: A remark on convex and starlike functions. J. Lond. Math. Soc. **21**, 287–290 (1980)
20. Whittaker, E.T., Watson, G.N.: A Course on Modern Analysis, An Introduction to the General Theory of Infinite Processess and of Analytic Functions: With an Account of the Principal Transcendental Functions, 4th edn. Cambridge University Press, (1927, Reprinted)
21. Patel, J.: On certain subclasses of multivalent functions involving Cho-Kwon-Srivastava operator, pp. 75–86. Ann. Univ. Mariae Curie-Sklodowska Sect. A, LX (2006)
22. Nehari, Z.: Conformal Mapping. MacGraw Hill Book Company, New York (1952)

23. MacGregor, T.H.: Majorization by univalent functions. Duke Math. J. **34**, 95–102 (1967)
24. Miller, S.S., Mocanu, P.T.: Univalent solutions of Briot-Bouquet differential subordinations. J. Differ. Equ. **58**, 297–309 (1985)
25. Stankiewicz, J., Stankiewicz, Z.: Some applications of the Hadamard convolution in the theory of functions. Ann. Uinv. Mariae Curie-Sklodowska Sect. A 22-24, 175-181 (1968/1970)

Inequalities for m-Convex Functions via Ψ-Caputo Fractional Derivatives

Ahmet Ocak Akdemir[1(✉)], Hemen Dutta[2], Ebru Yüksel[1], and Erhan Deniz[3]

[1] Faculty of Science and Letters, Department of Mathematics, Ağrı İbrahim Çeçen University, 04100 Ağrı, Turkey
aocakakdemir@gmail.com, yuksel.ebru90@hotmail.com
[2] Department of Mathematics, Gauhati University, Guwahati, India
hemen_dutta08@rediffmail.com
[3] Faculty of Science and Letters, Department of Mathematics, Kafkas University, 36100 Kars, Turkey
edeniz@gmail.com

Abstract. Fractional analysis has become a more popular topic in recent years as researchers working in applied mathematics and other areas of science have found numerios applications of it. Several new derivative and integral operators have been defined and various properties of these operators have been established. These new operators in the fractional analysis have attracted the attention of mathematicians working in the field of inequality and with the help of these new operators, the inequality theory has moved towards a new trend. Numerous inequalities have been studied with the caputo derivative, which is one of the most used operators in fractional analysis. In this context, we have established some new integral inequalities for m-convex functions by using ψ-Caputo derivatives and some basic definitions and techniques in this article. We have also given some special cases of our results for convexity.

1 Introduction

We will begin with the concept of convexity and m-convexity, which has applications in many areas of mathematics, engineering sciences, statistics and programming, and has become the focus of interest for many researchers on inequality theory and convex analysis.

Definition 1 (See [6, 7]). Let I be an interval in $\mathbb{R}$. Then $f : I \subseteq \mathbb{R} \to \mathbb{R}$ is said to be convex if

$$f(tx + (1-t)y) \leq tf(x) + (1-t)f(y)$$

is valid for all $x, y \in I$ and $t \in [0, 1]$.

Definition 2 (See [7, 17]). For $f : [0, b] \to \mathbb{R}$ and $m \in (0, 1]$, if the following inequality

$$f(tx + m(1-t)y) \leq tf(x) + m(1-t)f(y)$$

M. Zeki Sarıkaya et al. (Eds.): ICMRS 2019, LNNS 123, pp. 215–224, 2020.
https://doi.org/10.1007/978-3-030-43002-3_17

is valid for all $x, y \in [0, b]$ and $t \in [0, 1]$, we say that $f(x)$ is a m-convex function on $[0, b]$.

Obviously, if we put $m = 1$ in Definition 2, then f is just an ordinary convex function.

In this article, we will try to briefly mention the fractional derivatives and fractional integrals, which are the main motivation point, which have led to a new trend in analysis and applied mathematics in recent years and which have created new approaches in solving problems and opened the door to stable solutions. What makes the new operators valuable in fractional analysis is that they provide stable solutions to problems as well as shedding light on real life problems.

Let us now look at the definition of the Riemann-Liouville fractional integral, which is one of the well-known fractional integral operators.

Definition 3 (See [19]). Let $f \in L_1[a, b]$. The Riemann-Liouville integrals $J_{a+}^{\alpha} f$ and $J_{b-}^{\alpha} f$ of order $\alpha > 0$ are defined by

$$J_{a+}^{\alpha} f(t) = \frac{1}{\Gamma(\alpha)} \int_a^t (t - x)^{\alpha - 1} f(x) dx, \ t > a$$

and

$$J_{b-}^{\alpha} f(t) = \frac{1}{\Gamma(\alpha)} \int_t^b (x - t)^{\alpha - 1} f(x) dx, \ t < b$$

respectively where $\Gamma(\alpha) = \int_0^{\infty} e^{-t} t^{\alpha - 1} dt$. Here $J_{a+}^{0} f(t) = J_{b-}^{0} f(t) = f(t)$.

In the case of $\alpha = 1$, the fractional integral reduces to classical integral.

With the help of this operator and later fractional derivative and integral operators, many researchers have made impressive studies in the field of inequality theory, engineering and applied mathematics. Some of these are presented to readers in the form of [1–5, 8–16].

In [20], Khalil et al. gave a new definition that is called "conformable fractional derivative". They not only proved further properties of these definitions but also gave the differences with the other fractional derivatives. Besides, another considerable study have presented by Abdeljawad to discuss the basic concepts of fractional calculus. In [21], Abdeljawad gave the following definitions of right-left conformable fractional integrals.

Definition 4 (See [21]). Let $\alpha \in (n, n + 1]$, $n = 0, 1, 2, \ldots$ and set $\beta = \alpha - n$. Then the left conformable fractional integral of any order $\alpha > 0$ is defined by

$$\left(I_{\alpha}^{a} f\right)(t) = \frac{1}{n!} \int_a^t (t - x)^n (x - a)^{\beta - 1} f(x) dx, \ t > a$$

Definition 5 (See [21]). Analogously, the right conformable fractional integral of any order $\alpha > 0$ is defined by

$$\left({}^{b}I_{\alpha} f\right)(t) = \frac{1}{n!} \int_t^b (x - t)^n (b - x)^{\beta - 1} f(x) dx, \ b > t$$

Notice that if $\alpha = n+1$ then $\beta = \alpha - n = n+1-n = 1$, hence $\left(I^{\alpha}_{n+1} f\right)(t) = \left(J^{n+1}_{a+} f\right)(t)$ and $\left({}^{b}I_{n+1} f\right)(t) = \left(J^{n+1}_{b-} f\right)(t)$.

We will now recall an important derivative operator which is the basis for the main conclusions of the article and allows us to obtain inequalities for m-convex functions.

Definition 6 (See [18]). Let $\alpha > 0, n \in \mathbb{N}$, I is the interval $-\infty \le a < b \le \infty$, $f, \psi \in C^n(I)$ two functions such that ψ is increasing and $\psi'(x) \neq 0$, for all $x \in I$. The left ψ-Caputo fractional derivative of f of order α is given by

$$^{C}D^{\alpha,\psi}_{a^+} f(x) := I^{n-\alpha,\psi}_{a^+} \left(\frac{1}{\psi'(x)} \frac{d}{dx}\right)^n f(x),$$

and the right ψ-Caputo fractional derivative of f by

$$^{C}D^{\alpha,\psi}_{b^-} f(x) := I^{n-\alpha,\psi}_{b^-} \left(-\frac{1}{\psi'(x)} \frac{d}{dx}\right)^n f(x),$$

where $n = [\alpha] + 1$ for $\alpha \notin \mathbb{N}$, $n = \alpha$ for $\alpha \in \mathbb{N}$.

To simplify the notation, we will use the abbreviated symbol

$$f^{[n]}_{\psi} f(x) := \left(\frac{1}{\psi'(x)} \frac{d}{dx}\right)^n f(x).$$

From the definition, it is clear that, given $\alpha = m \in \mathbb{N}$, $^{C}D^{\alpha,\psi}_{a^+} f(x) = f^{[m]}_{\psi}(x)$ and $^{C}D^{\alpha,\psi}_{b^-} f(x) = (-1)^m f^{[m]}_{\psi}(x)$ and if $\alpha \notin \mathbb{N}$, then

$$^{C}D^{\alpha,\psi}_{a^+} f(x) = \frac{1}{\Gamma(n-\alpha)} \int_a^x \psi'(t)(\psi(x) - \psi(t))^{n-\alpha-1} f^{[n]}_{\psi}(t)dt$$

and

$$^{C}D^{\alpha,\psi}_{b^-} f(x) = \frac{1}{\Gamma(n-\alpha)} \int_x^b \psi'(t)(\psi(t) - \psi(x))^{n-\alpha-1} (-1)^n f^{[n]}_{\psi}(t)dt.$$

After this basic information and reminders, in the next part of our study, various integral inequalities for m-convex functions will be obtained with the help of ψ-Caputo fractional derivative operator which is a new derivative operator in fractional analysis.

2 Main Results

Theorem 1. Suppose that $n \in \mathbb{N}$, $f : I \subseteq \mathbb{R} \to \mathbb{R}$ be n-times differentiable, strictly increasing and real valued function such that $\psi' \in L([\mu, \xi])$. If $f^{(n)}$ is m-convex, then the following inequality for ψ-Caputo fractional derivative holds:

$$\Gamma(n-\alpha+1)\left({}^{c}D^{\alpha-1,\psi}_{\mu^+} f\right)(\kappa) + \Gamma(n-\beta+1)\left({}^{c}D^{\beta-1,\psi}_{\xi^-} f\right)(\kappa)$$
$$\le \frac{(\psi(\kappa) - \psi(\mu))^{n-\alpha}}{m\kappa - \mu}\Big[f^{(n)}(\mu)\left(\psi(\kappa)(m\kappa - \kappa) - \psi(\mu)(m\kappa - \mu)\right)$$

$$+mf^{(n)}(\kappa)\psi(\kappa)(\kappa-\mu)-\left(mf^{(n)}(\kappa)-f^{(n)}(\mu)\right)\int_{\mu}^{\kappa}\psi(x)dx\Big]$$

$$+\frac{(\psi(\xi)-\psi(\kappa))^{n-\beta}}{m\xi-\kappa}\Big[f^{(n)}(\kappa)\left(\psi(\xi)(m\xi-\xi)-\psi(\kappa)(m\xi-\kappa)\right)$$

$$+mf^{(n)}(\xi)\psi(\xi)(\xi-\kappa)-\left(mf^{(n)}(\xi)-f^{(n)}(\kappa)\right)\int_{\kappa}^{\xi}\psi(x)dx\Big]$$

where $m\in(0,1]$, $\mu,\xi\in I$, $\mu<\xi$ and $\alpha,\beta\geq 1$.

Proof. Let $\kappa\in[\mu,\xi]$ and $x\in[\mu,\kappa]$. Since ψ is a strictly increasing and differentiable function with $\psi'(x)>0$, we can write

$$\psi'(x)(\psi(\kappa)-\psi(x))^{n-\alpha}\leq\psi'(x)(\psi(\kappa)-\psi(\mu))^{n-\alpha}. \tag{2.1}$$

By using m-convexity of $f^{(n)}$, we have

$$f^{(n)}(x)\leq\frac{m\kappa-x}{m\kappa-\mu}f^{(n)}(\mu)+\frac{x-\mu}{m\kappa-\mu}mf^{(n)}(\kappa). \tag{2.2}$$

If we multiply the inequalities (2.1) and (2.2), then integrating the resulting inequality on $[\mu,\kappa]$, we get

$$\int_{\mu}^{\kappa}(\psi(\kappa)-\psi(x))^{n-\alpha}f^{(n)}(x)\psi'(x)dx$$
$$\leq\frac{(\psi(\kappa)-\psi(\mu))^{n-\alpha}}{m\kappa-\mu}\left(f^{(n)}(\mu)\int_{\mu}^{\kappa}(m\kappa-x)\psi'(x)dx+mf^{(n)}(\kappa)\int_{\mu}^{\kappa}(x-\mu)\psi'(x)dx\right) \tag{2.3}$$

By taking into account the definition of ψ-Caputo derivative, we obtain

$$\Gamma(n-\alpha+1)\left({}^{c}D_{\mu^{+}}^{\alpha-1,\psi}f\right)(\kappa)$$
$$\leq\frac{(\psi(\kappa)-\psi(\mu))^{n-\alpha}}{m\kappa-\mu}\Big[f^{(n)}(\mu)\left(\psi(\kappa)(m\kappa-\kappa)-\psi(\mu)(m\kappa-\mu)\right)$$
$$+mf^{(n)}(\kappa)\psi(\kappa)(\kappa-\mu)-\left(mf^{(n)}(\kappa)+f^{(n)}(\mu)\right)\int_{\mu}^{\kappa}\psi(x)dx\Big] \tag{2.4}$$

By a similar argument, the following inequality holds:

$$\psi'(x)(\psi(x)-\psi(\kappa))^{n-\beta}\leq\psi'(x)(\psi(\xi)-\psi(\kappa))^{n-\beta} \tag{2.5}$$

for $\kappa\in[\mu,\xi]$ and $x\in[\kappa,\xi]$.

By using m-convexity of $f^{(n)}$ again, we can write

$$f^{(n)}(x)\leq\frac{m\xi-x}{m\xi-\kappa}f^{(n)}(\kappa)+\frac{x-\kappa}{m\xi-\kappa}mf^{(n)}(\xi) \tag{2.6}$$

By making the similar processes, we can easily have

$$\int_{\kappa}^{\xi} (\psi(x) - \psi(\kappa))^{n-\beta} f^{(n)}(x)\psi'(x)dx$$
$$\leq \frac{(\psi(\xi) - \psi(\kappa))^{n-\beta}}{m\xi - \kappa}\left(f^{(n)}(\kappa)\int_{\kappa}^{\xi} (m\xi - x)\psi'(x)dx + mf^{(n)}(\xi)\int_{\kappa}^{\xi} (x - \kappa)\psi'(x)dx\right) \tag{2.7}$$

Simplifying the inequality (2.7) with definition of ψ-Caputo fractional derivative, it is obvious that;

$$\Gamma(n - \beta + 1)\left({}^{c}D_{\xi^-}^{\beta-1,\psi} f\right)(\kappa)$$
$$\leq \frac{(\psi(\xi) - \psi(\kappa))^{n-\beta}}{m\xi - \kappa}\Big[f^{(n)}(\kappa)\left(\psi(\xi)(m\xi - \xi) - \psi(\kappa)(m\xi - \kappa)\right)$$
$$+mf^{(n)}(\xi)\psi(\xi)(\xi - \kappa) - \left(mf^{(n)}(\xi) + f^{(n)}(\kappa)\right)\int_{\kappa}^{\xi} \psi(x)dx\Big] \tag{2.8}$$

By adding the inequalities (2.4) and (2.8), we get the desired result.

Corollary 1. If we choose $\alpha = \beta$ in Theorem 1, we have the following new inequality:

$$\Gamma(n - \alpha + 1)\left[\left({}^{c}D_{\mu^+}^{\alpha-1,\psi} f\right)(\kappa) + \left({}^{c}D_{\xi^-}^{\alpha-1,\psi} f\right)(\kappa)\right]$$
$$\leq \frac{(\psi(\kappa) - \psi(\mu))^{n-\alpha}}{m\kappa - \mu}\Big[f^{(n)}(\mu)\left(\psi(\kappa)(m\kappa - \kappa) - \psi(\mu)(m\kappa - \mu)\right)$$
$$+mf^{(n)}(\kappa)\psi(\kappa)(\kappa - \mu) - \left(mf^{(n)}(\kappa) + f^{(n)}(\mu)\right)\int_{\mu}^{\kappa} \psi(x)dx\Big]$$
$$+ \frac{(\psi(\xi) - \psi(\kappa))^{n-\alpha}}{m\xi - \kappa}\Big[f^{(n)}(\kappa)\left(\psi(\xi)(m\xi - \xi) - \psi(\kappa)(m\xi - \kappa)\right)$$
$$+mf^{(n)}(\xi)\psi(\xi)(\xi - \kappa) - \left(mf^{(n)}(\xi) + f^{(n)}(\kappa)\right)\int_{\kappa}^{\xi} \psi(x)dx\Big]$$

Theorem 2. Suppose that $n \in \mathbb{N}$, $f : I \subseteq \mathbb{R} \to \mathbb{R}$ be n-times differentiable, strictly increasing and real valued function such that $\psi' \in L([\mu, \xi])$. If $\left|f^{(n+1)}\right|$ is m-convex, then the following inequality for ψ-Caputo fractional derivative holds:

$$\Big|\Gamma(n - \alpha + 1)\left({}^{C}D_{\mu^+}^{\alpha,\psi} f\right)(\kappa) + \Gamma(n - \beta + 1)\left({}^{C}D_{\xi^-}^{\beta,\psi} f\right)(\kappa)$$
$$-\left[f^{(n)}(\mu)(\psi(\kappa) - \psi(\mu))^{n-\alpha} + f^{(n)}(\xi)(\psi(\xi) - \psi(\kappa))^{n-\beta}\right]\Big|$$
$$\leq \frac{(\psi(\kappa) - \psi(\mu))^{n-\alpha}}{2(m\kappa - \mu)}\left(\left|f^{(n+1)}(\mu)\right|(\kappa - \mu)(2mk - \kappa - \mu) + \left|mf^{(n+1)}(\kappa)\right|(\kappa - \mu)^2\right)$$
$$+ \frac{(\psi(\xi) - \psi(\kappa))^{n-\beta}}{2(m\xi - \kappa)}\left(\left|f^{(n+1)}(\kappa)\right|(\xi - \kappa)(2m\xi - \xi - \kappa) + \left|mf^{(n+1)}(\xi)\right|(\xi - \kappa)^2\right)$$

where $m \in (0, 1]$, $\mu, \xi \in I$, $\mu < \xi$ and $\alpha, \beta \geq 0$.

Proof. From the definition of $f^{(n+1)}$, we can write

$$\left|f^{(n+1)}(x)\right| \leq \frac{m\kappa - x}{m\kappa - \mu}\left|f^{(n+1)}(\mu)\right| + \frac{x-\mu}{m\kappa - \mu}\left|mf^{(n+1)}(\kappa)\right| \tag{2.9}$$

Namely,

$$f^{(n+1)}(x) \leq \frac{m\kappa - x}{m\kappa - \mu}\left|f^{(n+1)}(\mu)\right| + \frac{x-\mu}{m\kappa - \mu}\left|mf^{(n+1)}(\kappa)\right| \tag{2.10}$$

Since ψ is strictly increasing and differentiable function for $\kappa \in [\mu, \xi]$ and $x \in [\mu, \kappa]$, we have

$$(\psi(\kappa) - \psi(x))^{n-\alpha} \leq (\psi(\kappa) - \psi(\mu))^{n-\alpha} \tag{2.11}$$

If we multiply (2.10) and (2.11), then integrating on $[\mu, \kappa]$, we deduce

$$\begin{aligned}
&\int_{\mu}^{\kappa} (\psi(\kappa) - \psi(x))^{n-\alpha} f^{(n+1)}(x)dx \\
&\leq \frac{(\psi(\kappa) - \psi(\mu))^{n-\alpha}}{m\kappa - \mu}\left(\left|f^{(n+1)}(\mu)\right| \int_{\mu}^{\kappa} (m\kappa - x)dx + \left|mf^{(n+1)}(\kappa)\right| \int_{\mu}^{\kappa} (x-\mu)dx\right) \\
&= \frac{(\psi(\kappa) - \psi(\mu))^{n-\alpha}}{2(m\kappa - \mu)}\left(\left|f^{(n+1)}(\mu)\right| (\kappa - \mu)(2mk - \kappa - \mu)\right. \\
&\left. + \left|mf^{(n+1)}(\kappa)\right|(\kappa - \mu)^2\right)
\end{aligned} \tag{2.12}$$

Thus, it is clear to see that

$$\begin{aligned}
&\int_{\mu}^{\kappa} (\psi(\kappa) - \psi(x))^{n-\alpha} f^{(n+1)}(x)dx \\
&= -f^{(n)}(\mu)(\psi(\kappa) - \psi(\mu))^{n-\alpha} + \Gamma(n-\alpha+1)\left({}^{C}D_{\mu^+}^{\alpha,\psi} f\right)(\kappa)
\end{aligned} \tag{2.13}$$

Rewriting the inequality (2.12), we obtain

$$\begin{aligned}
&\Gamma(n-\alpha+1)\left({}^{C}D_{\mu^+}^{\alpha,\psi} f\right)(\kappa) - f^{(n)}(\mu)(\psi(\kappa) - \psi(\mu))^{n-\alpha} \\
&\leq \frac{(\psi(\kappa) - \psi(\mu))^{n-\alpha}}{2(m\kappa - \mu)}\left(\left|f^{(n+1)}(\mu)\right| (\kappa - \mu)(2mk - \kappa - \mu)\right. \\
&\left. + \left|mf^{(n+1)}(\kappa)\right|(\kappa - \mu)^2\right)
\end{aligned} \tag{2.14}$$

By using the properties of modulus, we have

$$f^{(n+1)}(x) \geq \frac{m\kappa - x}{m\kappa - \mu}\left|f^{(n+1)}(\mu)\right| + \frac{x-\mu}{m\kappa - \mu}\left|mf^{(n+1)}(\kappa)\right| \tag{2.15}$$

If we apply the similar argument to (2.15), we can easily obtain

$$\left|\Gamma(n-\alpha+1)\left({}^{C}D_{\mu^+}^{\alpha,\psi} f\right)(\kappa) - f^{(n)}(\mu)(\psi(\kappa) - \psi(\mu))^{n-\alpha}\right|$$

$$\leq \frac{(\psi(\kappa)-\psi(\mu))^{n-\alpha}}{2(m\kappa-\mu)}\Big(\left|f^{(n+1)}(\mu)\right|(\kappa-\mu)(2mk-\kappa-\mu)$$
$$+\left|mf^{(n+1)}(\kappa)\right|(\kappa-\mu)^2\Big) \tag{2.16}$$

We will proceed with m-convexity of $\left|f^{(n+1)}\right|$ and $\kappa\in[\mu,\xi]$,$x\in[\kappa,\xi]$ and $\beta>0$, we can prove

$$\left|\Gamma(n-\beta+1)\left({}^{C}D_{\xi^-}^{\beta,\psi}f\right)(\kappa)-f^{(n)}(\xi)(\psi(\xi)-\psi(\kappa))^{n-\beta}\right|$$
$$\leq \frac{(\psi(\xi)-\psi(\kappa))^{n-\beta}}{2(m\xi-\kappa)}\Big(\left|f^{(n+1)}(\kappa)\right|(\xi-\kappa)(2m\xi-\xi-\kappa)$$
$$+\left|mf^{(n+1)}(\xi)\right|(\xi-\kappa)^2\Big) \tag{2.17}$$

We omit the details. This completes the proof.

Corollary 2. If we choose $\alpha=\beta$ in Theorem 2 for ψ-Caputo derivative, we get the following inequality:

$$\Big|\Gamma(n-\alpha+1)\Big\{\left({}^{C}D_{\mu^+}^{\alpha,\psi}f\right)(\kappa)+\left({}^{C}D_{\xi^-}^{\alpha,\psi}f\right)(\kappa)\Big\}$$
$$-\left[f^{(n)}(\mu)(\psi(\kappa)-\psi(\mu))^{n-\alpha}+f^{(n)}(\xi)(\psi(\xi)-\psi(\kappa))^{n-\alpha}\right]\Big|$$
$$\leq \frac{(\psi(\kappa)-\psi(\mu))^{n-\alpha}}{2(m\kappa-\mu)}\Big(\left|f^{(n+1)}(\mu)\right|(\kappa-\mu)(2mk-\kappa-\mu)+\left|mf^{(n+1)}(\kappa)\right|(\kappa-\mu)^2\Big)$$
$$+\frac{(\psi(\xi)-\psi(\kappa))^{n-\alpha}}{2(m\xi-\kappa)}\Big(\left|f^{(n+1)}(\kappa)\right|(\xi-\kappa)(2m\xi-\xi-\kappa)+\left|mf^{(n+1)}(\xi)\right|(\xi-\kappa)^2\Big)$$

Lemma 1. $f:[\mu,\xi]\to\mathbb{R}$ be convex function and symmetric for $\frac{\mu+\xi}{2}$. Then, the following inequality holds:

$$f\left(\frac{\mu+\xi}{2}\right)\leq f(\kappa), \tag{2.18}$$

for $\kappa\in[\mu,\xi]$.

Theorem 3. Suppose that $n\in\mathbb{N}$, $f:I\subseteq\mathbb{R}\to\mathbb{R}$ be n-times differentiable, strictly increasing and real valued function such that $\psi'\in L([\mu,\xi])$. If $f^{(n)}$ is m-convex and symmetric for $\frac{\mu+\xi}{2}$, then the following inequality for ψ-Caputo fractional derivative holds:

$$\frac{1}{2}\left(\frac{1}{n-\alpha+1}+\frac{1}{n-\beta+1}\right)f^{(n)}\left(\frac{\mu+\xi}{2}\right)$$

$$\leq \frac{\Gamma(n-\alpha+1)\left({}^{C}D_{\xi+}^{\alpha-1,\psi}f\right)(\mu)}{2(\psi(\xi)-\psi(\mu))^{n-\alpha+1}}+\frac{\Gamma(n-\beta+1)\left({}^{C}D_{\eta^-}^{\beta-1,\psi}f\right)(\mu)}{2(\psi(\xi)-\psi(\mu))^{n-\beta+1}}$$

$$\leq \frac{1}{(\psi(\xi)-\psi(\mu))(m\xi-\mu)}\Big[f^{(n)}(\mu)\left(\psi(\xi)(m\xi-\xi)-\psi(\mu)(m\xi-\mu)\right)$$

$$+mf^{(n)}(\xi)\psi(\xi)(\xi-\mu)-\left(mf^{(n)}(\xi)-f^{(n)}(\mu)\right)\int_{\mu}^{\xi}\psi(\kappa)d\kappa\Big] \tag{2.19}$$

where $m \in (0,1]$, $\mu,\xi \in I$, $\mu<\xi$ and $\alpha,\beta \geq 0$.

Proof. Since ψ is a differentiable and strictly increasing function with $\kappa \in [\mu,\xi]$ and $\psi'(\kappa)>0$, we can write

$$\psi'(\kappa)(\psi(\kappa)-\psi(\mu))^{n-\beta} \leq \psi'(\kappa)(\psi(\xi)-\psi(\mu))^{n-\beta}. \tag{2.20}$$

By using the m-convexity of $f^{(n)}$, we get

$$f^{(n)}(\kappa) \leq \frac{m\xi-\kappa}{m\xi-\mu}f^{(n)}(\mu)+\frac{\kappa-\mu}{m\xi-\mu}mf^{(n)}(\xi). \tag{2.21}$$

If we multiply the inequalities (2.20) and (2.21), then by integrating on $[\mu,\xi]$ with respect to κ, we have

$$\int_{\mu}^{\xi}(\psi(\kappa)-\psi(\mu))^{n-\beta}f^{(n)}(\kappa)\psi'(\kappa)d\kappa$$

$$\leq \frac{(\psi(\xi)-\psi(\mu))^{n-\beta}}{m\xi-\mu}\left(mf^{(n)}(\xi)\int_{\mu}^{\xi}(\kappa-\mu)\psi'(\kappa)d\kappa+f^{(n)}(\mu)\int_{\mu}^{\xi}(m\xi-\kappa)\psi'(\kappa)d\kappa\right) \tag{2.22}$$

By rewriting the (2.22) in terms of definition of ψ- Caputo fractional derivative, we obtain

$$\Gamma(n-\beta+1)\left({}^{C}D_{\xi^-}^{\beta-1,\psi}f\right)(\mu)$$

$$\leq \frac{(\psi(\xi)-\psi(\mu))^{n-\beta}}{m\xi-\mu}\Big[f^{(n)}(\mu)\left(\psi(\xi)(m\xi-\xi)-\psi(\mu)(m\xi-\mu)\right)$$

$$+mf^{(n)}(\xi)\psi(\xi)(\xi-\mu)-\left(mf^{(n)}(\xi)-f^{(n)}(\mu)\right)\int_{\mu}^{\xi}\psi(\kappa)d\kappa\Big] \tag{2.23}$$

For $\kappa \in [\mu,\xi]$ and $\alpha>0$, the following inequality holds:

$$\psi'(\kappa)(\psi(\xi)-\psi(\kappa))^{n-\alpha} \leq \psi'(\kappa)(\psi(\xi)-\psi(\mu))^{n-\alpha} \tag{2.24}$$

By making use of similar computations for (2.21) and (2.24), one can easily obtain

$$\Gamma(n-\alpha+1)\left({}^{C}D_{\mu^+}^{\alpha-1,\psi}f\right)(\xi)$$

$$\leq \frac{(\psi(\xi)-\psi(\mu))^{n-\alpha}}{m\xi-\mu}\Big[f^{(n)}(\mu)\left(\psi(\xi)(m\xi-\xi)-\psi(\mu)(m\xi-\mu)\right)$$
$$+mf^{(n)}(\xi)\psi(\xi)(\xi-\mu)-\left(mf^{(n)}(\xi)-f^{(n)}(\mu)\right)\int_{\mu}^{\xi}\psi(\kappa)d\kappa\Big]$$

By multiplying the inequality (2.18) of Lemma 1 by $\psi'(\kappa)(\psi(\kappa)-\psi(\mu))^{n-\beta}$, then integrating on $[\mu,\xi]$ with respect to κ, we can write

$$f^{(n)}\left(\frac{\mu+\xi}{2}\right)\int_{\mu}^{\xi}(\psi(\kappa)-\psi(\mu))^{\eta-\beta}\psi'(\kappa)d\kappa$$
$$\leq \int_{\mu}^{\xi}(\psi(\kappa)-\psi(\mu))^{\eta-\beta}\psi'(\kappa)f^{(n)}(\kappa)d\kappa \tag{2.25}$$

Therefore, from the definition of ψ-Caputo fractional derivative, we get

$$\frac{f^{(n)}\left(\frac{\mu+\xi}{2}\right)}{2(n-\beta+1)} \leq \frac{\Gamma(n-\beta+1)\left({}^{C}D_{\eta^-}^{\beta-1,\psi}f\right)(\mu)}{2(\psi(\xi)-\psi(\mu))^{n-\beta+1}} \tag{2.26}$$

By a similar way, we have

$$\frac{f^{(n)}\left(\frac{\mu+\xi}{2}\right)}{2(n-\alpha+1)} \leq \frac{\Gamma(n-\alpha+1)\left({}^{C}D_{\xi^+}^{\alpha-1,\psi}f\right)(\mu)}{2(\psi(\xi)-\psi(\mu))^{n-\alpha+1}} \tag{2.27}$$

The proof is immediately follows from addition of (2.26) and (2.27).

Corollary 3. If we choose $\alpha=\beta$ in Theorem 3, we have the following new result:

$$\left(\frac{1}{n-\alpha+1}\right)f^{(n)}\left(\frac{\mu+\xi}{2}\right)$$
$$\leq \frac{\Gamma(n-\alpha+1)\left\{\left({}^{C}D_{\xi^+}^{\alpha-1,\psi}f\right)(\mu)+\left({}^{C}D_{\eta^-}^{\alpha-1,\psi}f\right)(\xi)\right\}}{2(\psi(\xi)-\psi(\mu))^{n-\alpha+1}}$$
$$\leq \frac{1}{(\psi(\xi)-\psi(\mu))(m\xi-\mu)}\Big[f^{(n)}(\mu)\left(\psi(\xi)(m\xi-\xi)-\psi(\mu)(m\xi-\mu)\right)$$
$$+mf^{(n)}(\xi)\psi(\xi)(\xi-\mu)-\left(mf^{(n)}(\xi)-f^{(n)}(\mu)\right)\int_{\mu}^{\xi}\psi(\kappa)d\kappa\Big]$$

3 Conclusion

Within the scope of our study, we have performed some new Hadamard type integral inequalities for differentiable m-convex functions via ψ-Caputo fractional derivative. To prove the results, we have used some known definitions, basic techniques and conditions of Theorems. We also give some special cases of our results. Interested researchers can obtain similar results for different types of convex functions. In addition, more general results and iterations can be selected using various derivative operators as further study subjects.

References

1. Abdeljawad, T., Baleanu, D.: On fractional derivatives with exponential kernel and their discrete versions. Rep. Math. Phys. **80**(1), 11–27 (2017)
2. Abdeljawad, T.: Fractional operators with exponential kernels and a Lyapunov type inequality. Adv. Differ. Equ. **2017**(1), 313 (2017)
3. Baleanu, D., Diethelm, K., Scalas, E., Trujillo, J.J.: Fractional Calculus: Models and Numerical Methods, 2nd edn. World Scientific Publishing Company, Singapore (2016)
4. Caputo, M., Fabrizio, M.: A new definition of fractional derivative without singular kernal. Prog. Fract. Differ. Appl. **1**(2), 73–85 (2015)
5. Das, S.: Functional Fractional Calculus. Springer, Heidelberg (2011)
6. Dragomir, S.S., Agarwal, R.P.: Two inequalities for differentiable mappings and applications to special means of real numbers and to trapezoidal formula. Appl. Math. Lett. **11**(5), 91–95 (1998)
7. Dragomir, S.S., Pearce, C.: Selected topics on Hermite-Hadamard inequalities and applications (2003). https://rgmia.org/papers/monographs/Master.pdf
8. Kilbas, A., Srivastava, M.H., Trujillo, J.J.: Theory and Application of Fractional Differential Equations. Mathematics Studies, vol. 204. Amsterdam, New York (2006)
9. Set, E., Akdemir, A.O., Gürbüz, M.: Integral inequalities for different kinds of convex functions involving Riemann-Liouville fractional integrals. Karaelmas Sci. Eng. J. **7**(1), 140–144 (2017)
10. Sarikaya, M.Z., Set, E., Yaldiz, H., Basak, N.: Hermite-Hadamard's inequalities for fractional integrals and related fractional inequalities. Math. Comput. Model. **57**, 2403–2407 (2013)
11. Farid, G., Nazeer, W., Saleem, M.S., Mehmood, S., Kang, S.M.: Bounds of Riemann-Liouville fractional integrals in general form via convex functions and their applications. Mathematics **6**, 248 (2018)
12. Gorenflo, R., Mainardi, F.: Fractional Calculus: Integral and Differential Equations of Fractional Order, pp. 223–276. Springer, New York (1997)
13. Jarad, F., Abdeljawad, T., Alzabut, J.: Generalized fractional derivatives generated by a class of local proportional derivatives. Eur. Phys. J. Spec. **226**, 3457–3471 (2018)
14. Jensen, J.L.W.V.: Sur les fonctions convexes et les inegalites entre les voleurs mogernmes. Acta Math. **30**, 175–193 (1906)
15. Jleli, M., O'Regan, D., Samet, B.: On Hermite-Hadamard type inequalities via generalized fractional integrals. Turk. J. Math. **40**, 1221–1230 (2016)
16. Kannappan, P.I.: Functional Equations and Inequalities with Applications. Springer, Dordrecht (2009)
17. Toader, G.: Some generalizations of the convexity. In: Proceedings of the Colloquium On Approximation and Optimization, Univ. Cluj-Napoca, Cluj-Napoca, pp. 329–338 (1984)
18. Almeida, R.: A Caputo fractional derivative of a function with respect to another function. Commun. Nonlinear Sci. Numer. Simul. **44**, 460–481 (2017)
19. Podlubny, I.: Fractional Differential Equations. Mathematics in Science and Engineering. Academic Press, New York (1999)
20. Khalil, R., Al Horani, M., Yousef, A., Sababheh, M.: A new definition of fractional derivative. J. Comput. Appl. Math. **264**, 65–70 (2014)
21. Abdeljawad, T.: On conformable fractional calculus. J. Comput. Appl. Math. **279**, 57–66 (2015)

A Certain Group Acting on a C^*-Algebra Generated by Countable Infinitely Many Semicircular Elements

Ilwoo Cho(✉)

Department of Mathematics and Statistics, St. Ambrose University,
421 Ambrose Hall, 518 W. Locust St., Davenport, IA 52803, USA
choilwoo@sau.edu

Abstract. In this paper, we consider an operator-algebraic structure of a C^*-algebra generated by mutually free, countable-infinitely many semicircular elements, and then construct-and-study a certain abelian infinite cyclic group acting on the C^*-algebra. We show how the group algebra generated by the group deform the free probability on the C^*-algebra.

Keywords: Free probability · Semicircular elements · Free-isomorphisms

1991 Mathematics Subject Classification: 46L10 · 46L54 · 47L55

1 Introduction

The main purposes of this paper are (i) to study a structure theorem of the C^*-*algebra* $\mathcal{X}$ generated by a set $X = \{x_n\}_{n=1}^{\infty}$ of mutually free, $|\mathcal{N}|$-many *semicircular elements* x_n's, (ii) to consider free-distributional data on $\mathcal{X}$, (iii) to construct a certain *group* λ, and the corresponding (pure-algebraic) *group algebra* Λ, acting on $\mathcal{X}$, and (iv) to show how the elements of Λ affect the *free probability* on $\mathcal{X}$.

1.1 Background

The study of *semicircular elements* are not only used in operator algebra theory (e.g., [20,21,29] and [30]), but also applicable to the related fields (e.g., [5–8,10,11] and [12]). The *free distributions* of semicircular elements are well-known, and well-characterized in free probability. These free distributions are called *the semicircular law* (e.g., [1,17,18,21,28,29] and [30]).

Studying semicircular elements is playing a key role in free-probabilistic operator algebra theory (including quantum statistical physics) by the (*free*) *central limit theorem*(*s*), e.g., see [2,17,19,28,29] and [30]. From combinatorial

M. Zeki Sarıkaya et al. (Eds.): ICMRS 2019, LNNS 123, pp. 225–259, 2020.
https://doi.org/10.1007/978-3-030-43002-3_18

approaches (e.g., [17,22] and [23]), the free distributions of such semicircular elements are universally characterized by the *Catalan numbers* c_n,

$$c_n = \frac{1}{n+1}\binom{2n}{n} = \frac{1}{n+1}\left(\frac{(2n)!}{n!(2n-n)!}\right) = \frac{(2n)!}{n!(n+1)!}, \tag{1.1}$$

for all $n \in \mathcal{N}_0 = \mathcal{N} \cup \{0\}$. i.e., the semicircular law is characterized by the *free-moment sequence*,

$$\left(\omega_n c_{\frac{n}{2}}\right)_{n=1}^{\infty} = (0, c_1, 0, c_2, 0, c_3, \ ...),$$

with

$$\omega_n = \begin{cases} 1 \text{ if } n \text{ is even} \\ 0 \text{ if } n \text{ is odd}, \end{cases} \tag{1.2}$$

for all $n \in \mathcal{N}$, where $\{c_k\}_{k=1}^{\infty}$ in (1.2) are the Catalan numbers (1.1).

1.2 Motivation

Recently, the author showed that, from the analysis on p-adic number fields, one can construct semicircular elements (e.g., [5] and [12]). It demonstrates that semicircular elements act well on quantum statistical physics with "very small" distance, too (e.g., [26] and [27]). Motivated by the constructions and results of [5] and [12], it is shown that semicircular elements are well-constructed whenever there are $|\mathcal{Z}|$-many orthogonal projections in a C^*-algebra (e.g., [6–8,10] and [11]), different from earlier works (e.g., [20,25,29] and [30]). In this new approach, the semicircular elements are understood as *Banach-space operators* acting on the given C^*-algebra, by regarding the C^*-algebra as a *Banach space* equipped with its C^*-norm (e.g., [13] and [14]).

Such results of [6,8] and [11] will be generalized abstractly in this paper. Also, to study free-distributional data on $\mathfrak{X}$, we consider *joint free distributions* of multi semicircular elements, too. To do that, the characterizations and estimations of [9] will be re-considered and applied (See Sects. 3 and 4 below).

1.3 Overview

Our main results show that (I) the free-distributional data induced by mutually free, multi semicircular elements are formulated and estimated canonically, and asymptotically (See Sects. 3 and 4, or [9]); (II) the C^*-algebra $\mathfrak{X}$ is not only $*$-isomorphic to the C^*-algebra $\mathcal{S}$ generated by mutually free, $|\mathcal{Z}|$-many semicircular elements $\{s_n\}_{n\in\mathcal{Z}}$, but also, *free-isomorphic to* $\mathcal{S}$ in the sense that: the free probability on $\mathfrak{X}$ is preserved to be that on $\mathcal{S}$ (See Sect. 5); (III) there are well-defined $*$-isomorphisms on $\mathcal{S}$ (and hence, on $\mathfrak{X}$), preserving the free probability on $\mathcal{S}$, and these $*$-isomorphisms induce a well-defined group λ acting on $\mathcal{S}$, preserving the free probability on $\mathcal{S}$ (See Sects. 6.1 and 6.2); (V) the pure-algebraic group algebra $\Lambda = \mathcal{C}[\lambda]$ of λ distorts the free probability on $\mathfrak{X}$, and some interesting distortions are characterized (See Sect. 6.3).

2 Preliminaries

In this section, we briefly introduce concepts used in text.

2.1 Free Probability

For more about free probability, e.g., see [3, 15–17, 28, 29] and [30]. *Free probability* is the noncommutative operator-algebraic version of classical *measure theory* (well-covering the cases where given (sub-)algebras are commutative, e.g., [5] and [12]). It is not only an important branch of operator algebra theory (e.g., [9, 17, 20–22, 28] and [29]), but also an interesting application in related fields (e.g., [4–8, 10–12, 24] and [25]).

In this paper, we use combinatorial free probability of Speicher (e.g., [17, 22] and [23]). Without introducing detailed definitions, or combinatorial backgrounds, the (joint) free moments and (joint) free cumulants are computed. Also, *free product probability spaces* are used without precise introduction.

Let B be a pure-algebraic, or topological (unital, or non-unital) $*$-algebra over $\mathbb{C}$, and let ψ be a (unbounded, or bounded) linear functional on B. Then, the mathematical pair (B, ψ) is said to be a *free $*$-probability space.* In particular, if B is unital containing its *unity*, or the multiplication-identity, 1_B in B, and if $\psi(1_B) = 1$, then the free probability space (B, ψ) is said to be a *unital free probability space.* However, in many cases, free probability spaces are not unital in the above sense. i.e., free "probability" theory is not only an operator-algebraic extension of classical "probability" theory, but also an operator-algebraic analogue of classical "bounded-or-unbounded-measure" theory. In text, our free probability spaces are "not" necessarily assumed to be unital.

2.2 Semicircular Elements

Let (A, φ) be a *topological $*$-probability space* (C^*-probability space, or W^*-probability space, or Banach $*$-probability space, etc.), consisting of a topological $*$-algebra A (C^*-algebra, resp., W^*-algebra, resp., Banach $*$-algebra, etc.), and bounded *linear functional* φ on A. Operators a of A are called *free random variable*, if one regards a as elements of (A, φ).

As in usual operator theory (e.g., [14]), we say a free random variable $a \in (A, \varphi)$ is *self-adjoint*, if a is self-adjoint in A as an operator, i.e., $a^* = a$, where a^* is the *adjoint of a*. Note that the free distribution of a self-adjoint free random variable a is fully characterized by

$$\text{the free-moment sequence } \left(\varphi(a^n)\right)_{n=1}^{\infty},$$

or,

$$\text{the free-cumulant sequence } \left(k_n(a, \ ..., \ a)\right)_{n=1}^{\infty}, \tag{2.2.1}$$

by [17] and [21], where $k_\bullet(.)$ is the *free cumulant on A in terms of φ*, under the *Möbius inversion* of [17] and [21].

Definition 2.1. *A "self-adjoint" free random variable* $x \in (A, \varphi)$ *is said to be semicircular, if*

$$\varphi(x^n) = \omega_n c_{\frac{n}{2}}, \text{ for all } n \in \mathcal{N}, \tag{2.2.2}$$

where ω_n *are in the sense of (1.2), and* c_k *are the* k*-th Catalan numbers (1.1) for all* $k \in \mathcal{N}_0$.

By the Möbius inversion, a self-adjoint free random variable x is semicircular in (A, φ), if and only if

$$k_n(x, \ldots, x) = \begin{cases} 1 & \text{if } n = 2 \\ 0 & \text{otherwise,} \end{cases} \tag{2.2.3}$$

for all $n \in \mathcal{N}$.

Since free-moment sequence and the free-cumulant sequence of x provide equivalent free-distributional data of x in (A, φ), one can use the definition (2.2.2) and the characterization (2.2.3) alternatively, as semicircularity, by (2.2.1).

i.e., the free distribution of a semicircular element $x \in (A, \varphi)$ is characterized by the free-moment sequence

$$(0,\ c_1,\ 0, c_2,\ 0,\ c_3,\ 0,\ c_4,\ \ldots), \tag{2.2.4}$$

or, equivalently, by the free-cumulant sequence

$$(0,\ 1,\ 0,\ 0,\ 0,\ 0,\ \ldots), \tag{2.2.5}$$

by (2.2.2) and (2.2.3), respectively.

By the universality (2.2.4) and (2.2.5), the free distributions of all semicircular elements are called "*the*" *semicircular law*. Equivalently, the free distributions of all semicircular elements are *identically free-distributed* to the semicircular law.

3 Relations on Catalan Numbers

In this section, we introduce some relations known in [9], used in our future works. For $k \in \mathcal{N}_0$, let c_k be the k-th Catalan number,

$$c_k = \frac{1}{k+1}\binom{2k}{k} = \frac{(2k)!}{k!(k+1)!}.$$

Among Catalan numbers, one can obtain the following relation.

Lemma 3.1. *Let* $k_1 > k_2$ *in* $\mathcal{N}_0$. *Then there exists a quantity* $\beta_{k_1>k_2}$ *in the set* $\mathcal{R}^+$ *of all positive real numbers, such that*

$$c_{k_1} = \beta_{k_1>k_2} c_{k_2}. \tag{3.1}$$

In particular,

$$\beta_{k_1>k_2} = 2^{k_1-k_2}\left(\frac{2k_2+1}{k_1+1}\right)\left(\prod_{l=1}^{k_1-k_2-1}\left(2 - \frac{1}{(k_1+1)-l}\right)\right), \tag{3.2}$$

in $\mathcal{R}^+$, with axiomatization:

$$\prod_{l=1}^{0}\left(2-\frac{1}{(k_1+1)-l}\right)=1.$$

Proof. If $k_1 > k_2$ in $\mathcal{N}_0$, then

$$\begin{aligned}
\frac{c_{k_1}}{c_{k_2}} &= \left(\frac{(2k_1)!}{(k_1)!(k_1+1)!}\right)\left(\frac{(k_2)!(k_2+1)!}{(2k_2)!}\right)\\
&= \left(\frac{(2k_1)!}{(2k_2)!}\right)\left(\frac{(k_2)!}{(k_1)!}\right)\left(\frac{(k_2+1)!}{(k_1+1)!}\right)\\
&= \frac{((2k_1)(2k_1-2)(2k_1-4)...(2k_2+2))\,((2k_1-1)(2k_1-3)(2k_1-5)...(2k_2+1))}{(k_2+1)(k_1+1)\left(k_1(k_1-1)...(k_2+2)\right)^2}\\
&= \frac{(2(k_1)2(k_1-1)2(k_1-2)...2(k_2+1))\,((2k_1-1)(2k_1-3)...(2k_2+1))}{(k_1+1)(k_2+1)\left(k_1(k_1-1)...(k_2+2)\right)^2}\\
&= \frac{2^{k_1-k_2}\,(k_1(k_1-1)...(k_2+2)(k_2+1))\,((2k_1-1)(2k_1-3)...(2k_2+1))}{(k_1+1)(k_2+1)\left(k_1(k_1-1)...(k_2+2)\right)^2}\\
&= \left(2^{k_1-k_2}\right)\left(\frac{1}{(k_1+1)}\right)\left(\frac{(2k_1-1)(2k_1-3)...(2k_2+3)}{k_1(k_1-1)...(k_2+2)}\right)(2k_2+1)\\
&= 2^{k_1-k_2}\left(\frac{1}{k_1+1}\right)\left(\frac{2k_1-1}{k_1}\right)\left(\frac{2k_1-3}{k_1-1}\right)\cdots\left(\frac{2k_2+3}{k_2+2}\right)(2k_2+1)\\
&= 2^{k_1-k_2}\left(\frac{2k_2+1}{k_1+1}\right)\left(\prod_{l=1}^{k_1-k_2-1}\left(2-\frac{1}{(k_1+1)-l}\right)\right).
\end{aligned}$$

Therefore, one obtains that

$$\frac{c_{k_1}}{c_{k_2}} = \beta_{k_1>k_2} \Longleftrightarrow c_{k_1} = \beta_{k_1>k_2}c_{k_2},$$

where $\beta_{k_1>k_2}$ is in the sense of (3.2). So, the relation (3.1) holds. ■

By (3.1), it is not hard to check that if $k_1 > k_2$ in $\mathcal{N}_0$, then

$$c_{k_1}c_{k_2} = c_{k_2}c_{k_1} = \beta_{k_1>k_2}c_{k_2}^2, \tag{3.3}$$

where $\beta_{k_1>k_2}$ is in the sense of (3.2). Inductive to (3.3), we obtain the following recurrence relation.

Theorem 3.2. *Let $k_1 > k_2 > ... > k_N$ in $\mathcal{N}_0$, for some $N \in \mathcal{N} \setminus \{1\}$, and take k_l-th Catalan numbers c_{k_l}, for all $l = 1, ..., N$. Now, take n_l-many c_{k_l}'s, for all $l = 1, \ldots, N$, and hence, choose $\left(s = \sum_{l=1}^{N} n_l\right)$-many total Catalan numbers with repetition. For convenience, let's denote these totally s-many Catalan numbers by*

$$c_{j_1}, c_{j_2}, ..., c_{j_s}.$$

Then, for every permutation α of the symmetric group S_X over $X = \{j_1, ..., j_s\}$, we have that

$$\prod_{l=1}^{s} c_{\alpha(j_l)} = \left(\prod_{i=1}^{N-1} \beta_{k_i > k_{i+1}}^{\Sigma_{l=1}^{i} n_l}\right)\left(c_{k_N}^{\Sigma_{l=1}^{N} n_l}\right),$$

with (3.4)

$$\beta_{k_i > k_{i+1}} = 2^{k_i - k_{i+1}} \left(\frac{2k_{i+1}+1}{k_i+1}\right)\left(\prod_{l=1}^{k_i - k_{i+1}-1}\left(2 - \frac{1}{(k_i+1)-l}\right)\right),$$

for all $i = 1, \ldots, N-1$.

Proof. Under hypothesis, for every permutation α of the symmetric group S_X over $X = \{j_1, ..., j_s\}$, one has that

$$\prod_{l=1}^{s} c_{\alpha(j_l)} = \prod_{l=1}^{N} c_{k_l}^{n_l} = c_{k_1}^{n_1} c_{k_2}^{n_2} ... c_{k_N}^{n_N}$$

$$= \left(\beta_{k_1 > k_2} c_{k_2}\right)^{n_1} c_{k_2}^{n_2} c_{k_3}^{n_3} \cdots c_{k_N}^{n_N}$$

by (3.1)

$$= \beta_{k_1 > k_2}^{n_1} c_{k_2}^{n_1+n_2} c_{k_3}^{n_3} \cdots c_{k_N}^{n_N}$$

$$= \beta_{k_1 > k_2}^{n_1} \left(\beta_{k_2 > k_3}^{n_1+n_2} c_{k_3}^{n_1+n_2}\right) c_{k_3}^{n_3} c_{k_4}^{n_4} \cdots c_{k_N}^{n_N}$$

by (3.1)

$$= \cdots$$

$$= \left(\beta_{k_1 > k_2}^{n_1} \beta_{k_2 > k_3}^{n_1+n_2} \cdots \beta_{k_{N-1} > k_N}^{n_1+n_2+...+n_{N-1}}\right) c_{k_N}^{n_1+n_2+...+n_{N-1}+n_N},$$

i.e.,

$$\prod_{l=1}^{s} c_{\alpha(j_l)} = \prod_{l=1}^{N} c_{k_l}^{n_l} = \left(\prod_{i=1}^{N-1} \beta_{k_i > k_{i+1}}^{\Sigma_{l=1}^{i} n_l}\right)\left(c_{k_N}^{\Sigma_{l=1}^{N} n_l}\right),$$

where $\beta_{k_i > k_{i+1}}$ are in the sense of (3.2), for all $i = 1, \ldots, N-1$. So, the relation (3.4) holds. ∎

By (3.4), we obtain the following estimation.

Theorem 3.3. *Under the same hypotheses of Theorem 3.2, for every permutation α of the symmetric group S_X, we have that*

$$\beta_{\min}\left(c_{k_N}^{\Sigma_{l=1}^{N} n_l}\right) \leq \prod_{l=1}^{s} c_{\alpha(j_l)} \leq \beta_{\max}\left(c_{k_N}^{\Sigma_{l=1}^{N} n_l}\right), \tag{3.5}$$

with

$$\beta_{\min} = \prod_{i=1}^{N-1}\left(\left(2^{k_i - k_{i+1}}\left(\frac{2k_{i+1}+1}{k_i+1}\right)\right)\left(2 - \frac{1}{k_{i+1}+2}\right)^{(k_i - k_{i+1}-1)}\right)^{\Sigma_{l=1}^{i} n_l},$$

and (3.6)

$$\beta_{\max} = \prod_{i=1}^{N-1}\left(\left(2^{k_i - k_{i+1}}\left(\frac{2k_{i+1}+1}{k_i+1}\right)\right)\left(2 - \frac{1}{k_i+1}\right)^{(k_i - k_{i+1}-1)}\right)^{\Sigma_{l=1}^{i} n_l}$$

in $\mathcal{R}^+$, for all $n_1, ..., n_N \in \mathcal{N}$.

Proof. Under the hypotheses, we have

$$\prod_{l=1}^{s} c_{\alpha(j_l)} = \left(\prod_{i=1}^{N-1} \beta_{k_i > k_{i+1}}^{\Sigma_{l=1}^{i} n_l}\right)\left(c_{k_N}^{\Sigma_{l=1}^{N} n_l}\right), \tag{3.7}$$

by (3.4). And, the quantity

$$\beta_{k_1,\ldots,k_N}^{n_1,\ldots,n_N} \overset{denote}{=} \prod_{i=1}^{N-1}\left(\beta_{k_i > k_{i+1}}^{\Sigma_{l=1}^{i} n_l}\right)$$

in (3.7) satisfies that

$$\beta_{\min} \leq \beta_{k_1,\ldots,k_N}^{n_1,\ldots,n_N} \leq \beta_{\max}, \tag{3.8}$$

where $\beta_{\min}$ and $\beta_{\max}$ are in the sense of (3.6).

Therefore, one obtains that

$$\beta_{\min}\left(c_{k_N}^{\Sigma_{l=1}^{N} n_l}\right) \leq \prod_{l=1}^{s} c_{j_{\alpha(l)}} \leq \beta_{\max}\left(c_{k_N}^{\Sigma_{l=1}^{N} n_l}\right),$$

by (3.8). Therefore, the estimation (3.5) holds. See [9] for more details. ∎

4 Joint Free Distributions of Multi Semicircular Elements

In this section, we consider free distributions of mutually free, multi semicircular elements in a C^*-probability space (A, φ). Without loss of generality, readers can regard (A, φ) as a W^*-probability space, or a Banach $*$-probability space etc. However, here, we will let (A, φ) be a C^*-probability space.

Let (A, φ) be a fixed C^*-probability space, and suppose there are N-many semicircular elements x_1, ..., x_N in (A, φ), for $N \in \mathcal{N}$. Assume further that they are free from each other in (A, φ). By the self-adjointness of x_1, ..., x_N in A, the free distribution, say

$$\rho \overset{denote}{=} \rho_{x_1,\ldots,x_N}, \tag{4.0.1}$$

of them are characterized by the joint free-moments

$$\bigcup_{n=1}^{\infty}\left(\bigcup_{(i_1,\ldots,i_n)\in\{1,\ldots,N\}^n} \{\varphi\left(x_{i_1}x_{i_2}\ldots x_{i_n}\right)\}\right) \tag{4.0.1$'$}$$

(e.g., [17,22] and [23]). More precisely, the free distribution ρ of (4.0.1), is characterized by the free-moments

$$\bigcup_{l=1}^{N} \{\varphi(x_l^n)\}_{n=1}^{\infty}, \tag{4.0.2}$$

and the "mixed" free-moments,

$$\bigcup_{s=2}^{\infty}\left\{\varphi\left(x_{i_1}^{n_1}x_{i_2}^{n_2}\ldots x_{i_s}^{n_s}\right) \middle| \begin{array}{l}(i_1,\ldots,i_s) \in \{1,\ldots,N\}^s \\ \text{are mixed in } \{1,\ldots,N\}\end{array}\right\}, \tag{4.0.3}$$

by (4.0.1)$'$.

In this section, by using the results of Sect. 3, we characterize, and estimate the free distribution ρ of (4.0.1).

4.1 Free-Distributional Data (4.0.2) of ρ

Let $\rho = \rho_{x_1,...,x_N}$ be the free distribution (4.0.1) of fixed N-many mutually free semicircular elements x_1, ..., x_N of (A, φ). Let's consider the free-distributional data (4.0.2). But these data are clear by the semicircularity (2.2.2) and (2.2.3), under the universality (2.2.4) and (2.2.5).

Corollary 4.1. *The free-distributional data (4.0.2) of the free distribution ρ of (4.0.1) are obtained by the semicircularity. i.e.,*

$$\varphi\left(x_l^n\right) = \omega_n c_{\frac{n}{2}}, \text{ for all } n \in \mathcal{N}, \tag{4.1.1}$$

for all $l = 1$, ..., N.

Proof. The formula (4.1.1) is proven by (2.2.2) and (2.2.4). ∎

4.2 Free-Distributional Data (4.0.3) of ρ

Let ρ be the free distribution (4.0.1) of mutually free, N-many multi semicircular elements x_1, ..., x_N of (A, φ). In this section, we concentrate on studying the free-distributional data (4.0.3) of ρ.

Throughout this section, for any $s \in \mathcal{N} \setminus \{1\}$, we fix an s-tuple I_s,

$$I_s \overset{denote}{=} (i_1, ..., i_s) \in \{1, ..., N\}^s, \tag{4.2.1}$$

which is mixed in $\{1, ..., N\}$. i.e., there exists at least one entry i_{k_0} of I_s, satisfying $i_{k_0} \neq i_k$, for some $k \neq k_0$ in $\{1, ..., s\}$.

For example,

$$I_8 = (1, 1, 3, 2, 4, 2, 2, 1),$$

is a mixed 8-tuple in $\{1, 2, 3, 4, 5\}^8$.

From the sequence I_s of (4.2.1), define a set

$$[I_s] = \{i_1, i_2, ..., i_s\}, \tag{4.2.2}$$

without considering repetition. For instance, if I_8 is as above, then

$$[I_8] = \{i_1, i_2, ..., i_8\},$$

with its cardinality 8, satisfying

$$\begin{gathered} i_1 = i_2 = i_8 = 1, \\ i_4 = i_6 = i_7 = 2, \\ i_3 = 3, and i_5 = 4. \end{gathered}$$

without considering repetition; for example, we regard all 1's in I_8 as different elements i_1, i_2 and i_8 in $[I_8]$.

Then from the set $[I_s]$ of (4.2.2), one can define a unique "noncrossing" partition $\pi_{(I_s)}$ of the noncrossing-partition lattice $NC([I_s])$ (e.g., [17,22] and [23]), such that: (i)

$$\forall V = \left(i_{j_1}, i_{j_2}, ..., i_{j_{|V|}}\right) \in \pi_{(I_s)},$$
$$\Longleftrightarrow \qquad \qquad (4.2.3)$$
$$\exists k \in \{1, ..., N\}, s.t., i_{j_1} = i_{j_2} = ... = i_{j_{|V|}} = k,$$

(ii) such a partition $\pi_{(I_s)}$ of (i) is "maximal" satisfying (4.2.3), under the partial ordering on $NC([I_s])$ (e.g., see [17,22] and [23]), and (iii) the very first maximal blocks of the partition in (ii) must contain i_1.

For example, if I_8 and $[I_8]$ are as above, then there exists a noncrossing partition

$$\begin{aligned}\pi_{(I_8)} &= \{(i_1, i_2, i_8), (i_3), (i_4, i_6, i_7), (i_5)\} \\ &= \{(1,1,1), (3), (2,2,2), (4)\},\end{aligned}$$

in $NC([I_8])$, satisfying the conditions (i), (ii) and (iii).

Now, suppose $\pi_{(I_s)} \in NC([I_s])$ is the noncrossing partition (4.2.3), and let

$$\pi_{(I_s)} = \{U_1, ..., U_t\},$$

where $t \leq s$ and $U_k \in \pi_{(I_s)}$ are the blocks of (ii), satisfying (i) and (iii), for $k = 1, ..., t$. For example,

$$\pi_{(I_8)} = \{U_1, U_2, U_3, U_4\},$$

with $t = 4 < 8 = s$, where

$$U_1 = \{i_1, i_2, i_8\}, U_2 = \{i_3\},$$

and

$$U_3 = \{i_4, i_6, i_7\}, U_4 = \{i_5\}.$$

Then the partition $\pi_{(I_s)}$ is regarded as the joint partition,

$$\pi_{(I_s)} = 1_{|U_1|} \vee 1_{|U_2|} \vee ... \vee 1_{|U_t|}, \qquad (4.2.4)$$

where $1_{|U_k|}$ are the maximal elements, the one-block partitions, of $NC(U_k)$, for all $k = 1, ..., t$.

Let I_s be in the sense of (4.2.1), and let $x_{i_1}, ..., x_{i_s}$ be the corresponding semicircular elements of (A, φ) induced by I_s, without considering repetition in the set $\{x_1, ..., x_N\}$. Define a free random variable $X[I_s]$ by

$$X[I_s] \overset{def}{=} \prod_{l=1}^{s} x_{i_l} \in (A, \varphi). \qquad (4.2.5)$$

If $X[I_s]$ is in the sense of (4.2.5), then

$$\varphi(X[I_s]) = \sum_{\pi \in NC([I_s])} k_\pi$$

by the Möbius inversion of [17] and [22], where

$$k_\pi = \prod_{V \in \pi} k_V$$

with

$$k_V = k_{|V|}\left(x_{i_{k_1}}, \ \ldots, \ x_{i_{k_{|V|}}}\right),$$

whenever $V = (i_{k_1}, \ \ldots, \ i_{k_{|V|}})$, where $|V|$ is the cardinality of V in π, and hence, it goes to

$$= \sum_{\pi \in NC([I_s]), \pi \leq \pi_{(I_s)}} k_\pi$$

by the mutual-freeness of $x_1, \ \ldots, \ x_N$ in (A, φ) in the sense of [17] and [22]

$$= \sum_{(\theta_1, \ldots, \theta_t) \in NC(U_1) \times \ldots \times NC(U_t)} k_{\theta_1 \vee \ldots \vee \theta_t}$$

by (4.2.4)

$$= \sum_{(\theta_1, \ldots, \theta_t) \in NC_2(U_1) \times \ldots \times NC_2(U_t)} k_{\theta_1 \vee \ldots \vee \theta_t}$$

$$= \sum_{(\theta_1, \ldots, \theta_t) \in NC_2(U_1) \times \ldots \times NC_2(U_t)} \left(\prod_{l=1}^{t} k_{\theta_l}\right), \tag{4.2.6}$$

by the semicircularity (2.2.3) and (2.2.5) of $x_{i_1}, \ \ldots, \ x_{i_s}$ in (A, φ), where $NC_2(X)$ is the subset of the noncrossing-partition lattice $NC(X)$,

$$NC_2(X) = \{\pi \in NC(X) : \forall V \in \pi, |V| = 2\}, \tag{4.2.7}$$

over countable finite sets X. i.e., $\pi \in NC_2(X)$, if and only if all blocks of π have two elements.

By (4.2.6) and (4.2.7), it can be checked that if there exists at least one $k_0 \in \{1, \ \ldots, \ t\}$, such that $|U_{k_0}|$ is odd in $\mathcal{N}$, then

$$\varphi\left(X[I_s]\right) = 0.$$

So, the formula (4.2.6) is non-zero, only if

$$|U_k| \in 2\mathcal{N}, \ \text{ for all } k = 1, \ldots, t, \tag{4.2.8}$$

where $2\mathcal{N} = \{2n : n \in \mathcal{N}\}$.

Moreover, if the condition (4.2.8) holds, then the summands $k_{\theta_1 \vee \ldots \vee \theta_t}$ of (4.2.6) satisfy that

$$k_{\theta_1 \vee \ldots \vee \theta_t} = \prod_{V \in \theta_1 \vee \ldots \vee \theta_t} k_V = \prod_{V \in \theta_1 \vee \ldots \vee \theta_t} \left(\prod_{i=1}^{t} 1^{\#(\theta_i)}\right) = 1, \tag{4.2.9}$$

by the semicircularity (2.2.3) and (2.2.5), where $\#(\theta_i)$ are the number of blocks of θ_i, for all $i = 1, \ \ldots, \ t$. Therefore, if the condition (4.2.8) holds, then

$$\varphi\left(X[I_s]\right) = \sum_{(\theta_1,...,\theta_t)\in NC_2([U_1])\times...\times NC_2([U_t])} 1$$
$$= \left|NC_2\left(U_1\right) \times ... \times NC_2\left(U_t\right)\right|, \tag{4.2.10}$$

by (4.2.6) and (4.2.9).

Lemma 4.2. *Let I_s be an s-tuple (4.2.1), and let $X[I_s] = \prod_{l=1}^{s} x_{i_l}$ be the corresponding free random variable (4.2.5) of (A, φ). If*

$$\pi_{(I_s)} = 1_{|U_1|} \vee ... \vee 1_{|U_t|},$$

in the sense of (4.2.3) and (4.2.4), then

$$\varphi\left(X[I_s]\right) = \begin{cases} \prod_{i=1}^{t} c_{\frac{|U_i|}{2}} & \text{if } |U_k| \in 2\mathcal{N}, \text{ for all } k = 1, ..., t \\ 0 & \text{otherwise.} \end{cases} \tag{4.2.11}$$

Proof. Under conditions,

$$\varphi\left(X[I_s]\right) = \begin{cases} \left|NC_2\left(U_1\right) \times ... \times NC_2\left(U_t\right)\right| & \text{if } |U_k| \in 2\mathcal{N}, \text{ for all } k = 1, ..., t \\ 0 & \text{otherwise,} \end{cases}$$

by (4.2.10).

Note that

$$\left|NC_2\left(U_k\right)\right| = \left|NC\left(\frac{|U_k|}{2}\right)\right|, \tag{4.2.12}$$

for all $k = 1, ..., t$ (e.g., see [17,22] and [23]).

So, the formula (4.2.10) goes to

$$\varphi\left(X[I_s]\right) = \begin{cases} \left|NC\left(\frac{|U_1|}{2}\right) \times ... \times NC\left(\frac{|U_t|}{2}\right)\right| & \text{if } |U_k| \in 2\mathcal{N}, \text{ for all } k = 1, ..., t \\ 0 & \text{otherwise,} \end{cases}$$

by (4.2.12)

$$= \begin{cases} \prod_{l=1}^{t} c_{\frac{|U_l|}{2}} & \text{if } |U_l| \in 2\mathcal{N}, \text{ for all } l = 1, ..., t \\ 0 & \text{otherwise,} \end{cases} \tag{4.2.13}$$

because $|NC(X)| = c_{|X|}$, for all finite sets X (e.g., [17,22] and [23]). Therefore, the formula (4.2.11) holds by (4.2.13). ∎

By the above lemma, we obtain the following free-distributional data, characterizing (4.0.3).

Theorem 4.3. *Let* I_s *be in the sense of (4.2.1), and* $X[I_s]$*, the corresponding free random variable (4.2.5) of* (A, φ)*, and assume that*

$$I_s = (U_1, \ ..., \ U_t),$$

satisfies (4.2.3) and (4.2.4). Also, let

$$k'_l = \frac{|U_l|}{2} \in \mathcal{R}^+, \text{for all } l = 1, ..., t,$$

and suppose there exist "mutually-distinct" $k'_{j_1}, \ ..., \ k'_{j_r}$ *among* $\{k'_1, \ ..., \ k'_t\}$*, where* $r \leq t$*, and assume that there are* n_{j_l}*-many such* k_{j_l}*'s, for* $l = 1, \ ..., \ r$*, satisfying*

$$\sum_{l=1}^{r} n_{j_l}(2k_{j_l}) = s.$$

Assume further that there is a permutation α *of the symmetric group* S_X *over* $X = \{k'_{j_1}, \ ..., \ k'_{j_r}\}$*, satisfying*

$$k_{j_1} > k_{j_2} > ... > k_{j_r} \text{ in } \mathcal{R}^+,$$

where

$$k_{j_l} = \alpha\left(k'_{j_l}\right) \in \mathcal{R}^+, \text{for all } l = 1, ..., r,$$

Then

$$\varphi\left(X[I_s]\right) = \begin{cases} \left(\prod_{i=1}^{r-1} \beta_{k_{j_i} > k_{j_{i+1}}}^{\Sigma_{l=1}^{i} n_{j_l}}\right)\left(c_{k_{j_r}}^{\Sigma_{l=1}^{N} n_{j_l}}\right) & \text{if } k_{j_1}, ..., k_{j_r} \in \mathcal{N} \\ \\ 0 & \text{otherwise} \end{cases} \tag{4.2.14}$$

with

$$\beta_{k_{j_i} > k_{j_{i+1}}} = 2^{k_{j_i} - k_{j_{i+1}}} \left(\frac{2k_{j_{i+1}}+1}{k_{j_i}+1}\right)\left(\prod_{l=1}^{k_{j_i}-k_{j_{i+1}}-1}\left(2 - \frac{1}{(k_{j_i}+1)-l}\right)\right),$$

for all $i = 1, \ ..., \ r-1$.

Proof. Under hypothesis, one has

$$\varphi\left(X[I_s]\right) = \begin{cases} \prod_{l=1}^{t} c_{k'_l} & \text{if } k'_l \in \mathcal{N}, \\ & \text{for all } l = 1, ..., t \\ 0 & \text{otherwise} \end{cases}$$

by (4.2.11)

$$= \begin{cases} \prod_{l=1}^{r} c_{k_{j_l}}^{n_{j_l}} & \text{if } k_{j_l} = \alpha\left(k'_{j_l}\right) \in \mathcal{N}, \\ & \text{for all } l = 1, ..., r \\ 0 & \text{otherwise} \end{cases}$$

by assumptions

$$= \begin{cases} \left(\prod_{i=1}^{r-1} \beta_{k_{j_i} > k_{j_{i+1}}}^{\Sigma_{l=1}^{i} n_{j_l}} \right) \left(c_{k_{j_r}}^{\Sigma_{l=1}^{r} n_{j_l}} \right) & \text{respectively,} \\ \\ 0 \end{cases}$$

by (3.4), where

$$\beta_{k_{j_i} > k_{j_{i+1}}} = 2^{k_{j_i} - k_{j_{i+1}}} \left(\frac{2k_{j_{i+1}} + 1}{k_{j_i} + 1} \right) \left(\prod_{l=1}^{k_{j_i} - k_{j_{i+1}} - 1} \left(2 - \frac{1}{(k_{j_i} + 1) - l} \right) \right),$$

for all $i = 1, ..., r - 1$. Therefore, the free-distributional data (4.2.14) holds. ■

The above formula (4.2.14) characterizes the free-distributional data (4.0.3) of our free distribution $\rho = \rho_{x_1, ..., x_s}$ of (4.0.1) in (A, φ).

Example 4.1. *Let x_1, x_2, x_3, x_4 be fixed mutually free semicircular elements of (A, φ), and let*

$$W = x_1^2 x_2^4 x_1^2 x_3^2 \in (A, \varphi)$$

induced by $\{x_1, x_2, x_3, x_4\}$. Then one can take

$$I_W = (1, 1, 2, 2, 2, 2, 1, 1, 3, 3) \overset{let}{=} (i_1, ..., i_{10}),$$

and

$$\pi_{(I_W)} = \{(i_1, i_2, i_7, i_8), (i_3, i_4, i_5, i_6), (i_9, i_{10})\},$$

with

$$U_1 = \{i_1, i_2, i_7, i_8\} = \{1, 1, 1, 1\}, \textit{having } k_1' = \frac{|U_1|}{2} = 2,$$

$$U_2 = \{i_3, i_4, i_5, i_6\} = \{2, 2, 2, 2\}, \textit{having } k_2' = \frac{|U_2|}{2} = 2,$$

and

$$U_3 = \{i_7, i_8\} = \{3, 3\}, \textit{having } k_3' = \frac{|U_3|}{2} = 1.$$

Therefore, one can take

$$k_1 = 2 > 1 = k_2, \textit{and hence}, r = 2,$$

and

$$n_1 = 2, \textit{and } n_2 = 1,$$

since there are two 2's (represented by k_1' and k_2'), and there is only one 1 (represented by k_3').

Therefore, we have

$$\varphi(W) = \left(\prod_{i=1}^{2-1} \beta_{k_i > k_{i+1}}^{\Sigma_{l=1}^{i} n_i} \right) \left(c_{k_r}^{\Sigma_{l=1}^{N} n_l} \right) = \beta_{k_1 > k_2}^{n_1} c_{k_2}^{n_1 + n_2}$$

$$= \left(\beta_{2>1}^2 \right) c_1^{2+1} = 4,$$

by (4.2.14).

By the above characterization (4.2.14) of (4.0.3), we obtain the following estimation.

Theorem 4.4. *Under the same conditions of Theorem 4.3, there exist* $\beta_{\min}^{(I_s)}$, $\beta_{\max}^{(I_s)} \in \mathcal{R}^+$, *such that*

$$\beta_{\min}^{(I_s)}\left(c_{k_r}^{\Sigma_{l=1}^r n_l}\right) \leq \varphi\left(X[I_s]\right) \leq \beta_{\max}^{(I_s)}\left(c_{k_r}^{\Sigma_{l=1}^r n_l}\right), \tag{4.2.15}$$

where

$$\beta_{\min}^{(I_s)} = \prod_{i=1}^{r-1}\left(2^{k_{j_i}-k_{j_{i+1}}}\left(\frac{2k_{j_{i+1}}+1}{k_{j_i}+1}\right)\left(2-\frac{1}{k_{j_{i+1}}+2}\right)^{k_{j_i}-k_{j_{i+1}}-1}\right)^{\Sigma_{l=1}^i n_l},$$

and

$$\beta_{\max}^{(I_s)} = \prod_{i=1}^{r-1}\left(2^{k_{j_i}-k_{j_{i+1}}}\left(\frac{2k_{j_{i+1}}+1}{k_{j_i}+1}\right)\left(2-\frac{1}{k_{j_i}+1}\right)^{k_{j_i}-k_{j_{i+1}}-1}\right)^{\Sigma_{l=1}^i n_l}.$$

Proof. The estimation (4.2.15) is obtained by (4.2.14), (3.5) and (3.6). ∎

The above theorem provides the estimation of the free-distributional data (4.0.3) of $\varphi\left(X[I_s]\right)$, by (4.2.15).

Example 4.2. *Let* x_1, x_2, x_3, x_4 *be fixed mutually free semicircular elements of* (A, φ), *and let*

$$W = x_1^2 x_2^4 x_1^2 x_3^2 \in (A, \varphi)$$

induced by $\{x_1, x_2, x_3, x_4\}$. *Then, by Example 4.1, one can take*

$$k_1 = 2 > 1 = k_2, \text{and hence}, r = 2,$$

and

$$n_1 = 2, \text{and } n_2 = 1.$$

Therefore, we have

$$\left(2^{2-1}\left(\frac{2\cdot 1+1}{2+1}\right)\left(2-\frac{1}{1+2}\right)^{2-1-1}\right)^2 (c_1^{2+1})$$
$$\leq \varphi(W)$$
$$\leq \left(2^{2-1}\left(\frac{2\cdot 1+1}{2+1}\right)^1\left(2-\frac{1}{2+1}\right)^0\right)^2 (c_1^{2+1})$$

$$\Longleftrightarrow$$

$$4 \leq \varphi(W) \leq 4.$$

by (4.2.15).

4.3 Asymptotic Properties of $\varphi(X[I_s])$

Now, by the main results of Sects. 4.1 and 4.2, we study asymptotic behaviors of the mixed free-distributional data $\varphi(X[I_s])$ for the free distribution $\rho = \rho_{x_1,...,x_N}$ of (4.0.1). For more details, see [9].

First, recall that if I_s and $X[I_s]$ are in the sense of (4.2.1), respectively, (4.2.5), and if $\varphi(X[I_s]) \neq 0$, then there exist

$$k_1 > k_2 > ... > k_r \in \mathcal{N}_0,$$

and

$$n_1, n_2, ..., n_r \in \mathcal{N},$$

such that

$$\Sigma_{i=1}^{r} n_i (2k_i) = s,$$

and

$$\varphi(X[I_s]) = \left(\prod_{i=1}^{r-1} \beta_{k_i > k_{i+1}}^{\Sigma_{l=1}^{i} n_l}\right)\left(c_{k_r}^{\Sigma_{l=1}^{r} n_l}\right), \tag{4.3.1}$$

by (4.2.14), where

$$\beta_{k_i > k_{i+1}} = 2^{k_i - k_{i+1}} \left(\frac{2k_{i+1}+1}{k_i+1}\right)\left(\prod_{l=1}^{k_i - k_{i+1} - 1}\left(2 - \frac{1}{(k_i+1)-l}\right)\right)$$

are in the sense of (3.4), for all $i = 1, \ldots, r-1$.

By (4.3.1), if $\varphi(X[I_s]) \neq 0$, then we have the following asymptotic behaviors.

Theorem 4.5. *Let $X[I_s] \in (A, \varphi)$ be a free random variable (4.2.5) induced by the s-tuple I_s of (4.2.1) for the fixed mutually free, N-many multi semicircular elements x_1, ..., $x_N \in (A, \varphi)$. Assume that $\varphi(X[I_s]) \neq 0$, and hence, it satisfies (4.3.1). If there exists $i_0 \in \{1, ..., r\}$, such that*

$$k_l \text{ are fixed in } \mathcal{N}_0, \text{for } l = i_0 + 1, ..., r,$$

and

$$k_l \to \infty \text{ in } \mathcal{N}_0, \text{for } l = 1, ..., i_0, \tag{4.3.2}$$

where $k_1 > k_2 > ... > k_r$ are in the sense of (4.2.14). Then there exist functions f and g on $\mathcal{R}^+$,

$$f(x) = 2^{-2k_{i_0+1}} (2k_{i_0+1} + 1)\left(\tfrac{2^{2x}}{x}\right),$$

and

$$g(x) = \tfrac{2^{2x}}{x}, \text{for all } x \in \mathcal{R}^+, \tag{4.3.3}$$

and a positive quantity $M_{i_0} \in \mathcal{R}^+$,

$$M_{i_0} = \left(\prod_{i=i_0+1}^{r-1} \beta_{k_i > k_{i+1}}^{\Sigma_{l=1}^{i} n_i}\right)\left(c_r^{\Sigma_{l=1}^{r} n_l}\right), \tag{4.3.4}$$

such that

$$\varphi(X[I_s]) \sim M_{i_0} (f(k_{i_0}))^{\Sigma_{l=1}^{i_0} n_l} \left(\prod_{i=1}^{i_0-1}\left(\frac{g(k_l)}{g(k_{l+1})}\right)^{\Sigma_{l=1}^{i} n_i}\right). \tag{4.3.5}$$

Proof. Let a free random variable $X[I_s]$ be given as above. Then, by (4.3.1), we have

$$\varphi\left(X[I_s]\right) = \left(\prod_{i=1}^{r-1} \beta_{k_i > k_{i+1}}^{N_i}\right)\left(c_{k_r}^{N_r}\right)$$

where

$$N_i = \sum_{l=1}^{i} n_l, \text{ for all } i = 1, ..., r,$$

and hence, it goes to

$$= \left(\beta_{k_1>k_2}^{N_1}\beta_{k_2>k_3}^{N_2} \cdots \beta_{k_{i_0}>k_{i_0+1}}^{N_{i_0}}\right)\left(\beta_{k_{i_0+1}>k_{i_0+2}}^{N_{i_0+1}} \cdots \beta_{k_{r-1}>k_r}^{N_{r-1}}\right)\left(c_{k_r}^{N_r}\right)$$

$$= \left(\beta_{k_1>k_2}^{N_1}\beta_{k_2>k_3}^{N_2} \cdots \beta_{k_{i_0-1}>k_{i_0}}^{N_{i_0-1}}\right)\left(\beta_{k_{i_0}>k_{i_0+1}}^{N_{i_0}}\right)$$

$$\cdot \left(\beta_{k_{i_0+1}>k_{i_0+2}}^{N_{i_0+1}} \cdots \beta_{k_{r-1}>k_r}^{N_{r-1}}\right)\left(c_{k_r}^{N_r}\right). \tag{4.3.6}$$

By the assumption that k_{i_0+1}, ..., k_r are fixed in $\mathcal{N}_0$, one can find the fixed quantity

$$M_{i_0} = \left(\beta_{k_{i_0+1}>k_{i_0+2}}^{N_{i_0+1}} \cdots \beta_{k_{r-1}>k_r}^{N_{r-1}}\right)\left(c_{k_r}^{N_r}\right) \in \mathcal{R}^+, \tag{4.3.7}$$

in (4.3.6), and hence, one has

$$\varphi\left(X[I_s]\right) = M_{i_0}\left(\beta_{k_1>k_2}^{N_1} \cdots \beta_{k_{i_0-1}>k_{i_0}}^{N_{i_0-1}}\right)\left(\beta_{k_{i_0}>k_{i_0+1}}^{N_{i_0}}\right), \tag{4.3.8}$$

by (4.3.7).

Also, by the assumption that $k_l \to \infty$, for $l = 1$, ..., i_0, there exists a function f on $\mathcal{R}^+$,

$$f(x) = \left(\frac{2k_{i_0+1}+1}{2^{2k_{i_0+1}}}\right)\left(\frac{2^{2x}}{x}\right), \text{ for all } x \in \mathcal{R}^+,$$

such that (4.3.9)

$$\beta_{k_{i_0}>k_{i_0+1}} \sim f(k_{i_0}),$$

by [9]. Thus, one can get that

$$\beta_{k_{i_0}>k_{i_0+1}}^{N_{i_0}} \sim f(k_{i_0})^{N_{i_0}}, \tag{4.3.10}$$

by (4.3.9).

Moreover, for all $l = 1$, ..., $i_0 - 1$, there exists a function g on $\mathcal{R}^+$,

$$g(x) = \frac{2^{2x}}{x}, \ \textit{for all } x \in \mathcal{R}^+,$$

such that (4.3.11)

$$\beta_{k_l>k_{l+1}} \sim \frac{g(k_{i+1})}{g(k_i)},$$

by [9], and hence,

$$\beta_{k_l>k_{l+1}}^{N_l} \sim \left(\frac{g(k_l)}{g(k_{l+1})}\right)^{N_l}, \tag{4.3.12}$$

in (4.3.8), for all $l = 1$, ..., $i_0 - 1$, by (4.3.11).

So,

$$\varphi\left(X[I_s]\right) \sim M_{i_0}\left(\frac{g(k_1)}{g(k_2)}\right)^{N_1}\left(\frac{g(k_2)}{g(k_3)}\right)^{N_2}\cdots\left(\frac{g(k_{i_0-1})}{g(k_{i_0})}\right)^{N_{i_0-1}}\left(f(k_{i_0})\right)^{N_{i_0}}, \tag{4.3.13}$$

by (4.3.11) and (4.3.12).

Therefore, the asymptotic estimation (4.3.5) holds true by (4.3.13). ■

4.4 Conclusion: The Free Distribution $\rho_{x_1,...,x_N}$ of (4.0.1)

Let $\rho = \rho_{x_1,..,x_N}$ be the joint free distribution (4.0.1) of mutually free, N-many semicircular elements x_1, ..., x_N of (A, φ). Then this free distribution ρ is characterized by the joint free moments of x_1, ..., x_N, and they consist of free moments (4.0.2) of each x_i, for $i = 1$, ..., N, and their mixed free moments (4.0.3). We summarize the main results of Sects. 4.2 and 4.3 by the following corollary.

Corollary 4.6. *The free distribution ρ is characterized by the free moments (4.1.1), and the mixed free moments (4.2.14). Moreover, the mixed free moments (4.2.14) have canonical estimations (4.2.15), and asymptotic estimations (4.3.5).* □

5 A C^*-Probability Space $(\mathcal{X}, \varphi)$ Generated by $|\mathcal{N}|$-Many Semicircular Elements

In this section, we consider a structure theorem of a C^*-algebra $\mathcal{X}$ generated by mutually free, $|\mathcal{N}|$-many semicircular elements

$$\{x_n\}_{n=0}^{\infty} = \{x_0, x_1, x_2, x_3, ...\}.$$

From now on, for notational convenience, the index for $|\mathcal{N}|$-many semicircular elements starts from 0 as above.

Let (A, φ) be a C^*-probability space, and assume that it contains a family $X = \{x_n\}_{n=0}^{\infty}$ of mutually free semicircular elements x_n's. In this section, we study C^*-probabilistic sub-structure $(\mathcal{X}, \varphi \mid_{\mathcal{X}})$, where $\mathcal{X}$ is the C^*-subalgebra $C^*(X)$ of A generated by the free family X, and $\varphi \mid_{\mathcal{X}}$ is the restriction of φ on $\mathcal{X}$.

5.1 Free-Isomorphic Relations

Let (A_1, φ_1) and (A_2, φ_2) be C^*-probability spaces. The C^*-probability space (A_1, φ_1) is said to be *free-homomorphic* to the C^*-probability space (A_2, φ_2), if there exists a $*$-homomorphism $\Omega : A_1 \to A_2$, such that

$$\varphi_2\left(\Omega(a)\right) = \varphi_1(a), \ \textit{for all } a \in (A_1, \varphi_1).$$

In such a case, the $*$-homomorphism Ω is called a *free-homomorphism from* (A_1, φ_1) *to* (A_2, φ_2). We write this free-homomorphic relation by

$$(A_1, \varphi_1) \overset{\text{free-homo}}{\subseteq} (A_2, \varphi_2). \tag{5.1.1}$$

Definition 5.1. *Suppose* $(A_1, \varphi_1) \overset{\text{free-homo}}{\subseteq} (A_2, \varphi_2)$ *in the sense of (5.1.1), via a free-homomorphism* $\Omega : A_1 \to A_2$. *If this free-homomorphism* Ω *is a* $*$-*isomorphism from* A_1 *onto* A_2, *then it is called a free-isomorphism. In this case,* (A_1, φ_1) *is said to be free-isomorphic to* (A_2, φ_2), *or* (A_1, φ_1) *and* (A_2, φ_2) *are free-isomorphic. This free-isomorphic relation is denoted by*

$$(A_1, \varphi_1) \overset{\text{free-iso}}{=} (A_2, \varphi_2). \tag{5.1.2}$$

By the definitions (5.1.1) and (5.1.2), if two C^*-probability spaces are free-isomorphic, then they are regarded as the same C^*-probability space.

5.2 A C^*-Probability Space $(\mathcal{X}, \varphi \mid_{\mathcal{X}})$

Let (A, φ) be a fixed C^*-probability space, and assume that the C^*-algebra A contains a family $X = \{x_n\}_{n=0}^{\infty}$, consisting of mutually free semicircular elements x_n's. Construct the C^*-subalgebra $\mathcal{X} = C^*(X)$ of A, generated by the family X. Then one can obtain a canonical C^*-probabilistic sub-structure

$$\mathcal{X}_\varphi \overset{denote}{=} (\mathcal{X}, \varphi = \varphi \mid_{\mathcal{X}}) \tag{5.2.1}$$

in (A, φ). From below, we consider the C^*-probability space $\mathcal{X}_\varphi$ of (5.2.1) as an independent C^*-probabilistic structure.

Now, let (B, ψ) be a C^*-probability space, containing a family $S = \{s_n\}_{n \in \mathbb{Z}}$ of mutually free $|\mathbb{Z}|$-many semicircular elements, and let

$$\mathcal{S}_\psi \overset{denote}{=} (\mathcal{S}, \psi = \psi \mid_{\mathcal{S}}) \tag{5.2.2}$$

be the corresponding C^*-probabilistic sub-structure of (B, ψ), where $\mathcal{S} = C^*(S)$ is the C^*-subalgebra of B generated by the family S. Also, one may/can understand $\mathcal{S}_\psi$ as an independent free-probabilistic structure.

In this section, we show that

$$\mathcal{X}_\varphi \overset{\text{free-iso}}{=} \mathcal{S}_\psi.$$

Remark 5.1. *Such a topological* $*$-*probability space* $\mathcal{S}_\psi$ *of (5.2.2) is well-determined naturally (e.g., [5] and [12]), or artificially (e.g., [6–8,10] and [11]).*

Consider the following structure theorem of the C^*-algebra $\mathcal{X}$ in (A, φ).

Proposition 5.1. *Let $\mathcal{X}$ be the C^*-subalgebra $C^*(X)$ in (A, φ). Then*

$$\mathcal{X} \overset{*\text{-}iso}{=} \underset{n \in \mathcal{N}_0}{\star} \overline{\mathbb{C}[\{x_n\}]} \overset{*\text{-}iso}{=} \overline{\mathbb{C}\left[\underset{n \in \mathcal{N}}{\star} \{x_n\}\right]}, \tag{5.2.3}$$

in (A, φ), where "$\overset{*\text{-}iso}{=}$" means "being $*$-isomorphic," and $\overline{\mathbb{C}[\{x_n\}]}$ are the C^*-subalgebras of A generated by $\{x_n\}$, where $\overline{Y}$ are the C^*-topology closures of subsets Y of A.

Here, the free product $(\star)$ in the first $*$-isomorphic relation of (5.2.3) is the free-probability-theoretic free product of [17, 22, 28] and [30], and the free product $(\star)$ in the second $*$-isomorphic relation of (5.2.3) is the pure-algebraic free product generating noncommutative free words in $X = \underset{n \in \mathcal{N}}{\cup} \{x_n\}$.

Proof. Let $\mathcal{X} = C^*(X)$ in a fixed C^*-algebra A. Then, by assumption,

$$\mathcal{X} \overset{def}{=} C^*(X) = \overline{\mathbb{C}[\{x_n\}_{n \in \mathcal{N}_0}]} \overset{*\text{-iso}}{=} \underset{n \in \mathcal{N}_0}{\star} \overline{\mathbb{C}[\{x_n\}]}, \tag{5.2.4}$$

since $X = \underset{n \in \mathcal{N}_0}{\cup} \{x_n\}$ is a free family of (A, φ) in the sense that: all elements x_n's of X are mutually free from each other in (A, φ) (e.g., [17] and [30]). Therefore, the first $*$-isomorphic relation of (5.2.3) holds. By (5.2.4), all elements T of $\mathcal{X}$ are the limits of linear combinations of free "reduced" words (under operator multiplication on A) in $\mathcal{X}$ by [17,22] and [30].

So, all (pure-algebraic) noncommutative free words in the family $X = \{x_n\}_{n \in \mathcal{N}_0}$ have their unique operator forms in $\mathcal{X}$, which are the free reduced words up to operator multiplication inherited from that on A, by (5.2.4). It shows that the second $*$-isomorphic relation of (5.2.3) holds, too. ∎

The above proposition provides a structure theorem of $\mathcal{X}$ in (A, φ). So, by (5.2.3), one can understand the C^*-probability space $\mathcal{X}_\varphi$ of (5.2.1) as an independent free-probabilistic structure,

$$\mathcal{X}_\varphi = \left(\underset{n \in \mathcal{N}_0}{\star} C^*(\{x_n\}), \underset{n \in \mathcal{N}_0}{\star} \varphi \mid_{C^*(\{x_n\})} \right). \tag{5.2.5}$$

Under similar arguments, one can get the following structure theorem of the C^*-probability space $\mathcal{S}_\psi$ of (5.2.2)

Proposition 5.2. *Let $\mathcal{S}_\psi$ be the C^*-probability space (5.2.2) in (B, ψ). Then*

$$\mathcal{S}_\psi = \left(\underset{j \in \mathcal{Z}}{\star} C^*(\{s_j\}), \underset{j \in \mathcal{Z}}{\star} \psi \mid_{C^*(\{s_j\})} \right). \tag{5.2.6}$$

Proof. Similar to the proof of (5.2.3), the C^*-subalgebra $\mathcal{S} = C^*(S)$ of B satisfies that

$$\mathcal{S} \overset{*\text{-iso}}{=} \underset{j \in \mathcal{Z}}{\star} \overline{\mathbb{C}[\{s_j\}]} \overset{*\text{-iso}}{=} \overline{\mathbb{C}\left[\underset{j \in \mathcal{Z}}{\star} \{s_j\}\right]},$$

in (B, ψ). Therefore, like in (5.2.5), the free-probabilistic structure (5.2.6) for the C^*-probability space $\mathcal{S}_\psi$ of (5.2.2) is obtained. ∎

To find a free-isomorphic relation between $\mathcal{X}_\varphi$ and $\mathcal{S}_\psi$, we regard them as the free-product C^*-probability spaces (5.2.3) and (5.2.6), respectively.

First, let's define a bijective function $g : \mathcal{N}_0 \to \mathcal{Z}$. To do that, we partition $\mathcal{N}_0$ and $\mathcal{Z}$ as follows:

$$\mathcal{N}_0 = \{0\} \sqcup (2\mathcal{N}) \sqcup (2\mathcal{N} - 1),$$

and (5.2.7)

$$\mathcal{Z} = (-\mathcal{N}) \sqcup \{0\} \sqcup \mathcal{N},$$

where

$$2\mathcal{N} = \{2n : n \in \mathcal{N}\}, 2\mathcal{N} - 1 = \{2n - 1 : n \in \mathcal{N}\},$$

and

$$-\mathcal{N} = \{-n : n \in \mathcal{N}\}.$$

Define a function $g : \mathcal{N}_0 \to \mathcal{Z}$ by

$$g(n) = \begin{cases} 0 & \text{if } n = 0 \\ \frac{n+1}{2} & \text{if } n \in 2\mathcal{N} - 1 \\ -\frac{n}{2} & \text{if } n \in 2\mathcal{N}, \end{cases} \tag{5.2.8}$$

in $\mathcal{Z}$, for all $n \in \mathcal{N}_0$. For instance,

$$g(0) = 0, g(1) = 1, (2) = -1,$$

and

$$g(3) = 2, g(4) = -2,$$

etc.

Then the function g of (5.2.8) is a well-defined bijection. By this bijection g, one can construct a bijection

$$G : X \to S,$$

by (5.2.9)

$$G(x_n) = s_{g(n)}, \text{ for all } n \in \mathcal{N}_0.$$

where X is the free family generating $\mathcal{X}$ of (5.2.1), and S is the free family generating $\mathcal{S}$ of (5.2.2).

Since g is a bijection from $\mathcal{N}_0$ onto $\mathcal{Z}$, the function G of (5.2.9) is a bijection from the generator set X of $\mathcal{X}$ onto the generator set S of $\mathcal{S}$. Therefore, one can define the corresponding "multiplicative" linear transformation,

$$\Gamma : \mathcal{X} \to \mathcal{S}$$

by the morphism satisfying

$$\Gamma(x_n) = G(x_n) = s_{g(n)} \in S, \tag{5.2.10}$$

for all $x_n \in X$, where G is in the sense of (5.2.9).

More precisely, for an alternating N-tuple $(n_1, ..., n_N) \in \mathcal{N}^N$, satisfying

$$n_1 \neq n_2, n_2 \neq n_3, ..., n_{N-1} \neq n_N \text{ in } \mathcal{N},$$

if one has a free reduced word $T = \prod_{l=1}^{N} x_{n_l}^{k_l} \in \mathcal{X}_\varphi$, where $x_{n_1}, ..., x_{n_N} \in X$, for $k_1, ..., k_N \in \mathcal{N}$, for $N \in \mathcal{N}$, then

$$\Gamma(T) = \Gamma\left(\prod_{l=1}^{N} x_{n_l}^{k_l}\right) = \prod_{l=1}^{N} \Gamma\left(x_{n_l}^{k_l}\right)$$

by the multiplicativity of Γ

$$= \prod_{l=1}^{N} \left(\Gamma(x_{n_l})\right)^{k_l}$$

by the multiplicativity of Γ

$$= \prod_{l=1}^{N} s_{g(n_l)}^{k_l} \overset{let}{=} W_T, \tag{5.2.11}$$

in $\mathcal{S}_\psi$.

Since $(n_1, ..., n_N)$ is an alternating N-tuple in $\mathcal{N}^N$, the N-tuple

$$(g(n_1), \; ..., \; g(n_N)) \in \mathcal{N}^N$$

is an alternating N-tuple, too, by the bijectivity (5.2.8) of g.

Thus, the images $\Gamma(T)$ of all free reduced words $T \in \mathcal{X}_\varphi$ with their lengths-N become free reduced words $W_T \in \mathcal{S}_\psi$ with the same lengths-N, by (5.2.11). i.e., the multiplicative linear transformation Γ of (5.2.10) preserves the free structures of $\mathcal{X}_\varphi$ to those of $\mathcal{S}_\psi$. Also, it is not hard to check that the bijectivity of the function G of (5.2.9) guarantees the bijectivity of Γ by (5.2.3) and (5.2.6), because of the freeness-preserving property (5.2.11).

Lemma 5.3. *Let $\Gamma : \mathcal{X} \to \mathcal{S}$ be the multiplicative linear transformation (5.2.10). Then it is a $*$-isomorphism. i.e.,*

$$\mathcal{X} \overset{*\text{-}iso}{=} \mathcal{S}. \tag{5.2.12}$$

Proof. By the discussions in the very above paragraphs, the morphism Γ is a bijective, freeness-preserving, multiplicative linear transformation. Remark again that all elements of $\mathcal{X}$ (or, of $\mathcal{S}$) are the limits of linear combinations of free reduced words in X (resp., in S) by (5.2.3) (resp., by (5.2.6)). And the morphism Γ satisfies (5.2.11). It implies indeed that Γ is bijective (and hence, bounded) and freeness-preserving.

Now, observe that, for any $x_n \in X \subset \mathcal{X}$, and $t \in \mathcal{C}$,

$$\Gamma\left((tx_n)^*\right) = \Gamma\left(\bar{t}x_n\right)$$

since $x_n^* = x_n$, by the semicircularity on the generator set X of $\mathcal{X}$

$$= \bar{t}\;\Gamma(x_n) = \bar{t}\; s_{g(n)} = \bar{t}\; s_{g(n)}^*$$

since $s_{g(n)}^{*} = s_{g(n)}$, by the semicircularity on the generator set S of $\mathcal{S}$

$$= \left(ts_{g(n)}\right)^{*} = (\Gamma(tx_n))^{*},$$

in $\mathcal{S}$.

Therefore, by (5.2.3), (5.2.6), and the linearity of Γ,

$$\Gamma\left(T^{*}\right) = (\Gamma(T))^{*} \text{ in } \mathcal{S}, \tag{5.2.13}$$

for all $T \in \mathcal{X}$.

Therefore, the morphism Γ of (5.2.10) is a $*$-isomorphism from $\mathcal{X}$ onto $\mathcal{S}$ by (5.2.13). Equivalently, the relation (5.2.12) holds. ∎

By the above lemma, we obtain the following free-isomorphic relation.

Theorem 5.4. *Let $\mathcal{X}_{\varphi}$ and $\mathcal{S}_{\psi}$ be the C^*-probability spaces (5.2.3) and (5.2.6), respectively. Then*

$$\mathcal{X}_{\varphi} \overset{\text{free-iso}}{=} \mathcal{S}_{\psi}. \tag{5.2.14}$$

Proof. By (5.2.12), there exists a $*$-isomorphism Γ of (5.2.10) from $\mathcal{X}$ onto $\mathcal{S}$. By (5.2.3) and (5.2.6), it suffices to show that the $*$-isomorphism Γ preserves the free distributions of generators of $\mathcal{X}_{\varphi}$ to those of generators of $\mathcal{S}_{\psi}$.

Let $x_n \in X \subset \mathcal{X}_{\varphi}$. Then

$$\psi\left((\Gamma(x_n))^{k}\right) = \psi\left(s_{g(n)}^{k}\right) = \left(\omega_k c_{\frac{k}{2}}\right) = \varphi\left(x_n^{k}\right),$$

for all $k \in \mathcal{N}$, by the semicircularity of $X \cup S$.

It shows that Γ preserves the free probability on $\mathcal{X}_{\varphi}$ to that on $\mathcal{S}_{\psi}$, and hence, it is a free-isomorphism by (5.1.2). Therefore, two C^*-probability space $\mathcal{X}_{\varphi}$ and $\mathcal{S}_{\psi}$ are free-isomorphic. ∎

The above free-isomorphic relation (5.2.14) illustrates that the study of free probability on $\mathcal{X}_{\varphi}$ is to study that on $\mathcal{S}_{\psi}$. So, one can use the known results from [6–8] and [11].

Assumption and Notation. From below, we will identify $\mathcal{X}_{\varphi}$ and $\mathcal{S}_{\psi}$ as the same C^*-probability space, and denote it by $\mathcal{X}_{\varphi}$. □

5.3 Free-Distributional Data on $\mathcal{X}_{\varphi}$

Let $\mathcal{A}$ be a C^*-probability space generated by mutually free, $|\mathcal{N}|$-many semicircular elements. Then it is free-isomorphic to the C^*-probability space $\mathcal{X}_{\varphi}$ of (5.2.5), and it is free-isomorphic to the C^*-probability space $\mathcal{S}_{\psi}$ of (5.2.6), by (5.2.14). So, without loss of generality, we let $\mathcal{X}_{\varphi}$ be "the" C^*-probability space generated by a free semicircular family $X = \{x_j\}_{j \in \mathcal{Z}}$, as a candidate of all such C^*-probability spaces.

Corollary 5.5. *Let $I_s = (i_1, ..., i_s)$ be an arbitrary s-tuple in $\mathcal{Z}^s$, for $s \in \mathcal{N}$, like in (4.2.1), and let $\pi_{(I_s)} \in NC(\{i_1, ..., i_s\})$ be in the sense of (4.2.4) for I_s. If $X[I_s]$ be a free random variable of $\mathcal{X}_\varphi$, in the sense of (4.2.5). If $\varphi(X[I_s]) \neq 0$, then there exist $r \leq s$ in $\mathcal{N}$, and $i_{l_1}, ..., i_{l_r} \in \{i_1, ..., i_s\}$, such that*

$$k_{l_1} > k_{l_2} > ... > k_{l_r} \text{ in } \mathcal{N}_0,$$

and

$$n_1, n_2, ..., n_r \in \mathcal{N},$$

such that

$$\sum_{l=1}^{r} n_l (2k_l) = s \text{ in } \mathcal{N},$$

and

$$\varphi(X[I_s]) = \left(\prod_{i=1}^{r-1} \beta_{k_{l_i} > k_{l_{i+1}}}^{\Sigma_{l=1}^{i} n_i}\right)\left(c_{k_r}^{\Sigma_{l=1}^{r} n_l}\right),$$

with (5.3.1)

$$\beta_{k_{l_i} > k_{l_{i+1}}} = 2^{k_{l_i} - k_{l_{i+1}}} \left(\frac{2k_{l_{i+1}}+1}{k_{l_i}+1}\right)\left(\prod_{u=1}^{k_{l_i} - k_{l_{i+1}} - 1} \left(2 - \frac{1}{(k_{l_i}+1)-u}\right)\right),$$

for all $i = 1, ..., r-1$.

Proof. The free-distributional data (5.3.1) on $\mathcal{X}_\varphi$ are obtained by (4.1.1) and (4.2.14). ■

The above corollary characterizes the general free-distributional data on $\mathcal{X}_\varphi$, because all elements of $\mathcal{X}_\varphi$ are the limits of linear combinations of free reduced words in $X = \{x_j\}_{j \in \mathcal{Z}}$, by (5.2.3), (5.2.6) and (5.2.14).

By (5.3.1), the following estimation on $\mathcal{X}_\varphi$ holds, too.

Corollary 5.6. *Under the same hypotheses with Corollary 5.5, there exists $\beta_{\min}$ and $\beta_{\max}$ in $\mathcal{R}^+$, such that*

$$\beta_{\min}\left(c_{k_r}^{\Sigma_{l=1}^{r} n_l}\right) \leq \varphi(X[I_s]) \leq \beta_{\max}\left(c_{k_r}^{\Sigma_{l=1}^{r} n_l}\right), \tag{5.3.2}$$

where $\beta_{\min}$ and $\beta_{\max}$ are in the sense of (4.2.15).

Proof. The estimation (5.3.2) on $\mathcal{X}_\varphi$ is obtained by (5.3.1) and (4.2.15). ■

Also, by (5.3.1) and (5.3.2), one has the following asymptotic properties of free-distributional data on $\mathcal{X}_\varphi$.

Corollary 5.7. *Under the same hypotheses with Corollary 5.5, assume that there exists $l_{i_0} \in \{l_1, ..., l_r\}$, such that $\left|j_{l_{i_0} - n}\right| \to \infty$, for all $n = 1, ..., r - l_{i_0}$, and $j_{l_{i_0}}, j_{l_{i_0 - 1}}, ..., j_{l_r}$ are fixed. Then there exist $M_{i_0} \in \mathcal{R}^+$, and two functions f and g on $\mathcal{R}^+$, such that*

$$\varphi(X[I_s]) \sim M_{i_0}\left(f(k_{j_{l_{i_0}}})\right)^{\Sigma_{l=1}^{i_0} n_l}\left(\prod_{i=1}^{l_{i_0}-1}\left(\frac{g(k_l)}{g(k_{l+1})}\right)^{\Sigma_{l=1}^{i} n_i}\right). \tag{5.3.3}$$

In particular, the quantity M_{i_0} is in the sense of (4.3.4), and the functions f and g are in the sense of (4.3.3).

Proof. The proof of the asymptotic estimation (5.3.3) on $\mathcal{X}_\varphi$ is done by (4.3.5). ∎

6 Certain Free-Isomorphisms on $\mathcal{X}_\varphi$

Let $\mathcal{X}_\varphi = (\mathcal{X}, \varphi)$ be a C^*-probability space generated by a free semicircular family $X = \{x_j\}_{j \in \mathcal{Z}}$, which is free-isomorphic to all C^*-probability spaces generated by mutually free, $|\mathcal{N}|$-many semicircular elements.

6.1 Shifts on $\mathcal{Z}$

Define bijections h_+ and h_- on the set $\mathcal{Z}$ of all integers by

$$h_+(j) = j + 1, \text{ and } h_-(j) = j - 1, \tag{6.1.1}$$

for all $j \in \mathcal{Z}$.

Remark 6.1. *Remark that, by (6.1.1), one can define two bijective functions h_+^0 and h_-^0 on $\mathcal{N}_0$ by*

$$h_\pm^0 = g^{-1} \circ h_\pm \circ g \text{ on } \mathcal{N}_0, \tag{6.1.1$'$}$$

where g is the bijection (5.2.8) from $\mathcal{N}_0$ onto $\mathcal{Z}$. Thus, the existence of the bijections $h_\pm$ of (6.1.1) on $\mathcal{Z}$ guarantees the existence of bijections $h_\pm^0$ of (6.1.1)$'$ on $\mathcal{N}_0$.

From the bijections $h_\pm$ of (6.1.1), one can define the bijections $h_\pm^{(n)}$ on $\mathcal{Z}$, by

$$h_\pm^{(n)} \overset{def}{=} \underbrace{h_\pm \circ h_\pm \circ \ldots\ldots \circ h_\pm}_{n\text{-times}}, \tag{6.1.2}$$

for all $n \in \mathcal{N}_0$, with axiomatization:

$$h_\pm^{(0)} = \text{the identity map } id_{\mathcal{Z}} \text{ on } \mathcal{Z},$$

where $(\circ)$ is the usual functional composition.

By (6.1.1) and (6.1.2), it is easy to check that

$$h_\pm^{(n)}(j) = j \pm n, \text{ for all } n \in \mathcal{N}_0.$$

Definition 6.1. *We call the bijections $h_\pm^{(n)}$ of (6.1.2), the n-($\pm$)-shifts on $\mathcal{Z}$, for all $n \in \mathcal{N}_0$.*

6.2 Integer Shifts on $\mathcal{X}_\varphi$

Let $h_\pm^{(n)}$ be the n-($\pm$)-shifts (6.1.2) on $\mathcal{Z}$. In this section, we construct certain $*$-isomorphisms on $\mathcal{X}_\varphi$ induced by the shifts $h_\pm^{(n)}$. For convenience, we let

$$\mathcal{N}_0^{\pm} \overset{denote}{=} \{\pm\} \times \mathcal{N}_0.$$

Let $(e, k) \in \mathcal{N}_0^{\pm}$, and define a multiplicative linear transformation λ_e^k acting on $\mathcal{X}_\varphi$ by the morphism satisfying

$$\lambda_e^k(x_j) = x_{jek}, \text{for all } x_j \in X \subset \mathcal{X}_\varphi,$$

where

$$jek = \begin{cases} j+k & \text{if } e = + \\ j-k & \text{if } e = -, \end{cases} \tag{6.2.1}$$

in $\mathcal{Z}$.

By the multiplicativity (6.2.1) of the morphism λ_e^k, if $T = \prod_{l=1}^{N} x_{j_l}^{n_l}$ is a free reduced word of $\mathcal{X}_\varphi$ with its length-N, then

$$\lambda_e^k(T) = \prod_{l=1}^{N} \lambda_e^k(x_{j_l})^{n_l} = \prod_{l=1}^{N} x_{j_l ek}^{n_l}, \tag{6.2.2}$$

which is a free reduced word with the length-N in $\mathcal{X}_\varphi$, since

$$(j_1, ..., j_N) \text{is alternating in } \mathcal{Z}^N,$$

if and only if

$$(j_1 ek, \; ..., \; j_N ek) \text{ is alternating in } \mathcal{Z}^N,$$

for all $N \in \mathcal{N}$.

So, the computation (6.2.2) shows that the morphism λ_e^k of (6.2.1) assign free reduced words to free reduced words with the same lengths in $\mathcal{X}_\varphi$.

Note that, for any $t \in \mathcal{C}$, and $x_j \in X \subset \mathcal{X}$,

$$\lambda_e^k\left((tx_j)^*\right) = \bar{t}x_{jek} = \bar{t}x_{jek}^* = \left(\lambda_e^k(tx_j)\right)^*,$$

implying that

$$\lambda_e^k(T^*) = \left(\lambda_e^k(T)\right)^*, \text{for all } T \in \mathcal{X}_\varphi, \tag{6.2.3}$$

in $\mathcal{X}_\varphi$, by (6.2.2).

Theorem 6.1. *A multiplicative linear transformation λ_e^k of (6.2.1) is a free-isomorphism on $\mathcal{X}_\varphi$, for all $(e, k) \in \mathcal{N}_0^{\pm}$.*

Proof. By (6.2.2) and (6.2.3), the morphism λ_e^k of (6.2.1) is a well-defined $*$-homomorphism on $\mathcal{X}_\varphi$. And, by the bijectivity of the k-(e)-shift $h_e^{(k)}$ on $\mathcal{Z}$, the restriction $\lambda_e^k \mid_X$ is a bijection on the free-generator set X of $\mathcal{X}_\varphi$. Thus, by (6.2.2) and (5.2.6), it is bijective, and hence, it is a $*$-isomorphism on $\mathcal{X}_\varphi$.

Observe now that

$$\varphi\left(\left(\lambda_e^k(x_j)\right)^n\right) = \varphi\left(x_{jek}^n\right) = \omega_n c_{\frac{n}{2}} = \varphi\left(x_j^n\right), \tag{6.2.4}$$

for all $n \in \mathcal{N}$, for all $x_j \in X$.

Therefore, for all s-tuple $I_s \in \mathcal{Z}^s$,

$$\varphi(X[I_s]) = \varphi\left(\lambda_e^k(X[I_s])\right) \text{ in } \mathcal{X}_\varphi,$$

by (6.2.4) and (5.3.1), where $X[I_s]$ are in the sense of (5.3.1). It guarantees that

$$\varphi(T) = \varphi\left(\lambda_e^k(T)\right), \text{for all } T \in \mathcal{X}_\varphi,$$

in $\mathcal{X}_\varphi$, by (5.2.6) and (6.2.2). Therefore, this $*$-isomorphism λ_e^k preserves the free probability on $\mathcal{X}_\varphi$ to that on $\mathcal{X}_\varphi$, i.e., it is a free-isomorphism. ■

Let $Aut(\mathcal{X}_\varphi)$ be the *automorphism group of* $\mathcal{X}_\varphi$,

$$Aut(\mathcal{X}_\varphi) \overset{def}{=} \left(\left\{\alpha \,\middle|\, \begin{array}{c}\alpha \text{ is a} \\ *\text{-isomorphism} \\ \text{on } \mathcal{X}_\varphi\end{array}\right\}, \cdot\right),$$

where $(\cdot)$ is the product (or composition) on $*$-isomorphisms.

Define now a subset λ of $Aut(\mathcal{X}_\varphi)$ by

$$\lambda = \{\lambda_e^k : (e, k) \in \mathcal{N}_0^{\pm}\}, \tag{6.2.5}$$

where λ_e^k are the free-isomorphisms (6.2.1) on $\mathcal{X}_\varphi$.

Theorem 6.2. *The subset λ of (6.2.5) is an abelian subgroup of $Aut(\mathcal{X}_\varphi)$.*

Proof. Let $\lambda_{e_1}^{k_1}, \lambda_{e_2}^{k_2} \in \lambda$. Then

$$\lambda_{e_1}^{k_1}\lambda_{e_2}^{k_2} = \lambda_{sgn(e_1k_1e_2k_2)}^{|e_1k_1e_2k_2|} \text{in } \lambda,$$

where sgn is the sign map on $\mathcal{Z}$,

$$sgn(j) = \begin{cases} + & \text{if } j \geq 0 \\ - & \text{if } j < 0, \end{cases}$$

for all $j \in \mathcal{Z}$, and $|.|$ is the absolute value on $\mathcal{Z}$. So, the algebraic structure $(\lambda, \cdot)$ is well-determined in $Aut(\mathcal{X}_\varphi)$.

One can check that

$$\left(\lambda_{e_1}^{k_1}\lambda_{e_2}^{k_2}\right)\lambda_{e_3}^{k_3} = \lambda_{sgn(e_1k_1e_2k_2)}^{|e_1k_1e_2k_2|}\lambda_{e_3}^{k_3}$$

$$= \lambda_{sgn(e_1k_1e_2k_2e_3k_3)}^{||e_1k_1e_2k_2|e_3k_3|} = \lambda_{sgn(e_1k_1e_2k_2e_3k_3)}^{|e_1k_1|e_2k_2e_3k_3||}$$

$$= \lambda_{e_1}^{k_1}\left(\lambda_{e_2}^{k_2}\lambda_{e_3}^{k_3}\right),$$

in λ. So, the algebraic sub-structure $(\lambda, \cdot)$ forms a semigroup.

This subset λ contains

$$\lambda_e^0 = 1_{\mathcal{X}_\varphi}, \text{the identity map on } \mathcal{X},$$

by (6.1.1) and (6.2.1), satisfying

$$1_{\mathcal{X}} \cdot \lambda_e^k = \lambda_e^k = \lambda_e^k \cdot 1_{\mathcal{X}} \text{ on } \mathcal{X}_\varphi,$$

for all $(e,k) \in \mathcal{N}_0^{\pm}$. Therefore, the semigroup $(\lambda, \cdot)$ is a monoid.

Note that, for any $(e,k) \in \mathcal{N}_0^{\pm}$,

$$\lambda_e^k \lambda_{-e}^k = \lambda_{sgn(0)}^0 = 1_{\mathcal{X}_\varphi} = \lambda_{-e}^k \lambda_e^k, \text{ in } \lambda.$$

It shows that every element $\lambda_e^k \in \lambda$ has its unique $(\cdot)$-inverse λ_{-e}^k, i.e.,

$$\left(\lambda_e^k\right)^{-1} = \lambda_{-e}^k, \text{ in } \lambda,$$

where y^{-1} mean the group-inverses of y. So, the monoid $(\lambda, \cdot)$ is a subgroup of $Aut(\mathcal{X}_\varphi)$.

It is not difficult to check that

$$\lambda_{e_1}^{k_1} \lambda_{e_2}^{k_2} = \lambda_{sgn(e_1k_1e_2k_2)}^{|e_1k_1e_2k_2|} = \lambda_{sgn(e_2k_2e_1k_1)}^{|e_2k_2e_1k_1|} = \lambda_{e_2}^{k_2} \lambda_{e_1}^{k_1},$$

in λ. Therefore, the subgroup $(\lambda, \cdot)$ is commutative in $Aut\left(\mathcal{X}_\varphi\right)$. ∎

The above theorem characterizes the algebraic structure of the subset λ of (6.2.5) in the automorphism group $Aut(\mathcal{X}_\varphi)$, i.e., it is an abelian group. More precisely, we have the following structure theorem of λ.

Theorem 6.3. *Let* $\lambda \overset{denote}{=} (\lambda, \cdot)$ *be the abelian subgroup (6.2.5) of* $Aut(\mathcal{X}_\varphi)$. *Then*

$$\lambda \overset{Group}{=} (\mathcal{Z}, +), \tag{6.2.6}$$

where "$\overset{Group}{=}$" *means "being group-isomorphic."*

Proof. Let λ be the subgroup (6.2.5) of $Aut(\mathcal{X}_\varphi)$. Define a map $\Phi : \mathcal{Z} \to \lambda$ by

$$\Phi\left(j\right) = \lambda_{sgn(j)}^{|j|}, including \Phi(0) = \lambda_e^0 = 1_{\mathcal{X}}.$$

Then one can check that Φ is a bijection from $\mathcal{Z}$ onto λ, because

$$j \longmapsto (sgn(j), |j|)$$

is bijective from $\mathcal{Z}$ onto $\mathcal{N}_0^{\pm}$. Also, we have

$$\Phi\left(j_1 + j_2\right) = \lambda_{sgn(j_1+j_2)}^{|j_1+j_2|} = \lambda_{sgn(j_1)}^{|j_1|} \lambda_{sgn(j_2)}^{|j_2|} = \Phi(j_1)\Phi(j_2),$$

in λ, for all $j_1, j_2 \in \mathcal{Z}$.

Therefore, the group-isomorphic relation (6.2.6) holds. ∎

The above theorem shows that the subgroup λ of (6.2.5) is an infinite abelian cyclic group embedded in $Aut(\mathcal{X}_{\varphi})$.

Since λ is a well-defined group, one can construct the corresponding (pure-algebraic) *group algebra* Λ over $\mathbb{C}$,

$$\Lambda \overset{def}{=} \mathbb{C}[\lambda], \tag{6.2.7}$$

generated by the group λ.

Definition 6.2. *Let λ be the group (6.2.5) and Λ, the corresponding group algebra (6.2.7) of λ over $\mathbb{C}$. Then λ is called the integer-shift group on $\mathcal{X}_{\varphi}$, and Λ is called the integer-shift-group algebra on $\mathcal{X}_{\varphi}$. All elements of λ are said to be integer shifts, and all elements of Λ are said to be integer-shift operators on $\mathcal{X}_{\varphi}$.*

By (6.2.7), all elements U of the integer-shift-group algebra Λ have their forms,

$$U = \sum_{(e,k)\in\mathcal{N}_0^{\pm}} t_{(e,k)}\lambda_e^k \text{ in } \Lambda, \text{ with } t_{(e,k)} \in \mathbb{C}, \tag{6.2.7$'$}$$

where $\sum$ in (6.2.7)$'$ is the *finite sum*.

By the very definitions (6.2.5) and (6.2.7), the group λ and the algebra Λ act naturally on the C^*-probability space $\mathcal{X}_{\varphi}$. i.e., there exists a well-defined *group-action α of λ acting on $\mathcal{X}_{\varphi}$*,

$$\alpha\left(\lambda_e^k\right)(T) = \lambda_e^k(T), \forall \lambda_e^k \in \lambda, \tag{6.2.8}$$

and there is a well-defined *algebra-action α_o of Λ acting on $\mathcal{X}_{\varphi}$*,

$$\alpha_o(U)(T) = U(T), \forall U \in \Lambda, \tag{6.2.9}$$

in $\mathcal{X}_{\varphi}$, for all $T \in \mathcal{X}_{\varphi}$.

As we have seen above, all integer shifts $\lambda_e^k \in \lambda$ are free-isomorphisms on $\mathcal{X}_{\varphi}$. It means that the group-action α of (6.2.8) preserves the free probability on $\mathcal{X}_{\varphi}$, characterized by (4.1.1) and (4.2.14). However, we cannot guarantee all integer-shift operators $U \in \Lambda$ are free-isomorphisms on $\mathcal{X}_{\varphi}$ (or not, in general).

6.3 Deformed Free Probability on $\mathcal{X}_{\varphi}$ by Λ

In this section, we study how the algebra-action α_o of (6.2.9) affects the free probability on $\mathcal{X}_{\varphi}$, by investigating how the operators of the integer-shift-operator algebra Λ of (6.2.7) deform the free-distributional data on $\mathcal{X}_{\varphi}$.

Theorem 6.4. *Let λ be the integer-shift group, and let α be the group-action (6.2.8) of λ, acting on $\mathcal{X}_{\varphi}$. Then the free probability on $\mathcal{X}_{\varphi}$ is preserved by α. i.e.,*

$$\varphi\left(\alpha\left(\lambda_e^k\right)(T)\right) = \varphi(T), \text{ for all } T \in \mathcal{X}_{\varphi}. \tag{6.3.1}$$

It implies that the integer-shift operators $\alpha_o\left(\lambda_e^k\right)$ preserves the free probability on $\mathcal{X}_{\varphi}$, for all $\lambda_e^k \in \lambda$ in Λ.

Proof. By (6.2.8), for any $T \in \mathcal{X}_\varphi$,

$$\alpha\left(\lambda_e^k\right)(T) = \lambda_e^k(T) \textit{ in } \mathcal{X}_\varphi.$$

And all integer-shifts $\lambda_e^k \in \lambda$ are free-isomorphisms on $\mathcal{X}_\varphi$, by (6.2.4). Therefore, for any $T \in \mathcal{X}_\varphi$,

$$\varphi\left(\alpha\left(\lambda_e^k\right)(T)\right) = \varphi\left(\lambda_e^k(T)\right) = \varphi(T).$$

Therefore, the action α of λ preserves the free probability on $\mathcal{X}_\varphi$ in the sense of (6.3.1). ■

By (6.3.1), we concentrate on the cases where integer-shift operators U of (6.2.7)′ contained in $\Lambda \setminus \lambda$.

Lemma 6.5. *Let $U = t\lambda_e^k \in \Lambda$, for $t \in \mathcal{C}^\times$, and $\lambda_e^k \in \lambda \subset \Lambda$. Then, for any semicircular element $x_j \in X \subset \mathcal{X}_\varphi$, we have that*

$$\varphi\left((\alpha_o(U)(x_j))^n\right) = t^n \varphi\left(x_j^n\right), \tag{6.3.2}$$

for all $n \in \mathcal{N}$.

Proof. Let $U = t\lambda_e^k$ be an integer-shift operator of Λ, for $t \in \mathcal{C}^\times$, and $\lambda_e^k \in \lambda$, and let $x_j \in X \subset \mathcal{X}_\varphi$. Then

$$(\alpha_o(U)(x_j))^n = \left(t\lambda_e^k(x_j)\right)^n = t^n x_{jek}^n,$$

in $\mathcal{X}_\varphi$, for all $n \in \mathcal{N}$. Thus,

$$\varphi\left(\alpha_o(U)\left(x_j^n\right)\right) = t^n \varphi\left(x_{jek}^n\right) = t^n\left(\omega_n c_{\frac{n}{2}}\right),$$

for all $n \in \mathcal{N}$, by the semicircularity of $x_{jek} \in X$ in $\mathcal{X}_\varphi$.

Therefore, the free-distributional data (6.3.2) is obtained. ■

Consider the following semicircular-like concept.

Definition 6.3. *Let y be a self-adjoint free random variable of a topological $*$-probability space (B, ψ). It is said to be weighted-semicircular with its weight $t_y \in \mathcal{C}^\times = \mathcal{C} \setminus \{0\}$ (or, in shift, t_y-semicircular) in (B, ψ), if*

$$k_n^B\left(\underbrace{y, y, \dots\dots, y}_{n\text{-times}}\right) = \begin{cases} t_y & \text{if } n = 2 \\ 0 & \text{otherwise,} \end{cases} \tag{6.3.3}$$

for all $n \in \mathcal{N}$, where $k_\bullet^B(\ldots)$ is the free cumulant on B in terms of ψ.

The free distribution of such a t_y-semicircular element y is called the weighted-semicircular law with its weight t_y (or, in short, the t_y-semicircular law).

By the definition (6.3.3), and by the Möbius inversion, it is not hard to check that: a self-adjoint free random variable y is t_y-semicircular in a topological $*$-probability space (B, ψ), if and only if

$$\psi(y^n) = \omega_n (t_y)^{\frac{n}{2}} c_{\frac{n}{2}}, \tag{6.3.4}$$

for all $n \in \mathcal{N}$.

So, one can re-define the weighted-semicircularity (6.3.3) by the free-moment characterization (6.3.4).

Theorem 6.6. *Let $U = t\lambda_e^k$ be an integer-shift operator of Λ, for $t \in \mathcal{R}^{\times} = \mathcal{R} \setminus \{0\}$, and $\lambda_e^k \in \lambda \subset \Lambda$. Then, for any semicircular elements $x_j \in X$ of $\mathcal{X}_\varphi$, the corresponding free random variables $\alpha_o(U)(x_j)$ are t^2-semicircular in $\mathcal{X}_\varphi$. i.e.,*

$$t \in \mathcal{R}^{\times} \Longrightarrow \alpha_o\left(t\lambda_e^k(x_j)\right) \textit{ are } t^2 - \textit{semicircular in } \mathcal{X}_\varphi. \tag{6.3.5}$$

Proof. Let $U \in \Lambda$ be the given integer-shift operator, and let $x_j \in X$ be an arbitrary semicircular element of $\mathcal{X}_\varphi$. Then, by (6.3.2),

$$\varphi\left((\alpha_o(U)(x_j))^n\right) = \omega_n t^n c_{\frac{n}{2}} = \omega_n \left(t^2\right)^{\frac{n}{2}} c_{\frac{n}{2}}, \tag{6.3.6}$$

for all $n \in \mathcal{N}$.

Since $t \in \mathcal{R}^{\times}$, one has

$$(\alpha_o(U)(x_j))^* = \bar{t}x_{jek}^* = tx_{jek} = \alpha_o(U)(x_j),$$

because $x_{jek} \in X$ is self-adjoint in $\mathcal{X}$.

So, the self-adjoint free random variable $\alpha_o(U)(x_j)$ is t^2-semicircular by (6.3.4) and (6.3.6). Therefore, the statement (6.3.5) holds. ■

The above theorem illustrates that the integer-shift operators $t\lambda_e^k$ of Λ deform the semicircular law for $x_j \in X \subset \mathcal{X}_\varphi$ to the t^2-semicircular law for $tx_{jek} \in \mathcal{X}_\varphi$, whenever $t \in \mathcal{R}^{\times}$, by (6.3.5). Clearly, if $t = 1$, then the corresponding t^2-semicircular law becomes the semicircular law. So, if $t \neq 1$ in $\mathcal{C}^{\times}$, then the integer-shift operators $t\lambda_e^k \in \Lambda$ distort the semicircular law for X to the t^2-semicircular laws in $\mathcal{X}_\varphi$.

Notation. From below, if there is no confusions, then we denote $\alpha_o(U)$ simply by U, for all $U \in \Lambda$. □

Let's now observe general cases where a given integer-shift operator U of (6.2.7)′ has multi summands. Note here that the integer-shift algebra Λ is a pure algebraic algebra over $\mathcal{C}$, so every element U of Λ has finite summands. For an integer-shift operator U of (6.2.7)′, define the "finite" subset $Supp(U)$ of $\mathcal{N}_0^{\pm}$, by

$$Supp(U) \overset{def}{=} \{(e,k) \in \mathcal{N}_0^{\pm} : t_{(e,k)} \neq 0 in (6.2.7)'\}, \tag{6.3.7}$$

making U be

$$U = \sum_{(e,k) \in Supp(U)} t_{(e,k)} \lambda_e^k \in \Lambda, \tag{6.3.7$'$}$$

as a finite sum.

We call the set $Supp(U)$ of (6.3.7), the *support of the integer-shift operatorU* $\in \Lambda$. The non-zero quantities

$$\{t_{(e,k)} : (e,k) \in Supp(U)\}$$

in (6.3.7)′ are called the *coefficients of* U.

Let U be a multi-sum element (6.3.7)′ of the integer-shift-operator algebra Λ, and let $x_j \in X$ be an arbitrary semicircular element of our C^*-probability space $\mathcal{X}_\varphi$. Then

$$U(x_j) = \sum_{(e,k)\in Supp(U)} t_{(e,k)} x_{jek} \in \mathcal{X}_\varphi. \tag{6.3.8}$$

By (6.3.8), one can get that

$$(U(x_j))^n = \sum_{((e_l,k_l))_{l=1}^n \in Supp(U)^n} \left(\prod_{l=1}^n t_{(e_l,k_l)}\right)\left(\prod_{l=1}^n x_{je_lk_l}\right), \tag{6.3.9}$$

for $n \in \mathcal{N}$.

For $n \in \mathcal{N}$, and for $U \in \Lambda$ of (6.3.7)′, let

$$\begin{array}{c} \mathbf{J}_{((e_l,k_l))_{l=1}^n} = (je_1k_1, je_2k_2, \ldots, je_nk_n) \in \mathcal{Z}^n, \\ \text{where} \\ ((e_l,k_l))_{l=1}^n \in Supp(U)^n = \underbrace{Supp(U) \times \ldots \times Supp(U)}_{n\text{-times}}. \end{array} \tag{6.3.10}$$

Then we denote a product $\prod_{l=1}^n x_{je_lk_l}$ in the summand of (6.3.9) by

$$X\left[\mathbf{J}_{((e_l,k_l))_{l=1}^n}\right] \overset{denote}{=} \prod_{l=1}^n x_{je_lk_l} \in \mathcal{X}_\varphi, \tag{6.3.11}$$

where $\mathbf{J}_{((e_l,k_l))_{l=1}^n}$ is in the sense of (6.3.10).

i.e., by (6.3.11), the formula (6.3.9) can be re-written by

$$(U(x_j))^n = \sum_{((e_l,k_l))_{l=1}^n \in Supp(U)^n} \left(\prod_{l=1}^n t_{(e_l,k_l)}\right) X\left[\mathbf{J}_{((e_l,k_l))_{l=1}^n}\right], \tag{6.3.12}$$

for $n \in \mathcal{N}$.

By (5.3.1), (6.3.9) and (6.3.12), the following proposition is obtained.

Proposition 6.7. *Let* $U \in \Lambda$ *be an integer-shift operator (6.3.7)′, and* $x_j \in X$*, an arbitrary semicircular element of* $\mathcal{X}_\varphi$*. Then*

$$\varphi\left((U(x_j))^n\right) = \sum_{((e_l,k_l))_{l=1}^n \in Supp(U)^n} \left(\prod_{l=1}^n t_{(e_l,k_l)}\right) \beta_{((e_l,k_l))_{l=1}^n}, \tag{6.3.13}$$

for all $n \in \mathcal{N}$*, where the quantities* $\beta_{((e_l,k_l))_{l=1}^n}$ *satisfy*

$$\beta_{((e_l,k_l))_{l=1}^n} = \varphi\left(X\left[\mathbf{J}_{((e_l,k_l))_{l=1}^n}\right]\right), \tag{6.3.14}$$

and they are determined by (5.3.1), by regarding $X\left[\mathbf{J}_{((e_l,k_l))_{l=1}^n}\right]$ *as* $X[I_n]$ *of (4.2.5), by understanding* $\mathbf{J}_{((e_l,k_l))_{l=1}^n}$ *as* I_n *of (4.2.1).*

Proof. The proof of the equality (6.3.14) is done by (5.3.1). Therefore, the free distributional data (6.3.13) holds by (6.3.11) and (6.3.12). ■

It is true that the formula (6.3.13) is hard to be simplified, since the computation is done case-by-case up to supports. However, it is meaningful because we at least have a tool (6.3.14) (and (5.3.1)) to compute (6.3.13).

Now, let $U \in \Lambda$ be an integer-shift operator (6.3.7)$'$ with its finite support $Supp(U)$ of (6.3.7). Then one can construct a new integer-shift operator $U^N \in \Lambda$, for $N \in \mathcal{N}$;

$$U^N = \left(\sum_{(e,k) \in Supp(U)} t_{(e,k)} \lambda_e^k \right)^N$$

$$= \sum_{((e_l,k_l))_{l=1}^N \in Supp(U)^N} \left(\prod_{l=1}^N t_{(e_l,k_l)} \right) \left(\prod_{l=1}^N \lambda_{e_l}^{k_l} \right) \quad (6.3.15)$$

$$= \sum_{((e_l,k_l))_{l=1}^N \in Supp(U)^N} \left(\prod_{l=1}^N t_{(e_l,k_l)} \right) \lambda_{sgn(\Sigma_{l=1}^N e_l k_l)}^{|\Sigma_{l=1}^N e_l k_l|},$$

by (6.2.6).

Now, let $W = ((e_l, k_l))_{l=1}^N \in Supp(U)^N$ in (6.3.15). Then we write the corresponding sign of $\sum_{l=1}^N e_l k_l$, and the absolute value $\left|\sum_{l=1}^N e_l k_l\right|$ by $\mathbf{e}_W$, and $\mathbf{k}_W$, respectively, i.e.,

$$\mathbf{e}_W = sgn\left(\sum_{l=1}^N e_l k_l\right) \in \{\pm\},$$

and (6.3.16)

$$\mathbf{k}_W = \left|\sum_{l=1}^N e_l k_l\right| \in \mathcal{N}_0.$$

By using the new notations (6.3.16), one can re-write (6.3.15) as follows;

$$U^N = \sum_{W \in Supp(U)^N} \mathbf{t}_W \lambda_{\mathbf{e}_W}^{\mathbf{k}_W},$$

with (6.3.17)

$$\mathbf{t}_W = \prod_{(e,k) \hookrightarrow W} t_{(e,k)} \in \mathcal{C}, \forall W \in Supp(U)^N,$$

for $N \in \mathcal{N}$, where "$(e,k) \hookrightarrow W$" means "(e,k) is an entry of $W \in Supp(U)^N$."

Theorem 6.8. *Let* $U = \sum_{(e,k) \in Supp(U)} t_{(e,k)} \lambda_e^k \in \Lambda$ *be an integer-shift operator (6.3.7)$'$, and let* $x_j \in X$ *be a semicircular element of* $\mathcal{X}_\varphi$. *Then*

$$\varphi\left(U^N(x_j^n)\right) = \left(\omega_n c_{\frac{n}{2}}\right) \left(\sum_{((e_l,k_l))_{l=1}^N \in Supp(U)^N} \left(\prod_{l=1}^N t_{(e_l,k_l)} \right) \right), \quad (6.3.18)$$

for all $N, n \in \mathcal{N}$.

Proof. Let $U \in \Lambda$ be a given integer-shift operator. Then, for any $N \in \mathcal{N}$,

$$U^N = \sum_{W \in Supp(U)^N} \mathbf{t}_W \lambda_{\mathbf{e}_W}^{\mathbf{k}_W} \in \Lambda,$$

by (6.3.17), where $\mathbf{k}_W, \mathbf{e}_W$ are in the sense of (6.3.16), and $\mathbf{t}_W$ are in the sense of (6.3.17).

So, if $x_j \in X \subset \mathcal{X}_\varphi$, then

$$U^N\left(x_j^n\right) = \sum_{W \in Supp(U)^N} \mathbf{t}_W x^n_{j\mathbf{e}_W \mathbf{k}_W}, \tag{6.3.19}$$

by (6.3.8), for all $n \in \mathcal{N}$.

Thus, one has that

$$\begin{aligned}\varphi\left(U^N\left(x_j^n\right)\right) &= \sum_{W \in Supp(U)^N} \mathbf{t}_W \left(\omega_n c_{\frac{n}{2}}\right) \\ &= \left(\omega_n c_{\frac{n}{2}}\right)\left(\sum_{W \in Supp(U)^N} \mathbf{t}_W\right),\end{aligned} \tag{6.3.20}$$

by (6.3.19), and by the semicircularity of $x_{j\mathbf{e}_W \mathbf{k}_W} \in X$ in $\mathcal{X}_\varphi$. Therefore, the free-distributional data (6.3.18) holds by (6.3.20). ■

The above free-probabilistic information (6.3.18) also characterizes how finite-sum integer-shift operators distort the free probability on $\mathcal{X}_\varphi$. The following corollary is a direct consequence of (6.3.18).

Corollary 6.9. *Let* $U_1 = \sum_{(e,k) \in Supp(U_1)} \lambda_e^k$ *be an integer-shift operator of* Λ, *whose coefficients are all* 1*'s. Then*

$$\varphi\left(U_1^N\left(x_j^n\right)\right) = \left(\omega_n c_{\frac{n}{2}}\right)\left|Supp(U_1)\right|^N, \tag{6.3.21}$$

for all $N, n \in \mathcal{N}$.

Proof. By (6.3.18), we have

$$\varphi\left(U_1^N\left(x_j^n\right)\right) = \left(\omega_n c_{\frac{n}{2}}\right)\left(\sum_{W \in Supp(U_1)^N} \mathbf{t}_W\right),$$

where $\mathbf{t}_W$ are in the sense of (6.3.17). And, by the definition of U_1,

$$\mathbf{t}_W = 1, \text{for all } W \in Supp(U_1)^N.$$

Therefore, the formula (6.3.21) holds, for all $N, n \in \mathcal{N}$. ■

References

1. Ahsanullah, M.: Some inferences on semicircular distribution. J. Stat. Theory Appl. **15**(3), 207–213 (2016)
2. Bercovici, H., Voiculescu, D.: Superconvergence to the central limit and failure of the cramer theorem for free random variables. Probab. Theory Relat. Fields **103**(2), 215–222 (1995)
3. Bozejko, M., Ejsmont, W., Hasebe, T.: Noncommutative probability of type D. Internat. J. Math. **28**(2), 1750010, 30 (2017)
4. Bozheuiko, M., Litvinov, E.V., Rodionova, I.V.: An extended anyon fock space and non-commutative meixner-type orthogonal polynomials in the infinite-dimensional case. Uspekhi Math. Nauk. **70**(5), 75–120 (2015)
5. Cho, I.: Semicircular families in free product banach $*$-algebras induced by p-adic number fields over primes p. Compl. Anal. Oper. Theory **11**(3), 507–565 (2017)
6. Cho, I.: Semicircular-like laws and the semicircular law induced by orthogonal projections. Complex Anal. Oper. Theory **12**, 1657–1695 (2018)
7. Cho, I.: Free stochastic integrals for weighted-semicircular motion induced by orthogonal projections. Adv. Top. Math. Anal. (2019, to appear)
8. Cho, I.: Acting semicircular elements induced by orthogonal projections on von neumann algebras. Mathematics **5**, 74 (2017). https://doi.org/10.3390/math5040074
9. Cho, I., Dong, J.: Numbers, Catalan, distributions, free, of mutually free multi semicircular elements (2018). Submitted to J. Probab. Relat. Fields
10. Cho, I., Jorgensen, P.E.T.: Semicircular elements induced by projections on separable Hilbert spaces. In: Operator Theory: Advances & Applications. Monograph Series. Birkhauser, Basel (2019, to appear)
11. Cho, I., Jorgensen, P.E.T.: Banach $*$-algebras generated by semicircular elements induced by certain orthogonal projections. Opuscula Math. **38**(4), 501–535 (2018)
12. Cho, I., Jorgensen, P.E.T.: Semicircular elements induced by p-adic number fields. Opuscula Math. **35**(5), 665–703 (2017)
13. Connes, A.: Geometry, Noncommutative. Academic Press, San Diego (1994). ISBN 0-12-185860-X
14. Halmos, P.R.: Hilbert Space Problem Books. Graduate Texts in Mathematics, vol. 19. Springer, Cham (1982). ISBN 978-0387906850
15. Kaygorodov, I., Shestakov, I.: Free generic Poisson fields and algebras. Commun. Algebra **46**(4) (2018). https://doi.org/10.1080/00927872.2017.1358269
16. Makar-Limanov, L., Shestakov, I.: Polynomials and Poisson dependence in free Poisson algebras and free Poisson fields. J. Algebra **349**(1), 372–379 (2012)
17. Nica, A., Speicher, R.: Lectures on the Combinatorics of Free Probability. London Mathematical Society Lecture Note Series, 1st edn, p. 335. Cambridge University Press (2006). ISBN-13:978-0521858526
18. Nourdin, I., Peccati, G., Speicher, R.: Multi-dimensional semicircular limits on the free wigner chaos. Progr. Probab. **67**, 211–221 (2013)
19. Pata, V.: The central limit theorem for free additive convolution. J. Funct. Anal. **140**(2), 359–380 (1996)
20. Radulescu, F.: Random matrices, amalgamated free products and subfactors of the C^*-algebra of a free group of nonsingular index. Invent. Math. **115**, 347–389 (1994)
21. Radulescu, F.: Free group factors and Hecke operators, notes taken by N. Ozawa. In: Proceedings of the 24th Conference in Operator Theory, Theta Advanced Series in Math. Theta Foundation (2014)

22. Speicher, R.: Combinatorial Theory of the Free Product with Amalgamation and Operator-Valued Free Probability Theory, vol 132, no. 627. Memoirs of the American Mathematical Society (1998)
23. Speicher, R.: A conceptual proof of a basic result in the combinatorial approach to freeness. Infinite Dimension. Anal. Quant. Probab. Relat. Top. **3**, 213–222 (2000)
24. Speicher, R., Kemp, T.: Strong Haagerup inequalities for free R-diagonal elements. J. Funct. Anal. **251**(1), 141–173 (2007)
25. Speicher, R., Haagerup, U.: Brown's spectrial distribution measure for R-diagonal elements in finite von neumann algebras. J. Funct. Anal. **176**(2), 331–367 (2000)
26. Vladimirov, V.S.: p-adic quantum mechanics. Commun. Math. Phys. **123**(4), 659–676 (1989)
27. Vladimirov, V.S., Volovich, I.V., Zelenov, E.I.: p-Adic Analysis and Mathematical Physics. Series on Soviet and East European Mathematics, vol. 1. World Scientific (1994). ISBN 978-981-02-0880-6
28. Voiculescu, D.: Aspects of free analysis. Jpn. J. Math. **3**(2), 163–183 (2008)
29. Voiculescu, D.: Free probability and the von neumann algebras of free groups. Rep. Math. Phys. **55**(1), 127–133 (2005)
30. Voiculescu, D., Dykemma, K., Nica, A.: Free Random Variables. CRM Monograph Series, vol. 1. American Mathematical Society (1992). ISBN-13: 978–0821811405

The manufacturer's authorised representative in the EU is Springer Nature Customer Service Centre GmbH, Europaplatz 3, 69115 Heidelberg, Germany. If you have any concerns regarding our products, please contact ProductSafety@springernature.com

Printed and bound by CPI Group (UK) Ltd, Croydon, CR0 4YY
30/09/2026
02225436-0004